AF689580

Biotechnology in Agriculture and Forestry

Springer
Berlin
Heidelberg
New York
Barcelona
Hong Kong
London
Milan
Paris
Tokyo

Volumes already published

Volume 1: Trees I (1986)
Volume 2: Crops I (1986)
Volume 3: Potato (1987)
Volume 4: Medicinal and Aromatic Plants I (1988)
Volume 5: Trees II (1989)
Volume 6: Crops II (1988)
Volume 7: Medicinal and Aromatic Plants II (1989)
Volume 8: Plant Protoplasts and Genetic Engineering I (1989)
Volume 9: Plant Protoplasts and Genetic Engineering II (1989)
Volume 10: Legumes and Oilseed Crops I (1990)
Volume 11: Somaclonal Variation in Crop Improvement I (1990)
Volume 12: Haploids in Crop Improvement I (1990)
Volume 13: Wheat (1990)
Volume 14: Rice (1991)
Volume 15: Medicinal and Aromatic Plants III (1991)
Volume 16: Trees III (1991)
Volume 17: High-Tech and Micropropagation I (1991)
Volume 18: High-Tech and Micropropagation II (1992)
Volume 19: High-Tech and Micropropagation III (1992)
Volume 20: High-Tech and Micropropagation IV (1992)
Volume 21: Medicinal and Aromatic Plants IV (1993)
Volume 22: Plant Protoplasts and Genetic Engineering III (1993)
Volume 23: Plant Protoplasts and Genetic Engineering IV (1993)
Volume 24: Medicinal and Aromatic Plants V (1993)
Volume 25: Maize (1994)
Volume 26: Medicinal and Aromatic Plants VI (1994)
Volume 27: Somatic Hybridization in Crop Improvement I (1994)
Volume 28: Medicinal and Aromatic Plants VII (1994)
Volume 29: Plant Protoplasts and Genetic Engineering V (1994)
Volume 30: Somatic Embryogenesis and Synthetic Seed I (1995)
Volume 31: Somatic Embryogenesis and Synthetic Seed II (1995)
Volume 32: Cryopreservation of Plant Germplasm I (1995)
Volume 33: Medicinal and Aromatic Plants VIII (1995)
Volume 34: Plant Protoplasts and Genetic Engineering VI (1995)
Volume 35: Trees IV (1996)
Volume 36: Somaclonal Variation in Crop Improvement II (1996)
Volume 37: Medicinal and Aromatic Plants IX (1996)
Volume 38: Plant Protoplasts and Genetic Engineering VII (1996)
Volume 39: High-Tech and Microprogation V (1997)
Volume 40: High-Tech and Microprogation VI (1997)
Volume 41: Medicinal and Aromatic Plants X (1998)
Volume 42: Cotton (1998)
Volume 43: Medicinal and Aromatic Plants XI (1999)
Volume 44: Transgenic Trees (1999)
Volume 45: Transgenic Medicinal Plants (1999)
Volume 46: Transgenic Crops I (1999)
Volume 47: Transgenic Crops II (2001)
Volume 48: Transgenic Crops III (2001)
Volume 49: Somatic Hybridization in Crop Improvement II (2001)
Volume 50: Cryopreservation of Plant Germplasm II (2002)
Volume 51: Medicinal and Aromatic Plants XII (2002)

Volume in preparation

Volume 52: Brassicas and Legumes: From Genome Structure to Breeding

Biotechnology in Agriculture and Forestry 51

Medicinal and Aromatic Plants XII

Edited by
T. Nagata and Y. Ebizuka

With 133 Figures, 3 in Color, and 44 Tables

Springer

Series Editors
Professor Dr. Toshiyuki Nagata
University of Tokyo
Graduate School of Science
Department of Biological Sciences
7-3-1 Hongo, Bunkyo-ku
Tokyo 113-0033, Japan

Professor Dr. Y.P.S. Bajaj †
New Delhi, India

Volume Editors

Professor Dr. Toshiyuki Nagata
(*address see above*)

Professor Dr. Yutaka Ebizuka
University of Tokyo
Graduate School of Pharmaceutical Sciences
7-3-1 Hongo, Bunkyo-ku
Tokyo 113-0033, Japan

ISSN 0934-943-X
ISSN 3-540-41686-2 Springer-Verlag Berlin Heidelberg New York

Library of Congress Cataloging-in-Publication Data. Medicinal and aromatic plants. (Biotechnology in agriculture and forestry; 4-). Includes bibliographies and index. 1. Medicinal plants – Biotechnology. 2. Aromatic plants – Biotechnology. 3. Plant cell culture. 4. Materia medica, Vegetable. I. Bajaj, Y.P.S., 1936– . II. Series. TP248.27.P55.M43 1998 660.62 88-3059. ISBN 3-540-41686-2

This work is subject to copyright. All rights reserved, whether the whole or part of the material is concerned, specifically the rights of translation, reprinting, reuse of illustrations, recitation, broadcasting, reproduction on microfilm or in any other way, and storage in data banks. Duplication of this publication or parts thereof is permitted only under the provisions of the German Copyright Law of September 9, 1965, in its current version, and permission for use must always be obtained from Springer-Verlag. Violations are liable for prosecution under the German Copyright Law.

Springer-Verlag Berlin Heidelberg New York
a member of BertelsmannSpringer Science+Business Media GmbH
http.//www.springer.de

© Springer-Verlag Berlin Heidelberg 2002
Printed in Germany

The use of general descriptive names, registered names, trademarks, etc. in this publication does not imply, even in the absence of a specific statement, that such names are exempt from the relevant protective laws and regulations and therefore free for general use.

Cover design: Design & Production GmbH, Heidelberg
Typesetter: SNP Best-Set Typesetter Ltd., Hong Kong
SPIN 10779423 31/3130 5 4 3 2 1 0-Printed on acid-free paper

This Volume is dedicated to the late
Professor Toshio Yamaki, Tokyo, Japan,
who introduced me to the field of plant sciences.

Toshiyuki Nagata

Preface

This volume on *Medicinal and Aromatic Plants XII* is the 51st volume in the series, *Biotechnology in Agriculture and Forestry* (BAF), which was founded by Professor Bajaj in 1986. At the onset, however, it is necessary to describe why and how the editing of this series has been transferred to me. To my surprise, I was informed of the decease of Professor Bajaj in the early months of 2000. Since his toils and efforts to continue this series were tremendous, we have to be thankful to him for his energy with a kind of cultural legacy of his contributions to plant biotechnology. An even more surprising matter was an inquiry by Dr. Dieter Czeschlik, the executive director for books and journals at Springer Verlag, whether I could take charge of the succeeding series of BAF. Initially, I was very hesitant; however, discussions with some of my colleagues and communication with the staff of Springer Verlag gradually convinced me to accept this honorable proposal. After my acceptance as Series Editor, the submitted manuscripts for Volume 51 were sent to me, however, there were just ten of them. Soon I asked other possible contributors working in this field, who had exchanged letters with Professor Bajaj, to write respective chapters. Independently, I looked for other possible contributors as well. In the meantime, Professor Ebizuka of the University of Tokyo, who, as a professor for pharmaceutical chemistry, is more knowledgeable on this topic, very kindly joined me in editing this volume. Without his help, I would not have been able to finalize the editing of this volume. Although he has been, and is extremely busy, he helped me to proceed with the editorial work. Thus, I could finally send the completed manuscripts for this series to the publisher.

In fact, during this work, I became aware that the previous books in this series are considered to be resource books of "Medicinal and Aromatic Plants" throughout the world, and many colleagues of former contributors and people working in the pharmaceutical industries consistently commented on this fact. However, many people simultaneously felt the necessity of a change in perspective and offered some constructive comments to this series. Even in some of the contributions to this volume, hints of changes may be noted. This was due mainly to the determination of the whole genomic sequence of *Arbidopsis thaliana* towards the end of 2000 and we are also aware that the sequence of rice genomes will also be determined by the end of 2004. Such progress, together with established transformation techniques, should allow the manipulation of pharmaceutical plants with specific genes, which will certainly change the features of medicinal and aromatic plants greatly. Thus, the contributions in the Series, *Medicinal and Aromatic Plants XII*, will focus on these points.

To close these introductory remarks, I am very pleased to inform readers that the forthcoming volumes of the BAF Series will be edited by an international editorial team, in which I remain the Series Editor, while two co-editors, Professor Horst Lörz of Hamburg University, Germany, and Professor Jack Widholm of the University of Illinois, USA, join me. The effects of this change to an international editorial team will appear as of 2001 and intensive collaboration among them will make the new features of the BAF Series quite useful.

Tokyo, January 2002 TOSHIYUKI NAGATA

Contents

List of Contributors

M.S. Al-Said, Department of Pharmacognosy, College of Pharmacy, King Saud University, P.O. Box 2457, Riyadh 11451, Saudi Arabia

L. Alvárez, Centro de Investigaciones Quìmicas. Universidad Autónoma del Estado de Morelos, México

T. Bahorun, Department of Biological Sciences, Faculty of Science, University of Mauritius, Republic of Mauritius

W. Barz, Institut für Biochemie und Biotechnologie der Pflanzen, Westfälische Wilhelms-Universität, Hindenburgplatz 55, 48143 Münster, Germany

J. Budzianowski, Department of Pharmaceutical Botany, Karol Marcinkowski University of Medical Sciences in Poznañ, ul. Sw. Marii Magdaleny 14, 61-861 Poznañ, Poland

P. Christen, University of Geneva, Laboratory of Pharmaceutical Analytical Chemistry, 20, Bd D'Yvoy, 1211 Geneva 4, Switzerland

B. Dános, Research Institute for Medicinal Plants, 2011 Budakalász, Hungary

H.M. Elhag, Department of Pharmacognosy, College of Pharmacy, King Saud University, P.O. Box 2457, Riyadh 11451, Saudi Arabia

M.M. El-Olemy, Department of Pharmacognosy, College of Pharmacy, King Saud University, P.O. Box 2457, Riyadh 11451, Saudi Arabia

J.F.S. Ferreira, Aventis CropScience, P.O. Box 12014, RTP, North Carolina 27709, USA

R. Handayani, Faculty of Pharmaceutical Science, Nagasaki University, 1-14 Bunkyo-machi, Nagasaki 852-8521, Japan

T. Ikenaga, Faculty of Environmental Sciences, Nagasaki University, 1-14 Bunkyo-machi, Nagasaki 852-8521, Japan

A. IKUTA, Research Institutes for Science and Technology, Science University of Tokyo, 2669 Yamazaki, Noda, Chiba 278-8510, Japan

I. IONKOVA, Department of Pharmacognosy, Faculty of Pharmacy, Medical University, Sofia, 2 Dunav Str., 1000 Sofia, Bulgaria

K. ISHIMARU, Department of Applied Biological Sciences, Faculty of Agriculture, Saga University, 1 Honjo, Saga 840-8502, Japan

J. JANICK, Department of Horticulture and Landscape Architecture, Purdue University, West Lafayette, Indiana 47907-1165, USA

É. KERTÉSZ-DOBOS, Research Institute for Medicinal Plants, 2011 Budakalász, Hungary

S. KIKUTA, Faculty of Pharmaceutical Science, Nagasaki University, 1-14 Bunkyo-machi, Nagasaki 852-8521, Japan

S.W. KIM, Plant Cell Biotechnology Laboratory, Korea Research Institute Bioscience and Biotechnology, 52-Oun-dong, Yusong, Taejon, 305-333 Korea

S.E. KINTZIOS, Agricultural University of Athens, Faculty of Agricultural Biotechnology, Laboratory of Plant Physiology, Iera Odos 75, 11855 Athens, Greece

A. KORTENBUSCH, Institut für Biochemie und Biotechnologie der Pflanzen, Westfälische Wilhelms-Universität, Hindenburgplatz 55, 48143 Münster, Germany

K. KUKULCZANKA, Botanical Garden, University of Wrocław, ul. Sienkiewicza 23, 50-335 Wrocław, Poland

Á. LÁSZLÓ-BENCSIK, Research Institute for Medicinal Plants, 2011 Budakalász, Hungary

J.R. LIU, Plant Cell Biotechnology Laboratory, Korea Research Institute Bioscience and Biotechnology, 52-Oun-dong, Yusong, Taejon, 305-333 Korea

S. MEDEROS-MOLINA, Department of Plant Physiology, Facultad de Farmacia, Universidad de La Laguna, 38200 La Laguna, Tenerife, Canary Islands, Spain

U. MÜHLENBECK, Institut für Biochemie und Biotechnologie der Pflanzen, Westfälische Wilhelms-Universität, Hindenburgplatz 55, 48143 Münster, Germany

Y. Murakami, Department of Applied Biological Sciences, Faculty of Agriculture, Saga University, 1 Honjo, Saga 840-8502, Japan

S.C. Oh, Plant Cell Biotechnology Laboratory, Korea Research Institute Bioscience and Biotechnology, 52-Oun-dong, Yusong, Taejon, 305-333 Korea

T. Oyama, Faculty of Pharmaceutical Science, Nagasaki University, 1-14 Bunkyo-machi, Nagasaki 852-8521, Japan

R. Quintero, Centro de Investigación en Biotecnología. Universidad Autónoma del Estado de Morelos, México

K. Shimomura, Tsukuba Medicinal Plant Research Station, National Institute of Health Sciences, 1 Hachimandai, Tsukuba, Ibaraki 305-0843, Japan

M.A.L. Smith, University of Illinois, Department of Natural Resources and Environmental Sciences, Urbana, Illinois 61801, USA

Y. Sugimoto, Arid Land Research Center, Tottori University, 1390 Hamasaka, Tottori 680-0001, Japan

F. Trotin, Laboratoire de Pharmacognosie, Faculté des Sciences Pharmaceutiques et Biologiques, B.P. 83, 59006 Lille Cedex, France

J. Vasseur, Laboratoire de Physiologie et de Morphogénèse Végétales, Université des Sciences et Technologies de Lille, 59655 Villeneuve d'Ascq Cedex, France

M.L. Villarreal, Centro de Investigación en Biotecnología, Universidad Autónoma del Estado de Morelos, México

M. Wink, University of Heidelberg, Institute for Pharmaceutical Biology, Im Neuenheimer Feld 364, 69120 Heidelberg, Germany

1 Production of Artemisinin from in Vitro Cultures of *Artemisia annua* L.

JORGE F.S. FERREIRA[1] and JULES JANICK[2]

1 Introduction

Malaria, one of the oldest known diseases, was referred to in Egyptian writings of the 16th century B.C. In the 17th century, Italians believed that breathing bad air (*mal aria*) arising from swamps was responsible for the disease, and the term malaria first entered the English medical literature in the first half of the 19th century. Each year, this disease afflicts over 300 million people worldwide, killing up to 2.7 million, mostly children. Most of these cases occur in Africa, but large areas of Asia, Central, and South America have high incidences of the disease (Nussenzweig and Long 1994). Out of 37 countries and territories, which are members of the Pan American Health Organization (PAHO), World Health Organization (WHO), 21 still have active malaria transmission (PAHO/WHO 1998).

Malaria has been treated for over 40 years with quinine-derived drugs. However, *Plasmodium falciparum* has developed resistance against these drugs in several areas of the world. Artemisinin (*qinghaosu*) (Fig. 1), a sesquiterpene lactone belonging to the cadinane series, is an antimalarial compound first isolated from *Artemisia annua* L. (Fig. 2) by Chinese scientists in 1972 (Anonymous 1979). In addition to a lactone group, artemisinin contains an endoperoxide bridge, which is rarely found in secondary metabolites. One of the few compounds, which contain a peroxide group, is 10, 12-peroxycalamanene. It is extracted from tubers of *Cyperus rotundus*, and is also effective against *Plasmodium falciparum* (Thebtaranonth et al. 1995). Schmid and Hofheinz (1983), Xu et al. (1986), Ravindranathan et al. (1990), and Avery et al. (1992) achieved complete chemical (de novo) synthesis of artemisinin. The procedures require several steps, and can start from different raw materials. A comprehensive review of the chemistry of artemisinin has been recently published (Ziffer et al. 1997). However, low yield, complexity, and high cost make extraction from plants the most economically feasible method for artemisinin production at present. *Artemisia annua* plants

[1]Department of Plant Biology, Southern Illinois University-Carbondale, Carbondale, Illinois 62901-6509, USA

[2]Department of Horticulture and Landscape Architecture, Purdue University, West Lafayette, Indiana 47907-1165, USA

Biotechnology in Agriculture and Forestry, Vol. 51
Medicinal and Aromatic Plants XII (ed. by T. Nagata and Y. Ebizuka)
© Springer-Verlag Berlin Heidelberg 2002

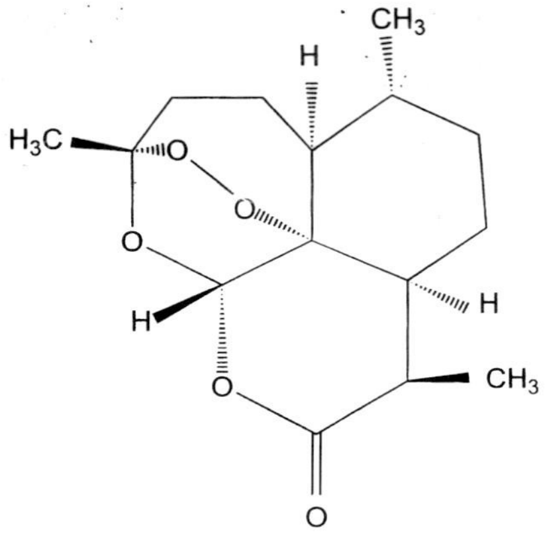

Fig. 1. Artemisinin (qinghaosu) molecule. A sesquiterpene lactone with a peroxide bridge, produced by aerial parts of *Artemisia annua* L.

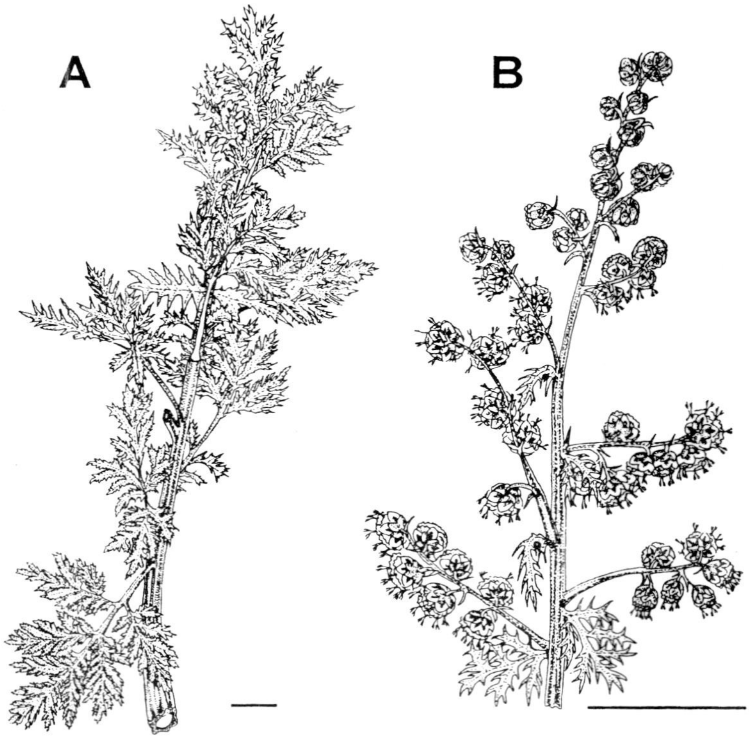

Fig. 2. A Vegetative and B flowering shoots of *Artemisia annua*. *Bar* = 1 cm. (Ferreira and Janick 1995)

normally produce from 0.01 to 0.5% (w/w) artemisinin. However, researchers in Switzerland have developed artemisinin hybrids which produce over 0.8% (w/w) artemisinin, in the field, in about 8 months from sowing to harvest (Delabays et al. 1993). Pure artemisinin (98%) presently has a value of approximately $300.00/g, while crude artemisinin sells for ca. $300.00/kg and costs $1,200.00/ton for extraction, plus the costs for purification and crystallization.

Artemisinin has been detected in leaves, small green stems, buds, flowers, and seeds of artemisia (Acton et al. 1985; Zhao and Zeng 1985; Liersch et al. 1986; Martinez and Staba 1988; Singh et al. 1988; Madhusudanan 1989; Ferreira et al. 1995a). Artemisinin has not been reported in roots of field-grown plants (Pras et al. 1991; Klayman 1993; Ferreira et al. 1995a) or pollen, and the detection of artemisinin from seeds appears to be due to the presence of floral debris (Ferreira et al. 1995a). The highest concentration of artemisinin is found in the inflorescence, which at anthesis may contain more than ten times as much artemisinin as leaves (Ferreira et al. 1995a). Artemisinin accumulates in glandular trichomes (Fig. 3), which are present in leaves, stems, and flowers of the plant (Duke et al. 1994; Ferreira et al. 1997).

Production of artemisinin under in vitro conditions has attracted the attention of several investigators due to the progress achieved in the production of other medicinal natural compounds. Examples include shikonins and related naphtoquinones, produced from cell cultures of *Lithospermum erythrorhizon* by Mitsui Petrochemicals in China and Japan (Tabata et al. 1974); sanguinarine, produced by elicited cell cultures of *Papaver somniferum* (Eilert et al. 1985); and taxol, the commercial production of which is being pursued by Phyton Catalytic, Inc. and ESCA Genetics (Gibson et al. 1995).

Attempts to produce artemisinin and related compounds by tissue culture systems have been reviewed by Woerdenbag et al. (1994), Ferreira and Janick (1996), and Ferreira et al. (1997). This review will focus on recent results on the search of artemisinin production by in vitro cultures of untransformed and transformed tissues of *A. annua*.

1.1 Distribution and Importance of *Artemisia annua*

Artemisia annua L. (Asteraceae or Compositae), also known as *qinghao* (Chinese), annual or sweet wormwood, or sweet Annie, is an annual herb native to Asia, most probably China (McVaugh 1984). *Artemisia annua* occurs naturally as part of the steppe vegetation in the northern parts of Chahar and Suiyuan provinces (40°N, 109°E) in China, at 1000–1500 m above sea level (Wang 1961). The plant now grows wild in many countries, such as Argentina, Bulgaria, France, Hungary, Romania (cultivated for its essential oil), Italy, Spain, the United States, and the former Yuguslavia (Klayman 1989, 1993).

Artemisinin, along with taxol, is considered one of the novel discoveries in recent medicinal plant research, and its isolation and characterization has increased interest in *A. annua* worldwide. Artemisinin is the base compound

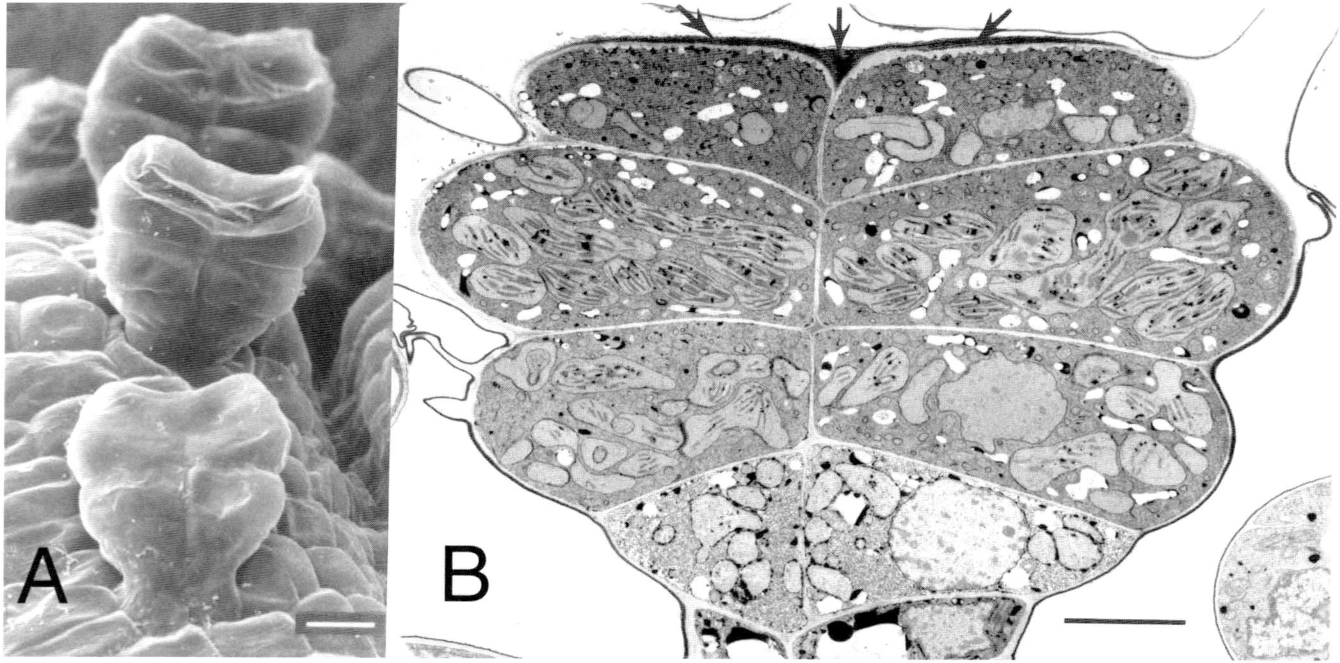

Fig. 3. Biseriate glandular trichomes of *A. annua*. **A** Scanning electron micrograph (SEM) of glandular trichome at the basal portion of the corolla with apical sac collapsed due to fixation process. (Ferreira and Janick 1995) **B** Transmission electron micrograph (TEM) of a ten-cell stage gland. The *arrows* indicate the dense layer of osmiophilic material that comprises a portion of the cell walls of the two apical cells. The apical cell pair has no chloroplasts, but the subapical two cell pairs (four cells) contain large, amorphous chloroplasts without starch grains. *Bar* = 5 μm. (Duke et al. 1994)

for the synthesis of more potent and stable antimalarial drugs with reduced toxicity for humans. Artemisinin is effective against *Plasmodium* species, including *P. vivax* and *P. falciparum*, two of the four species that cause human malaria; with *P. falciparum* being responsible for the often fatal cerebral malaria, an advanced stage of the disease.

1.2 Conventional Practices for the Propagation and Production of Artemisinin, and the Demand on the World Market

Propagation of artemisia is normally done by seeds. Seeds keep their vigor for at least 3 years if stored under dry and cool conditions. Several researchers transplant artemisia to the field at the five- to six-leaf stage, which requires 4–6 weeks of greenhouse growth. Vegetative propagation of artemisia is achieved easily from cuttings. The shoots can be taken from juvenile or adult plants and have a rooting rate of 95–100%. Cuttings will root in about 2 weeks in a mist chamber (Ferreira 1994).

Traditionally, artemisinin has been extracted from wild stands in China, with artemisinin concentrations varying from 0.01 to 0.5% (w/w). Material from Sichuan Province is reported to have the highest artemisinin levels (Trigg 1989). Macro- and micronutrients have minor influence on artemisinin production by field plants (Laughlin 1994). Applications of $50\,mg\,l^{-1}$ gibberellic acid (GA_3) to field grown plants increased artemisinin content from 0.77 to $1.10\,mg\,g^{-1}$; kinetin (10 and $20\,mg\,l^{-1}$) increased leaf yield and oil content, but decreased artemisinin content; and triacontanol had no effect on artemisinin content (Farooqi et al. 1996). The levels of artemisinin increased from 0.77 to 1.3% when $80\,mg\,l^{-1}$ GA_3 was applied to field crops, but artemisinin levels were not correlated to the levels of GA_3 applied to the crop (Siyapata-Ntakirutimana et al. 1996). Salinity stress did not influence artemisinin production (Prasad et al. 1997).

China and Vietnam are the main producers of artemisinin and its derivatives either for oral or parenteral use. Malaria control program officials have distributed, between 1991–1998, 31.6 million tablets of artemisinin, 10.5 million of artesunate, and 793,500 vials of injectable artesunate in Vietnam. Although recent data from China are not available, sales of artesunate tablets rose from 185,000 to 2,545,000 between 1991 and 1995. In Thailand, consumption of artesunate rose from 2880 tablets in 1993 to 653,199 tablets in 1997.

Artemisinin, artemether, arteether, artesunate, and dihydroartemisinin (the last four being artemisinin semi-synthetic derivatives) can all be purchased as drug substances from producers in China, while artemisinin, artemether, and artesunate can be purchased from Vietnam (WHO/MAL 1998). According to the World Health Organization, artemisinin and its derivatives are widely registered as antimalarial drugs in countries where malaria is endemic. Paluther, a commercial brand of artemether (Aventis Pharma), is available in over 100 countries, including 20 countries from Asia and Africa. Combined sales of Paluther in Peru and Brazil reached $200,000.00 in 1999. Currently, no formulation has been registered in Europe or North America, but intramuscular artemether can be made available in France and Denmark upon request. Because countries such as Bangladesh and the Philippines have no problem with multi-drug resistant malaria, artemisinin-derived drugs are unavailable. However, countries such as Myanmar and Vietnam require the use of artemisinin drugs due to the existence of multi-drug resistant *Plasmodium* strains. The development and spread of multi-drug resistant *Plasmodium falciparum* currently dictate the demand of artemisinin-derived drugs in the world.

2 In Vitro Approaches

2.1 In Vitro Culture of *A. annua*

Several investigators, cited by Ferreira et al. (1997), report successful in vitro propagation of *A. annua* via shoot cultures. Shoots are easily micropropagated and cultured using standard protocols and cytokinin supplementation. Plants grown in vitro readily acclimate to soil, but many plants derived from shoot cultures display cytokinin abnormalities producing highly branched plants that lack apical dominance (Ferreira 1994; Paniego and Giuliette 1996). Although inflorescence tissue will proliferate in vitro (Mathur and Kumar 1996), cuttings from underdeveloped inflorescences take longer to root because they need to revert to vegetative growth. A constant supply of vegetative plants, for cuttings, can be maintained under photoperiods of 16 h in a greenhouse; the critical photoperiod for flowering of *A. annua* is between 13 and 14 h (Ferreira et al. 1995a). There is evidence that artemisinin production in the field can be predicted from greenhouse performance. Artemisinin contents of 14 greenhouse-grown clones were highly correlated ($r = 0.954$) with the same clones field-grown the same year (Ferreira et al. 1995b). However, when 24 clones, ranging from low to high artemisinin, were micropropagated for 2 years the correlation was lower ($r = 0.61$), and even lower ($r = 0.50$) when the clones were regenerated and grown in a greenhouse under a 16-h photoperiod. Many of the clones lost their apical dominance and never flowered in the field, unless they regained their normal growth habit. The low correlation in artemisinin among the same clones, analyzed 2 years apart, indicated that epigenetic changes, or somaclonal variations, might hinder long-term maintenance of stock cultures. Ideally, clones selected for high artemisinin in the greenhouse and stored in vitro should be regenerated and reevaluated every year or two with new cultures reestablished. Alternatively, clones pre-screened for artemisinin content can be maintained vegetatively in a greenhouse, under a 16-h photoperiod. Because artemisinin levels from vegetative plants are positively correlated with artemisinin content of the same clones after full bloom (Jain et al. 1996), it is unnecessary to induce flowering in the evaluation process. Artemisinin is found in in vitro differentiated shoots at levels similar to those of field crops (Ferreira et al. 1997). Treatments of shoot cultures with kinetin, chlormequat chloride (CCC), or daminozide did not affect artemisinin content but benzyl adenine (BA) reduced root growth and artemisinin content (Ferreira et al. 1997).

Ferreira and Janick (1996) reported a statistically significant (1% level) correlation ($r = 0.78$) between artemisinin content of shoots in vitro and the number of roots, but not with the number of shoots. These results indicate that the presence of roots enhances the biosynthesis of artemisinin by shoots. Because artemisia roots are devoid of artemisinin, those results suggest that roots may be the source of artemisinin precursors.

2.2 Establishment of Callus and Cell Cultures

Callus can be obtained on media supplemented with combinations of auxins and cytokinins, but non-friable callus is usually obtained. Ferreira and Janick (1996) obtained the highest yield of friable callus with a combination of BA at 4.44 μM and 2,4-dichlorophenoxy acetic acid (2,4-D) at 4.52 μM. Vitamin C reduced browning, and GA_3 improved friability. However, after 17 months in culture, only 3 out of 24 clones kept as shoot cultures generated callus. Cell cultures were established using the same callus-inducing medium, without agar. However, artemisinin was not detected from callus or cell cultures or from the liquid medium.

2.3 Agrobacteria-Mediated Transformed Cultures of *A. annua*

Cai et al. (1995) established root cultures of *A. annua* using *Agrobacterium rhizogenes* R1601, and detected artemisinin in cultures kept under illumination. Maximum artemisinin production was obtained with media supplemented with 0.3% sucrose. Enhancement of artemisinin occurred with the addition of 4.8 mgl^{-1} GA_3 to the medium. Low concentration of 1, α-naphthaleneacetic acid (0.025 mgl^{-1}) enhanced root growth, but inhibited artemisinin production. Teoh et al. (1996), working on cryopreservation of hairy root cultures, reported an average survival rate of 65%. The production of artemisinin (0.214 mgg^{-1} fresh weight) from frozen cultures was similar to that of unfrozen controls (0.246 mgg^{-1} fresh weight).

Rao et al. (1998) transformed several parts of artemisia seedlings with *A. rhizogenes* and produced hairy root cultures. Cultures kept in the dark produced no artemisinin, but artemisinin was detected under a 16-h photoperiod. Smith et al. (1997) reported that low concentrations of GA_3 increased artemisinin production in transformed root cultures, and that a maximum artemisinin production of 4.0 mg/flask was obtained with GA_3 (0.01 mg l^{-1}), but production on a dry weight basis, or after transformation, has not been presented. All root cultures produced arteannuin B and artemisinic acid at all levels of GA_3, but GA_3 levels greater than 0.01 mg l^{-1} had a detrimental effect on artemisinin production.

Paniego and Giulietti (1996) produced transformed shoot cultures with *A. tumesfasciens*, and root cultures with *A. rhizogenes*. Artemisinin content averaged 0.018% (dry weight) for transformed shoots, and 0.010% for transformed roots, whereas artemisinin content in the original flowering plants and in non-transformed cultures was ca. 0.1%. Addition of Ca^{2+}, Mg^{2+}, and PO_4^{3-} failed to increase artemisinin, but GA_3 (5–10 ppm) increased artemisinin in shoots by 300–400%. This success lasted only for one culture cycle because GA_3 caused vitrification and loss of the clones. The transformed shoots (20% success) retained normal morphology, aromaticity, and the capacity to produce artemisinin, but they lost apical dominance. Aromaticity in those shoot cultures suggests the presence of glandular trichomes, the normal site of accumulation of essential oils, artemisinin, and artemisitene (Duke et al. 1994; Ferreira and Janick 1995). Ghosh et al. (1997) transformed *A. annua* with *Agrobacterium tumefasciens* and detected artemisinin (0.06%) in shooty teratomas. However, transformed but undifferentiated gall callus did not produce artemisinin.

These reports indicate the feasibility of transforming artemisia with agrobacteria. The production of artemisinin in roots of these transformed plants was serendipitous and not the result of transformation with a cloned gene involved in artemisinin biosynthesis. A first step towards the understanding of genetic control of artemisinin biosynthesis has been reported by Matsushita et al. (1996) who cloned a cDNA encoding farnesyl diphosphate synthase (FPPS) from a cDNA library of *Artemisia annua*. Insertion of the DNA encoding for FPPS in *E. coli* resulted in measurable levels of farnesyl diphosphate.

2.4 Structure and Extraction of Artemisinin

Artemisinin (Fig. 1) is an odorless, non-volatile compound, which is purified as white crystals with a melting point of 156–157 °C. Its molecular weight is m/e 282.1742 M+ (Luo and Shen 1987). It has an empirical formula of $C_{15}H_{22}O_5$ without chromophores; thus no absorption in most of the UV range.

Artemisinin can be easily extracted with petroleum ether, which boils from 30 to 60 °C or another miscible solvent such as chloroform, acetonitrile, and ether, which have a boiling point lower than the critical temperature for artemisinin stability. Artemisinin was established to be stable up to 150 °C in neutral solvents (Lin et al. 1985). The first published laboratory procedure for isolation of artemisinin was described by the late Dr. Daniel Klayman

(Klayman et al. 1984). Artemisinin can be detected and quantified using several methods, including HPLC, GC, TLC, radio and enzyme immunoassays (Ferreira et al. 1997). Duke et al. (1994) demonstrated that artemisinin could be extracted by briefly dipping plant tissues (fresh or dry) in miscible solvents. Extraction of artemisinin and artemisinic acid is also achieved by an improved method of supercritical CO_2, with optimal pressure of 15 MPa, temperature of 50 °C, methanol (or ethanol) concentration of 3%, flow rate of 2 ml/min, and extraction time of 20 min (Kohler et al. 1997). It has been suggested that artemisinin is produced and sequestered in heart-shaped glandular trichomes (Fig. 3) present mainly in leaves and flowers of *A. annua*, thus readily available for extraction without the need for tissue grinding (Duke et al. 1994; Ferreira and Janick 1995).

3 Conclusions and Prospects

Secondary plant metabolites result from the activity of several enzymes in a multi-step pathway. These enzymes are encoded by genes, which are activated at a certain time and in a specific tissue during the plant life cycle. Currently, the de novo synthesis of most compounds has not yet been mastered with profitable results. While results have been promising for some classes of compounds, such as alkaloids, only a few cDNAs have been cloned, which regulate the production of enzymes needed early in the sesquiterpene pathway. Artemisinin, a sesquiterpene lactone, is an efficient herbicide and esquizontocide, which requires over ten steps to be produced from raw materials. An alternative approach for artemisinin synthesis would be to clone genes regulating late enzymes in the pathway, and then to provide bacterial cultures with immediate precursors, such as artemisinic acid and/or arteannuin B. However, it is not clear if these cultures could perform all the steps required for artemisinin biosynthesis, even if they contained all the genes coding for the needed enzymes. The positive association of artemisinin with light, both in in vivo and in vitro cultures, indicates that its biosynthesis is associated with chloroplasts, lacking in bacteria. Also, if successfully produced in large amounts, artemisinin could very well quench its own biosynthesis, or kill the cultures, due to its anti-microbial effects.

Previous attempts to produce artemisinin by undifferentiated in vitro cultures have yielded inconsistent results. Recent studies have concentrated on the transformation of organs of the artemisia plant with *Agrobaterium rhizogenes* to produce hairy roots, or with *A. tumefasciens* to produce shooty teratomas. However, we suggest that some key factors have been overlooked:

1. Past successes in producing secondary metabolites by hairy root cultures in a number of species have been limited to compounds which are originally

produced by roots of differentiated plants. Roots of *Artemisia annua* do not normally produce artemisinin.
2. Hairy roots exposed to light might behave as "pseudoshoots." The detection of artemisinin in green hairy roots suggests an association of artemisinin production with chloroplasts.
3. Artemisinin is synthesized and stored in leaves, stems, flower receptacles, and the corolla, all bearing glandular trichomes. This suggests that artemisinin production is best approached through differentiated cultures or shooty teratomas, which have the potential to bear glandular trichomes or, at least, to exert photosynthetic activity.
4. Elicitation has been a successful approach to trigger the production of certain metabolites, particularly antifungal or antibacterial compounds. However, elicitors have not succeeded in triggering artemisinin production either in vivo or in vitro. This suggests that artemisinin function to the plant might not be as an antifungal or bactericidal compound.
5. The only phytohormone that appears to influence artemisinin production is gibberellic acid (GA_3) which is also known to promote flowering. Artemisinin production is reported to reach its peak close to or during full flowering, and GA_3 is reported to promote flowering in shoot cultures of *A. annua* (Gulati et al. 1996). Thus, gibberellins might be one of the biochemical signals produced by the plant, which leads to an increased production of artemisinin. The influence of gibberellins could be verified by further investigating the effect of GA_3 antagonizers, such as abscisic acid (ABA), on artemisinin production.
6. *Artemisia annua* is a highly cross-pollinating species, and plants derived from seeds will have highly variable levels of artemisinin. Preferably, high artemisinin-producing clones should be used for in vitro studies.

Although field production of artemisia is presently the most commercially feasible approach to produce artemisinin and related compounds, molecular studies with green hairy roots and shooty teratomas could contribute to our understanding of artemisinin biosynthesis. In addition to research on the genetic potential of artemisinin biosynthesis by *Artemisia annua*, factors that affect temporal (when artemisinin reaches its maximum) or spatial (where it is stored) accumulation must not be ignored.

Acknowledgements. We acknowledge the help of Rawleen White and Jerry L. Neal with the figures.

References

Acton N, Klayman DL, Rollman IJ (1985): Reductive electrochemical HPLC assay for artemisinin (quinghaosu). Planta Med 51:445–446

Anonymous (1979) Qinghaosu Antimalarial Coordinating Research Group. Antimalarial studies on qinghaosu. Chin Med J 92:811–816

Avery MA, Chong WKM, Jennings-White C (1992) Stereoselective total synthesis of (+)-artemisinin, the antimalarial constituent of *Artemisia annua* L. J Am Chem Soc 114:974–979

Cai G, Li G, Ye H, Li G (1995) Hairy root culture of *Artemisia annua* L. by Ri plasmid transformation and biosynthesis of artemisinin. Chin J Biotech 11:227–235

Delabays N, Benakis A, Collet G (1993) Selection and breeding for high artemisinin (qinghaosu) yielding strains of *Artemisia annua*. Acta Hort 330:203–207

Duke MV, Paul RN, Elsohly HN, Sturtz G, Duke SO (1994) Localization of artemisinin and artemisitene in foliar tissues of glanded and glandless biotypes of *Artemisia annua* L. Int J Plant Sci 155:365–372

Eilert U, Kurz WGW, Contabel F (1985) Stimulation of sanguinarine accumulation in *Papaver somniferum* cell cultures by fungal elicitors. Plant Physiol 119:65–76

Farooqi AHA, Shukla A, Sharma S, Khan A (1996) Effect of plant age and GA_3 on artemisinin and essential oil yield in *Artemisia annua* L. J Herbs Spices Med Plant 4:73–80

Ferreira JFS (1994) Production and detection of artemisinin in *Artemisia annua* L. PhD Thesis, Purdue Univ, West Lafayette, IN

Ferreira JFS, Janick J (1995) Floral morphology of *Artemisia annua* with special reference to trichomes. Int J Plant Sci 156:807–815

Ferreira JFS, Janick J (1996) Roots as an enhancing factor for the production of artemisinin in shoot cultures of *Artemisia annua*. Plant Cell Tissue Organ Cult 44:211–217

Ferreira JFS, Simon JE, Janick J (1995a) Developmental studies of *Artemisia annua*: flowering and artemisinin production under greenhouse and field conditions. Planta Med 61:167–170

Ferreira JFS, Simon JE, Janick J (1995b) Relationship of artemisinin content of tissue-cultured, greenhouse-grown, and field-grown plants of *Artemisia annua*. Planta Med 61:351–355

Ferreira JFS, Simon JE, Janick J (1997) *Artemisia annua*: botany, horticulture, pharmacology. Hort Rev 19:319–371

Gibson DM, Ketchum REB, Hirasuna TJ, Shuler ML (1995) Potential of plant cell culture for taxane production. In: Suffness M (ed) Taxol: science and applications. CRC Press Boca Raton, pp 71–95

Ghosh BS, Mukherjee S, Jha S (1997) Genetic transformation of *Artemisia annua* by *Agrobacterium tumefasciens* and artemisinin synthesis in transformed cultures. Plant Sci 122:193–199

Gulati A, Bharel S, Jain SK, Abdin MZ, Srivastava PS (1996) In vitro micropropagation and flowering in *Artemisia annua*. J Biochem Biotech 5:31–35

Jain DC, Mathur AK, Gupta MM, Singh AK, Verma RK, Gupta AP, Kumar S (1996) Isolation of high artemisinin-yielding clones of *Artemisia annua*. Phytochemistry 43:993–1001

Klayman DL (1989) Weeding out malaria. Nat Hist Oct:18–26

Klayman DL (1993) *Artemisia annua*: From weed to respectable antimalarial plant. In: Kinghorn AD, Balandrin MF (eds) Human medicinal agents from plants. Am Chem Soc Symp Ser. Washington, DC, pp 242–255

Klayman DL, Lin AJ, Acton N, Scovill JP, Hoch JM, Milhous WK, Theorides AD (1984) Isolation of artemisinin (qinghaosu) from *Artemisia annua* growing in the United States. J Nat Prod 47:715–717

Kohler M, Haerdi W, Christen P, Veuthey J-L (1997) Extraction of artemisinin and artemisininc acid from *Artemisia annua* L. using supercritical carbon dioxide. J Chromat 785:353–360

Laughlin JC (1994) Agricultural production of artemisinin - a review. Trans R Soc Trop Med Hig 88 (Suppl 1):21–22

Liersch R, Soicke H, Stehr C, Tüllner H-U (1986) Formation of artemisinin in *Artemisia annua* during one vegetation period. Planta Med 52:387–390

Lin AJ, Klayman DL, Hoch JM (1985) Thermal rearrangement and decomposition products of artemisinin (qinghaosu) J Org Chem 50:4504–4508

Luo XD, Shen CC (1987) The chemistry, pharmacology, and clinical applications of qinghaosu (artemisinin) and its derivatives. Med Res Rev 7:29–52

Madhusudanan K (1989) Mass spectral studies on artemisinin, dehydroartemisinin, and arteether. Indian J Chem 28B:751–754

Martinez BC, Staba J (1988) The production of artemisinin in *Artemisia annua* L. tissue cultures. Adv Cell Cult 6:69–87

Mathur AK, Kumar S (1996) Micropropagation of *Artemisia annua* via the inflorescence. J Herbs Spices Med Plant 4:61–71
Matsushita Y, Kang WK, Charlwood BV (1996) Cloning and analysis of a cDNA encoding farnesyl diphosphate synthase from *Artemisia annua*. Gene 172:207–209
McVaugh R (1984)Compositae. In: Anderson WR (ed) Flora Novo-Galiciana: a descriptive account of the vascular plants of western Mexico, vol 12. Univ Michigan Press, Ann Arbor
Nussenzweig RS, Long CA (1994) Malaria vaccines: multiple targets. Science 265:1381–1383
Pan American Health Organization (PAHO)/World Health Organization (WHO) (1998) Status of the malaria programs in the Americas, XLVI Report. 25th Pan American Sanitary Conference; 50th session of the regional committee. PAHO/WHO, Washington, DC, 33 pp
Paniego NB, Giulietti AM (1996) Artemisinin production by *Artemisia annua* L.-transformed organ cultures. Enz Microb Technol 18:1–5
Pras N, Visser JF, Batterman S, Woerdenbag HJ, Malingré TM, Lugt CB (1991) Laboratory selection of *Artemisia annua* L. for high artemisinin yielding types. Phytochem Anal 2:80–83
Prasad A, Kumar D, Anwar M, Singh DV, Jain DC (1997) Response of *Artemisia annua* L. to soil salinity. J Herbs Spices Med Plants 5:49–55
Rao KV, Venkanna N, Narasu ML (1998) *Agrobacterium rhizogenes* mediated transformation of *Artemisia annua*. J Sci Ind Res 57:773–776
Ravindranathan T, Kumar MA, Menon RB, Hiremath SV (1990) Stereoselective synthesis of artemisinin+. Tetrahedron Lett 31:755–758
Schmid G, Hofheinz W (1983) Total synthesis of qinghaosu. J Am Chem Soc 105:624–625
Singh A, Kaul VK, Mahajan VP, Singh A, Misra LN, Thakur RS, Husain A (1988) Introduction of *Artemisia annua* in India and isolation of artemisinin, a promising antimalarial drug. Indian J Pharm Sci 48:137–138
Siyapata-Ntakirutmana E, Jaziri M, Homes J, Diallo B, Vanhaelen M (1996) Action des gibberellines sur la production de biomasse et d'artemisinine par *Artemisia annua* L. cultivée en champs. Belg J Bot 129:24–32
Smith TC, Weathers PJ, Cheetham RD (1997) Effects of gibberellic acid on hairy root cultures of *Artemisia annua*: growth and artemisin production. In Vitro Cell Dev Biol 33:75–79
Tabata M, Mizukami H, Hiraoka N, Konoshima M (1974) Pigment formation in callus cultures of *Lithospermum erythrorhizon*. Phytochemistry 13:927–932
Teoh KH, Weathers PJ, Cheetham RD, Walcerz DB (1996) Cryopreservation of transformed (hairy) roots of *Artemisia annua*. Cryobiology 33:106–117
Thebtaranonth C, Thebtaranonth Y, Wanauppathamkul S, Yuthavong Y (1995) Antimalarial sesquiterpenes from tubers of *Cyperus rotundus*: structure of 10,12-peroxycalamanene, a sesquiterpene endoperoxide. Phytochemistry 40:125–128
Trigg PI (1989) Qinghaosu (artemisinin) as an antimalarial drug. Econ Med Plant Res 3:19–55
Wang C-W (1961) The forests of China, with a survey of grassland and desert vegetations. In: Harvard Univ Maria Moors Cabot Foundation 5 Harvard Univ, Cambridge, MA, pp 171–187
World Health Organization (WHO)/Malaria Unit (MAL) (1998) The use of artemisinin and its derivatives as antimalarial drugs. In: WHO/MAL (ed) CTD/DMP/TDR. Geneva, Switzerland
Woerdenbag JH, Pras N, Uden WV, Wallaart TE, Beekman AC, Lugt CB (1994) Progress in the research of artemisinin-related antimalarials: an update. Pharmacy World Sci 16:169–180
Xu XX, Zhu J, Huang DZ, Zho WS (1986) Total synthesis of arteannuin and deoxyarteannuin. Tetrahedron 42:819–828
Zhao KC, Zeng MY (1985) Spektrometrische hochdruck-flussigkeits-chromatographische (HPLC) Untersuchungen zur Analytik von quinghaosu. Planta Med 51:233–237
Ziffer H, Highet RJ, Klayman DL (1997) Artemisinin: an endoperoxidic antimalarial from *Artemisia annua* L. Progr Chem Org Nat Prod 72:121–214

2 In Vitro Culture and the Production of Secondary Metabolites in *Coriandrum sativum* L. (Coriander)

J.R. Liu, S.W. Kim, and S.C. Oh[1]

1 Introduction

Coriander (*Coriandrum sativum* L.) is an annual herb that belongs to the Umbelliferae. Major producers are Morocco, Canada, India, Pakistan, Rumania, and the former Soviet Union. The coriander plant yields both the fresh green herb and the spice seed. The green herb is one of the most commonly used flavorings in the world and is a key component to curries. The seed has a distinctive sweet citrus, mint, musty aroma that has been valued over the centuries (Anonymous 1998; Grieve 1999).

The seed contains 13–18% dry weight of oil, of which up to 75% can be petroselenic acid. Petroselinic acid is the *cis* Δ^6 isomer of octadecenoic acid and the major fatty acid of the seed oil of most Umbelliferae, Araliaceae, and Garryaceae species. Although it shares a structural similarity with the widely occurring oleic acid (*cis* Δ^9 octadecenoic acid), petroselinic acid has distinct properties resulting from the novel position of its double bond. *cis* Δ^9 octadecenoic acid (oleic acid) is widely found in oil crops such as rapeseed, but one of its isomeric forms, *cis* Δ^6 octadecenoic acid (petroselenic acid), is rare (Cahoon et al. 1994). The single double bond is moved from the nine to the six position in this form and the difference in structure and the unique position of this double bond allows potential oxidative cleavage to form a C12 lauric acid with industrial usage in soaps and detergents and also a C6 dicarboxylic acid for use as a feedstock in the manufacture of nylon (Millam et al. 1997).

Despite its commercial importance, only a few in vitro culture studies have been published. These studies include reports of plant regeneration via somatic embryogenesis and production of petroselinic acid in suspension cultures (Kim et al. 1996a).

[1]Plant Cell Biotechnology Laboratory, Korea Research Institute Bioscience and Biotechnology, 52-Oun-dong, Yusong, Taejon, 305-333 Korea

Biotechnology in Agriculture and Forestry, Vol. 51
Medicinal and Aromatic Plants XII (ed. by T. Nagata and Y. Ebizuka)
© Springer-Verlag Berlin Heidelberg 2002

2 Plant Regeneration Via Somatic Embryogenesis

2.1 Plant Material and Callus Induction

Coriander seeds (*Coriandrum sativum* L.) were surface-sterilized with 70% ethanol for 1 min, and immersed in a 0.4% sodium hypochlorite solution for 10 min. They were then rinsed four times with sterile distilled water. The basal medium used throughout the experiments consisted of Murashige and Skoog's (1962) inorganic salts, 100 mg/l myo-inositol, 0.4 mg/l thiamine· HCl, 3% sucrose, and 0.4% Gelrite (MS medium). The pH of all media was adjusted to 5.8 before autoclaving. Twenty-five ml of medium was dispensed into 87 × 15 mm plastic petri dishes.

In each petri dish, ten seeds were placed onto MS basal medium and incubated at 25 °C in the dark. Two-week-old seedlings and zygotic embryos excised from imbibed seeds were used as the source of explants. Transversely sliced cotyledon, hypocotyl, and root segments (approximately 2.5 mm in length) of seedlings and intact zygotic embryos (approximately 1 mm in length) were placed onto MS medium supplemented with 0.1, 0.3, 1, or 3 mg/l 2,4-dichlorophenoxyacetic acid (2,4-D). Three petri dishes per treatment with ten explants each were cultured. After 4 weeks of culture, the number of explants with embryogenic calluses were counted to determine the frequency of embryogenic callus formation. Embryogenic calluses were distinguished from nonembryogenic calluses by the presence of somatic embryos on the surface at early developmental stages. Embryogenic calluses detached from the explants were subcultured on MS medium with 1 mg/l 2,4-D (MS1D) every 4 weeks. Unless mentioned otherwise, all cultures were maintained at 25 °C in the dark.

2.2 Initiation and Proliferation of Embryogenic Cell Suspension Cultures

Subcultured hypocotyl-derived embryogenic calluses (approximately 1 g) were placed in a 300-ml Erlenmeyer flask containing 50 ml of liquid MS1D medium. Cell suspension cultures were maintained on a gyratory shaker (100 rpm) and subcultured every 2 weeks.

2.3 Plant Regeneration

Cell aggregates (1–2 mm in diameter) maintained for 2 weeks in liquid MS1D medium were collected on stainless steel mesh (pore size: 1 mm) and rinsed with MS basal medium. Twenty cell aggregates were plated on MS basal medium in a petri dish. After 4 weeks of culture the regeneration frequency was determined from four petri dishes. To enhance the regeneration frequency, cell suspension cultures were transferred to MS medium supplemented with 4g/l KNO_3, 0.29g/l NH_4NO_3, 3mg/l thiamine HCl, 0.5mg/l pyridoxine HCl, 5mg/l nicotinic acid, 2% sucrose, and 0.1mg/l 2,4-D (SC-1) (Jeon et al. 1986) and cultured 2 weeks before plating onto MS basal medium. The regenerants were transplanted to potting soil and maintained in a phytotron (27°C day/22 °C night at 16-h photoperiods).

2.4 Results and Discussion of Plant Regeneration via Somatic Embryogenesis

After 2 weeks of culture, calluses began to form on the cut surface of all explants regardless of explant source and 2,4-D concentration. After 4 weeks of culture, white compact calluses were formed on cotyledon, hypocotyl, and zygotic embryos, but not on root segments. The calluses gave rise to numerous somatic embryos as the culture proceeded, indicating that they were embryogenic (Fig. 1A, B). The highest frequencies of embryogenic callus formation on hypocotyl, cotyledon, and zygotic embryo explants were 75, 30, and 77%, respectively (Fig. 2). However, root explants were incompetent and did

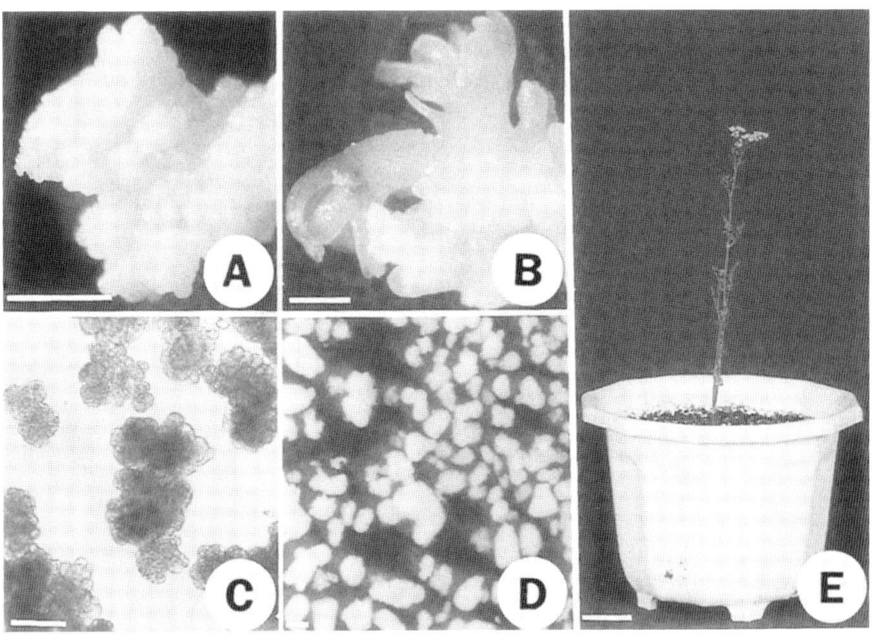

Fig. 1. Plant regeneration of coriander via somatic embryogenesis. **A** Hypocotyl-derived embryogenic callus cultured on MS medium with 1mg/l 2,4-D (*bar* 2mm). **B** Shoot differentiation from somatic embryos (*bar* 2mm). **C** Embryogenic cell suspension cultures of coriander (*bar* 100µm). **D** Somatic embryos developed from cell suspension cultures (*bar* 100µm). **E** Flowering plant regenerated from somatic embryo (*bar* 3mm)

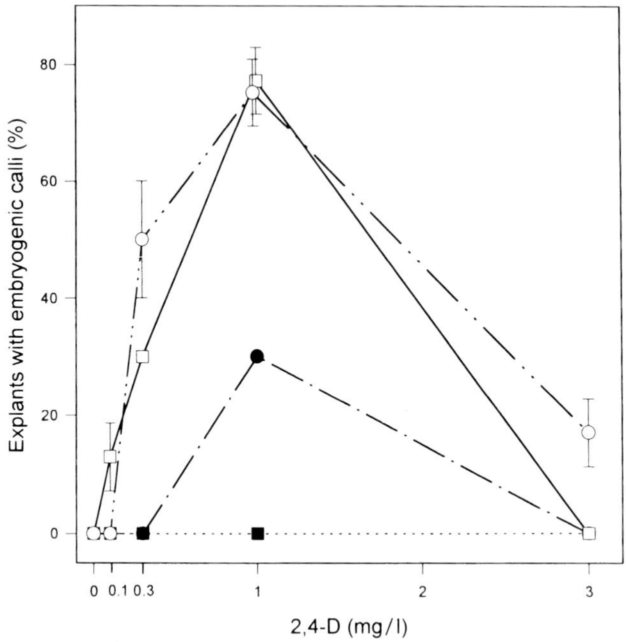

Fig. 2. Effects of 2,4-D concentration on embryogenic callus formation on cotyledon (*filled circles*), hypocotyl (*open circles*), root (*filled squares*), and zygotic embryo explants (*open squares*). Each treatment consisted of ten explants with three replicates. *Vertical bars* represent SD

not produce embryogenic calluses. The three different explants exhibited an optimum concentration of 2,4-D for embryogenic callus formation at 1 mg/l (Fig. 2).

Hypocotyl-derived embryogenic calluses were used to initiate cell suspension cultures. Embryogenic cell suspension cultures were primarily composed of small isodiametric meristematic cells (approximately 10 μm in diameter) containing a prominent nucleus and dense cytoplasm with numerous vacuoles (Fig. 1C). Of cell aggregates maintained in MS1D medium 19% gave rise to numerous globular to cotyledonary embryos when transferred to MS basal medium (Fig. 1D).

Cell aggregates precultured in SC-1 medium gave rise to numerous somatic embryos at an approximate frequency of 75%. Approximately 70% of somatic embryos developed into plantlets. Thus, preculture in SC-1 medium resulted in a fourfold increase in regeneration frequency when compared with the treatment without preculture. Also, the precultured cell aggregates produced plantlets approximately 2 weeks earlier. The major difference between MS1D medium and SC-1 medium is nitrogen ion content. It is uncertain how nitrogen sources influence cell growth and plant regeneration. However, similar effects of nitrogen source on plant regeneration were reported in rice. It has also been reported that the inorganic $NO_3^-:NH_4^+$ ratio alters cell sensitivity to auxin and regulation of hormone metabolism in rice (Grimes and Hodges 1991; Jeong et al. 1991).

Fifty plants were regenerated from somatic embryos. Ten of them were successfully transplanted to potting soil and grown to maturity (Fig. 1E).

3 Production of Petroselinic Acid from Cell Suspension Cultures

3.1 Plant Materials

Transversely sliced cotyledon, hypocotyl, and root segments (approximately 2.5 mm in length) of 2-week-old seedlings and intact zygotic embryos (approximately 1.5 mm in length) were placed onto Murashige and Skoog's (1962) medium containing 100 mg/l myo-inositol, 0.4 mg/l thiamine· HCl, 3% sucrose, 0.4% Gelrite, and 1 mg/l 2,4-D (MS1D). After 4 weeks of culture, calluses which formed on cut surfaces of explants were subcultured on MS medium with 1 mg/l 2,4-D (MS1D) every 4 weeks. To establish cell suspension cultures, subcultured hypocotyl-derived embryogenic calluses (approximately 1 g) were placed in a 300-ml Erlenmeyer flask containing 50 ml of liquid MS1D medium (Kim et al. 1996b). Cell suspension cultures were maintained on a gyratory shaker (100 rpm) and subcultured every 2 weeks. All cultures were maintained at 25 °C in the dark.

3.2 Analysis of Fatty Acids

The fatty acid pattern and the petroselinic acid content in calluses derived from various explants, including cotyledon, hypocotyl, root, zygotic embryo, and cell suspension cultures were analyzed by the method of Miller and Berger. Fatty acid methyl esters were prepared from calluses and somatic embryos after 4 weeks of culture on MS1D medium, and from 2-week-old cell suspension cultures (fresh weight: 200 mg) by heating at 100 °C for 30 min in a saponification solution (15% NaOH in 50% methanol). Methyl ester formation of the fatty acids was performed by addition of methanolic HCl (6 N HCl/methanol), followed by incubation at 80 °C for 10 min. Resulting methyl esters were transferred from the aqueous phase to an organic phase by solvent (hexane/MTBE) extraction, washed with a dilute NaOH solution, and analyzed by GC using a 25 × 0.2 mm Ultra 2 silica capillary column and oven temperature programming from 170 to 270 °C at 5 °C/min. Quantitative analysis of petroselinic acid was carried out by comparing the peak areas of the samples with that of the authentic petroselinic acid.

Table 1. Yield of petroselinic acid produced in calluses and cell suspension cultures and total content of saturated and unsaturated fatty acids

Callus origin[a]	Saturated fatty acid (%)	Unsaturated fatty acid (%)	Petroselinic acid (%)	Yield of petroselinic acid (mg/g)
ZE	19.0	81.0	0.29	0.20
R	18.2	81.8	0.33	0.19
H	21.4	78.6	0.46	0.26
C	20.0	80.0	0.25	0.15
SE	13.9	86.1	0.32	0.23
S	13.3	86.7	0.3	0.23

[a] ZE, Zygotic embryo; R, root; H, hypocotyl; C, cotyledon; SE, somatic embryo; S, cell suspension cultures.

3.3 Results and Discussion of Production of Petroselinic Acid in Suspension Cultures

Fatty acids composition and petroselinic acid content were shown in Fig. 1 and Table 1, respectively. The major fatty acids produced in coriander were palmitic acid, oleic acid, and α- and γ-linoleic acid. The patterns of fatty acid production in calluses derived from various explants, including cotyledon, hypocotyl, root and zygotic embryo, were all similar. However, the pattern of fatty acid production in cell suspension cultures was slightly different from that of calluses. Petroseliadic acid, elaidic acid, and *cis*-vaccenic acid, which were the common unsaturated fatty acids produced in calluses, were not detected in cell suspension cultures. Instead, 16:0 *anteiso*, 16:1w9c and unidentified fatty acid were produced. Furthermore, the unsaturated:saturated fatty acid ratio was approximately 80:20 in calluses. However, the unsaturated fatty acid content was increased in cell suspension cultures and somatic embryos. The petroselinic acid content was 0.15–0.23 mg/g fresh wt. in calluses, somatic embryo, and cell suspension cultures. The petroselinic acid content was not significantly changed, regardless of callus origin or culture conditions.

Petroselinic acid comprises as much as 85% of the total fatty acid content of Umbelliferae seeds, but it is virtually absent from leaves and other tissues of these plants (Kleiman and Spencer 1982). Petroselinic acid is metabolized and accumulated in the developing endosperm of some Umbelliferae species, including coriander and carrot (Cahoon and Ohlrogge 1994b). Petroselinic acid is the product of acyl-acyl carrier protein (ACP) desaturase. This polypeptide is highly expressed in seed. However, it is absent in tissues that do not synthesize petroselinic acid, including leaves and roots of coriander (Cahoon et al. 1993). In this study, however, a small amount (approximately 0.20 mg g^{-1} fresh wt.) of petroselinic acid was detected in calluses, somatic embryos and cell suspension cultures. The yield of petroselinic acid in calluses was much lower than that of seed endosperm in coriander (Cahoon and Ohlrogge 1994a).

This result indicates that petroselinic acid can be continuously synthesized at a low level in cultured cells. However, the petroselinic acid content between somatic and zygotic embryos was not compared in this study.

Recent increased interest in plant fatty acid metabolism has been stimulated by the potential to design new oilseed crops which produce higher value oils (Cahoon et al. 1994; Topfer et al. 1995). Modification of fatty acid composition in seeds of transgenic oil crops has been accomplished in soybean (Cahoon et al. 1999), canola (Facciotti et al. 1999), and tobacco (Cahoon et al. 1993; Cahoon and Ohlrogge 1994a; Hamada et al. 1998) by introduction of new genes or inhibition of existing enzyme activities using antisense RNA technology (Cahoon et al. 1993; Kinney et al. 1993). The plant regeneration system described in this study was successfully used for genetic transformation of coriander (Oh et al. 2002).

4 Summary and Conclusions

Hypocotyl segments and zygotic embryos of coriander formed embryogenic calluses at frequencies of up to 75% when cultured on MS medium supplemented with 1 mg/l 2,4-D. The calluses were transferred to MS liquid medium with 1 mg/l 2,4-D to initiate cell suspension cultures. Embryogenic cells became finely dispersible in the medium as the subculture proceeded. Cultures were transferred to a nitrogen compound-enriched liquid MS medium containing 2% sucrose and 0.1 mg/l 2,4-D, and cultured 2 weeks before plating on MS basal medium. Approximately 75% of cell aggregates (1–2 mm in diameter) underwent development into globular to cotyledonary somatic embryos after 2 weeks of plating. Most of the embryos were subsequently regenerated into plantlets. Regenerants were successfully transplanted to potting soil and grown to maturity in a phytotron.

The fatty acid pattern and petroselinic acid content in calluses, embryos and cell suspension cultures of *Coriander sativum* are described. The petroselinic acid content was 0.15–0.23 mg/g fresh wt. of calluses, somatic embryo and cell suspension cultures and varied little with callus origin and culture conditions. The ratio of unsaturated: saturated fatty acid was approximately 4:1 in calluses. However, the unsaturated fatty acid content was greater in somatic embryos and cell suspension cultures.

5 Protocol for Plant Regeneration Via Somatic Embryogenesis

1. For surface sterilization, soak seeds of coriander in 70% ethanol for 1 min, then in a 0.4% sodium hypochlorite solution for 10 min.
2. Rinse them four times with sterile distilled water.

3. Dispense (25 ml) the basal medium consisting of MS inorganic salts, 100 mg/l myo-inositol, 0.4 mg/l thiamine· HCl, 3% sucrose, and 0.4% Gelrite (MS medium) at pH 5.8 into a 15-mm petri dish (87 × 15 mm).
4. Place seeds into petri dishes containing MS basal medium (ten seeds per dish) and incubate them at 25 °C in the dark.
5. Two-week-old seedlings and zygotic embryos excised from imbibed seeds were used as the source of explants. Place transversely sliced cotyledon, hypocotyl, and root segments (approximately 2.5 mm in length) of seedlings and intact zygotic embryos (approximately 1 mm in length) onto MS medium supplemented with 0.1, 0.3, 1, or 3 mg/l 2,4-D. Three petri dishes per treatment with ten explants each were cultured.
6. After 4 weeks of culture, count the number of explants with embryogenic calluses to determine the frequency of embryogenic callus formation. Embryogenic calluses are distinguished from nonembryogenic calluses by the presence of somatic embryos on the surface at early developmental stages.
7. Detach embryogenic calluses from the explants with forceps and subculture them on MS medium with 1 mg/l 2,4-D (MS1D) every 4 weeks.
8. Place subcultured hypocotyl-derived embryogenic calluses (approximately 1 g) in a 300-ml Erlenmeyer flask containing 50 ml of liquid MS1D medium.
9. Subculture cell suspension cultures every 2 weeks and maintain them on a gyratory shaker (100 rpm).
10. Collect cell aggregates (1–2 mm in diameter) maintained for 2 weeks in liquid MS1D medium with a stainless steel mesh (pore size: 1 mm) and rinse them with liquid MS basal medium.
11. Transfer cell suspension cultures to MS medium supplemented with 4 g/l KNO_3, 0.29 g/l NH_4NO_3, 3 mg/l thiamine HCl, 0.5 mg/l pyridoxine· HCl, 5 mg/l nicotinic acid, 2% sucrose, and 0.1 mg/l 2,4-D (SC-1) (Jeon et al. 1986) and culture them for 2 weeks.
12. Plate 20 cell aggregates on MS basal medium in a petri dish.
13. After 4 weeks of culture, determine the regeneration frequency from four petri dishes.
14. Transplant the regenerants to potting soil and maintain them in a phytotron (27 °C day/22 °C night at 16-h photoperiods).

6 Protocol for Production of Petroselenic Acid from Cell Suspension Cultures

1. Place transversely sliced cotyledon, hypocotyl, and root segments (approximately 2.5 mm in length) of 2-week-old seedlings and intact zygotic embryos (approximately 1.5 mm in length) onto MS medium containing

100 mg/l myo-inositol, 0.4 mg/l thiamine· HCl, 3% sucrose, 0.4% Gelrite, and 1 mg/l 2,4-D (MS1D).
2. After 4 weeks of culture, detach calluses which formed on cut surfaces of explants with forceps and subculture on MS medium with 1 mg/l 2,4-D (MS1D) every 4 weeks.
3. To establish cell suspension cultures, place subcultured hypocotyl-derived embryogenic calluses (approximately 1 g) in a 300-ml Erlenmeyer flask containing 50 ml of liquid MS1D medium (Kim et al. 1996b).
4. Maintain cell suspension cultures on a gyratory shaker (100 rpm) at 25 °C in the dark and subculture every 2 weeks.
5. Analyze the fatty acid pattern and the petroselinic acid content in calluses derived from various explants, including cotyledon, hypocotyl, root, zygotic embryo, and cell suspension cultures by the method of Miller and Berger (1985).
6. Prepare fatty acid methyl esters from calluses and somatic embryos after 4 weeks of culture on MS1D medium, and from 2-week-old cell suspension cultures (fresh wt.: 200 mg) by heating at 100 °C for 30 min in a saponification solution (15% NaOH in 50% methanol).
7. Add methanolic HCl (6 N HCl/methanol) to the solution, then incubate it at 80 °C for 10 min.
8. Add solvent (hexane/MTBE) to the methyl esters (aqueous phase), then collect organic phase. Wash the organic phase with a dilute NaOH solution, and analyze the organic phase by GC using a 25 m × 0.2 mm Ultra 2 silica capillary column and oven temperature programming from 170 to 270 °C at 5 °C/min.
9. Analyze petroselinic acid quantitatively by comparing the peak areas of the samples with that of the authentic petroselinic acid.

References

Anonymous (1998) Coriander. http://www.agric.gov.ab.ca/crops/special/coriander.html

Cahoon EB, Ohlrogge JB (1994a) Metabolic evidence for the involvement of a Δ^4-palmitiyl-acyl carrier protein desaturase in petroselinic acid synthesis in coriander endorsperm and transgenic tobacco cells. Plant Physiol 104:827–837

Cahoon EB, Ohlrogge JB (1994b) Apparent role of phosphatidylcholine in the metabolism of petroselinic acid in developing Umbelliferae endosperm. Plant Physiol 104:845–855

Cahoon EB, Shanklin J, Ohlrogge JB (1993) Expression of a coriander desaturase results in petroselinic acid production in transgenic tobacco. Proc Natl Acad Sci USA 89:11184–11188

Cahoon EB, Dormann P, Ohlrogge JB (1994) Petroselinic acid biosynthesis and production in transgenic plants. Prog Lipid Res 33:155–163

Cahoon EB, Carlson TJ, Ripp KG, Schweiger BJ, Cook GA, Hall SE, Kinney AJ (1999) Biosynthetic origin of conjugated double bonds: production of fatty acid components of high-value drying oils in transgenic soybean embryos. Proc Natl Acad Sci USA 96:12935–12940

Facciotti MT, Bertain PB, Yuan L (1999) Improved stearate phenotype in transgenic canola expressing a modified acyl-acyl carrier protein thioesterase. Nat Biotechnol 17: 593–597

Grieve M (1999) Coriander. http://www.botanical.com/botanical/mgmh/c/corian99.html

Grim HD, Hodges TK (1991) The inorganic NO_3^-:NH_4^+ ratio influences plant regeneration and auxin sensitivity in primary callus derived from immature embryos of indica rice (*Oryza sativa* L.). J Plant Physiol 136:362–367

Hamada T, Kodama H, Takeshita K, Utsumi H, Iba K (1998) Characterization of transgenic tobacco with an increased linolenic acid level. Plant Physiol 118:591–598

Jeon JH, Liu JR, Yang SG, Lee HS, Joung H, Han MH (1986) Development of a model system for artificial seed production I. Encapsulation of somatic embryos by alginic acid. Korean J Plant Tissue Cult 18:209–214

Jeong WJ, Song NH, Min SR, Kim MK, Liu JR (1991) Effect of ABA and the total inorganic nitrogen content on plant regeneration from cultured cells of rice (*Oryza sativa* L. cv. Taebaegbyeo). Korean J Plant Tissue Cult 18:209–214

Kim SW, Park MK, Bae KS, Rhee MS, Liu JR (1996a) Production of petroselinic acid from cell suspension cultures of *Coriandrum sativum*. Phytochemistry 42:1581–1582

Kim SW, Park MK, Liu JR (1996b) High frequency plant regeneration via somatic embryogenesis in cell suspension cultures of coriander (*Coriandrum sativum* L.). Plant Cell Rep 15:751–753

Kinney AJ, Hitz WD, Yadav NS, Perez-Grau L (1993) Genes of fatty acid biosynthesis in developing oilseeds. Plant Physiol (Suppl 1) 102:1

Kleiman R, Spencer GF (1982) Search for new industrial oils. XVI. Umbelliflorae-seed oils rich in petroselinic acid. J Am Oil Chem Soc 59:29–38

Millam S, Mitchell S, Craig A, Paoli M, Moscheni E, Angelini L (1997) In vitro manipulation as a means for accelerated improvement of some new potential oil crop species. Ind Crops Prod 6:213–219

Miller LT, Berger T (1985) Hewlett-Packard application note 228. MID Inc, Delaware, USA

Murashige T, Skoog F (1962) A revised medium for rapid growth and bioassays with tobacco tissue culture. Physiol Plant 15:473–497

Oh SC, Kim SW, Suh MJ, Liu JR (2002) *Agrobacterium*-mediated genetic transformation of Coriander. Plant Cell Rep (submitted)

Topfer R, Martini N, Schell J (1995) Modification of plant lipid synthesis. Science 268:681–686

3 Polyphenol Production in *Crataegus* Tissue Cultures (Hawthorn)

T. BAHORUN[1], F. TROTIN[2,3], and J. VASSEUR[3]

1 Introduction

1.1 Distribution and Importance of the Plant

The genus *Crataegus* originates from northern temperate regions and comprises about 280 species depending upon botanical sources (Grisvard et al. 1964; Mabberley 1987). It is widespread in western Asia, North America and in Europe, where 21 species have been indexed (Tutin et al. 1968). Hybridization is frequent in this genus (Mabberley 1987), causing many identification difficulties. Among the best-known species are *Crataegus monogyna* Jacq. and *Crataegus laevigata* (Poiret) DC (= *C. oxyacantha* auct.= *C. oxyacanthoides* Thuill.) belonging to the Eurasian subgenus *Crataegus* which are most common and are of growing pharmaceutical use. They have lobed or divided leaves, while the northern American subgenus *americanae* is characterized by entire or weakly lobed leaves (Mabberley 1987).

These deciduous, ramified, thorny shrubs produce white-pinkish scented flowers in corymbs during midsummer and the resulting red fruits remain on the trees until eaten by birds in winter (Courtney and Manzur 1985). They show some differences in habitat: *C. monogyna* is found mainly in open fields, thickets and used as quickset hedges, while *C. laevigata* prefers woods or their borders. Ornamental pink- or red-flowered cultivars of *C. laevigata* such as "*Coccinea plena*", "*Rosea*", "*Rubra*" (Grisvard et al. 1964) are planted in avenues, parks and gardens.

C. laevigata is characterized by ovobate leaves, with 3–5 obtuse few profound lobes, 2–3 styled flowers, red anthers and a fruit with 2–3 seeds. *C. monogyna* shows 3–7 more acute lobes on the leaves, more pleasantly perfumed one-styled flowers with black anthers and a one-seeded fruit (Coste 1937). *C. monogyna*, considered as a Euro-Siberian species (Rameau et al.

[1]Department of Biological Sciences, Faculty of Science, University of Mauritius, Republic of Mauritius
[2]Laboratoire de Pharmacognosie, Faculté des Sciences Pharmaceutiques et Biologiques, B.P. 83, 59006 Lille Cedex, France
[3]Laboratoire de Physiologie et de Morphogénèse Végétales, Université des Sciences et Technologies de Lille, 59655 Villeneuve d'Ascq Cedex, France

Biotechnology in Agriculture and Forestry, Vol. 51
Medicinal and Aromatic Plants XII (ed. by T. Nagata and Y. Ebizuka)
© Springer-Verlag Berlin Heidelberg 2002

1989), covers a wide area in Europe (Oberdorfer 1983) from Sweden to Russia and Greece and was introduced into northern America, while *C. laevigata* is a sub-Atlantic species found in western and central Europe.

Unripe hawthorn fruits are astringent and unpleasant but ripe fruits are edible; generally insipid except those of *C. azarolus* which are acidulous and aromatic. Seeds have been found in European prehistoric lacustrine sites, while a flour made from the fruit (preferably cooked before being ground; Couplan and Styner 1994) was used in a kind of bread in central Europe until the mid-20th century (Fournier 1947).

Nevertheless, hawthorn, besides its wide horticultural use, should mainly be considered a pharmaceutical plant. Hawthorn teas and pharmaceutical preparations belong to the most commonly studied drugs of plant origin, which are widely prescribed or used in self-medication (Kurzmann and Schimmer 1996). Many Pharmacopoeias index indiscriminately both *C. monogyna* and *C. laevigata* parts, mainly flowers or leaves with flowers (flowering tops), and sometimes fruits. The third edition of the European Pharmacopoeia includes the following five *genera* as pharmaceutical sources: *C. laevigata* and *C. monogyna* are more commonly used, while *C. pentagyna* Waldst. et Kit. ex Willd., *C. nigra* Waldst. et Kit., and *C. azarolus* L. (Wichtl and Anton 1999) have less frequent utilizations.

Pharmacological properties mainly focus on the cardiovascular effects (Ammon and Kaul 1994a,b,c) and sedative actions (Wichtl and Anton 1999). Hawthorn is officially recognized by health authorities in different countries for its action against mild cardiac insufficiency not requiring digitalis glycosides (stages I–II of the New York Heart Association, NYHA) (Weihmayr and Ernst 1996), against mild bradyarrhythmias in the elderly, and also as a sedative in neurotony including sleep and common anxiety disorders.

Its action on the heart has received special attention in Germany. It is mainly characterized by a positive inotropic action (enhanced myocardial contractility) (Ammon and Kaul 1994b), partly due to inhibition of c-AMP and c-GMP phosphodiesterases (Weihmayr and Ernst 1996) but also to other mechanisms. This was demonstrated on heart preparations or in animals for extracts (water, alcoholic), plant powder, flavonoid-rich fractions, and isolated flavonoids such as vitexin-2′-*O*-rhamnoside, luteolin-7-*O*-glucoside, rutin, quercetin (Ammon and Kaul 1994b; Schüssler et al. 1995), while hyperoside gave either negative or positive effects according to the system used. It must be noted that flavanolic fractions, such as total, polymeric and more particularly oligomeric proanthocyanidins exhibit an important influence.

Other important cardiac properties include a neutral (Loew 1997), or more often, moderately positive chronotropic action (heart rate increase) (Blesken 1992), in addition to an enhancement of atrio-ventricular conductibility (positive dromotropic action (Weihmayr and Ernst 1996). A few cases of heart rhythm attenuation have been reported (Wichtl and Anton 1999).

Another well-known effect of *Crataegus* is the increase in coronary flow observed in the perfused heart. Other parameters, such as improvement of the heart ejection fraction (Schüssler et al. 1995; Weihmayr and Ernst 1996), the occurrence of a mild anti-arrhythmic effect and a decrease in vascular peripheral resistance, all result in better cardiovascular performance. Consequently, a better tolerance to oxygen deficiency and exercise in clinical studies is observed (Tauchert et al. 1994).

Moreover, oral administration of the standardized extract induces a significant decrease in mortality after ischemia reperfusion in animals (Chatterjee et al. 1997). An aqueous fraction of this extract, rich in low-molecular oligomeric procyanidins and poor in flavonoids, is significantly more active against ischemia-reperfusion. This fraction also proved to be a more efficient free radical scavenger and elastase inhibitor than the flavonoids alone.

These protective activities have frequently been attributed to polyphenols. Antioxidant or Free radical scavenging properties have been demonstrated for many flavonoids (Rice-Evans et al. 1995; Van Acker et al. 1996), for flavanols and procyanidins (Hatano et al. 1989; Ricardo Da Silva et al. 1991). We reported important anti-lipoperoxidative actions in extracts of fresh hawthorn organs, especially floral buds, young leaves and unripe fruits (Bahorun et al. 1994b), clearly correlated to high phenolic contents. This led us to study phenolics and their action in hawthorn tissue cultures.

1.2 Conventional Practices for the Propagation and Production of Pharmaceuticals, and Demand on the World Market

1.2.1 Propagation

Propagation of specific cultivars such as ornamental ones can be made by grafting, for example on *C. laevigata* (Grisvard et al. 1964). It should be considered that exposure to cold in winter is necessary for the very slow seed germination process, which can take up to 18 months (Morris 1997). This delay can be reduced by scarification, diluted sulfuric acid, fermentation treatments, or by the early sowing of unripe fruits.

Crataegus belongs to the Rosaceae, which together with other genera like *Chaenomeles*, *Sorbus*, *Pyrus*, *Malus*, *Cydonia*, *Cotoneaster*, and *Pyracantha* are susceptible to fire blight (*Erwinia amylovora*). This is responsible for spectacular damage to fruit trees or its transmission. In France, a decree (24 December 1994) officially prohibits "the importation, propagation, marketing and planting" of different species known as vectors of this disease. This includes "all species and cultivars of *Crataegus*" besides some varieties of *Cotoneaster*, *Pyracantha*, *Pyrus*, and *Malus*.

In north-eastern France, however, the decree of 1994 authorizes, by special demand, the sowing and planting of hawthorn by professionals in regions neighboring Benelux and Germany where its propagation is not prohibited.

1.2.2 Production of Pharmaceuticals, Demand on the World Market

Hawthorn is not a source of isolated individual pharmaceutical molecules. Moreover, the major therapeutically used flavonoids are ubiquitous molecules, rather easy to extract, found in high yields in other plant sources (rutin, citro-flavonoids, diosmin).

Flowers and flowering tops are often used in herbal teas, especially in Europe, the fruits are used less often – except in the USA. The plant powder (dried or freeze-dried) as well as spray-dried extracts are commonly used in capsules sold as OTC or under medical prescription. The dried plant powder is an extraction product used for the manufacture of galenical forms (tincture, fluid or concentrated extract) or other standardized preparations used either alone or more frequently in complex pharmaceuticals. In Europe, despite the extensive distribution of the plant, Italy, Yugoslavia, and Albania are the main producers. France, although the plant is very common, imports it from southern Europe.

In the USA, the market for medicinal botanicals, one of the fastest-growing pharmaco-economical segments, approached $4 billion in 1998, including nearly 270 million prescriptions (Brevoort 1998). The sales figures for hawthorn fruit closely follow those of the presently popular herbal medicines such as green tea, kava, ginseng, grape seed, echinacea, and ginkgo (Brevoort 1997).

In Europe, the top sales of herbal remedies are in Germany and France, followed to a lesser extent by Italy and the United Kingdom (Keller 1996). In Germany, 19% of the total cardiac drug prescriptions are phytotherapeutic and hawthorn is one of the major ones (Heide 1996). In France, an increased number of phytotherapeutical neurosedative prescriptions has been noted. Hawthorn, in association with *Valeriana*, *Passiflora*, and *Ballota*, is considered by an increasing number of practitioners as a potential "non-harmful" alternative to synthetic tranquilizers to be tested first in moderate neurotonic disorders.

In 1996, the number of hawthorn-specific preparations (alone or associated with other plants or pharmaceuticals) was 81 in Germany, 39 in France, 20 in Switzerland, 14 in Italy and 1 in the United Kingdom (Martindale 1996).

1.2.3 Active Principles

The main constituents of hawthorn flowers, leaves, and fruits are polyphenols (Fig. 1).

The *flavonoid* content ranges from 1 to 2% (leaves or leaves with flowers) up to nearly 2.5% (flowers) including flavone and flavanol glycosides (Nikolov et al. 1982; Lamaison and Carnat 1990, 1991; Ammon and Kaul 1994a; Wichtl and Anton 1999). Hyperoside is the major flavonoid (0.6–1%) in flowers of *C. monogyna* and *C. laevigata*. Vitexin-2″-O-rhamnoside is dominant in the

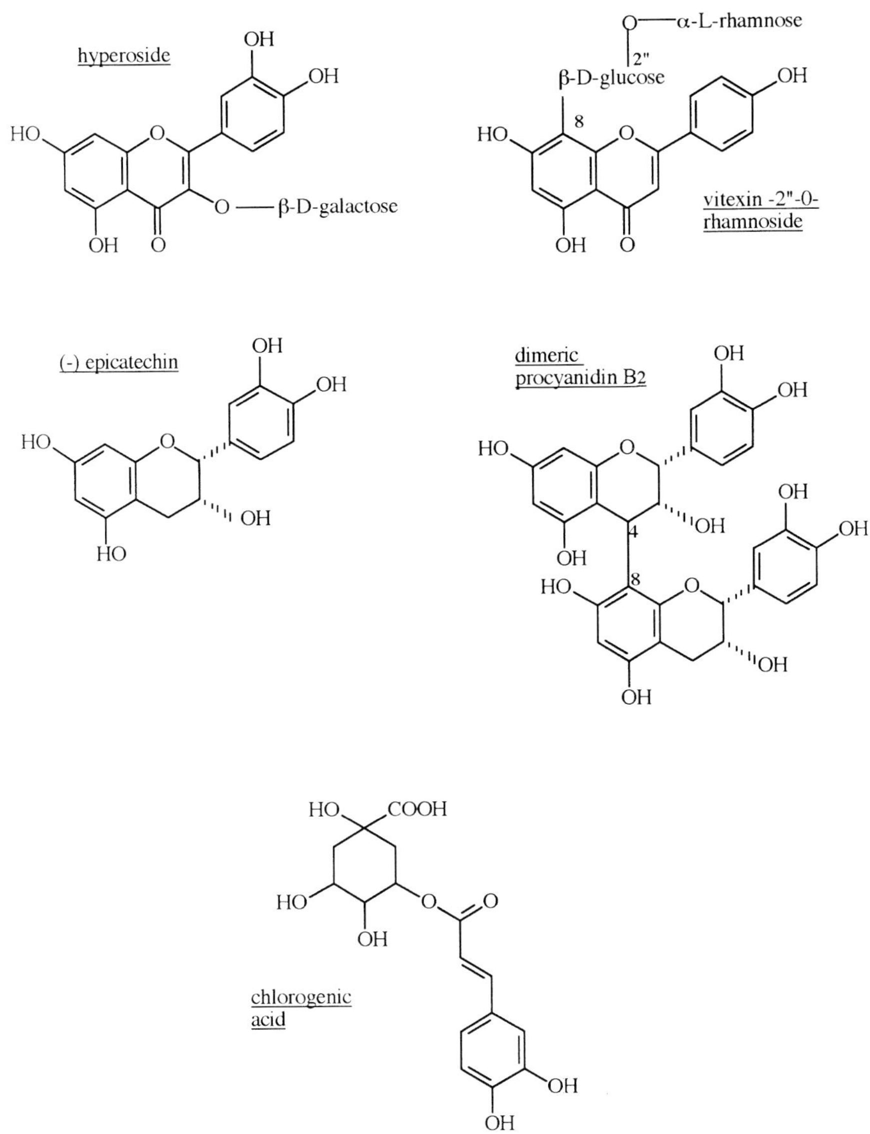

Fig. 1. Structures of main phenolics in *Crataegus*

leaves, but a 4′-acetyl-2′-rhamnoside of vitexin is only present in *C. monogyna* leaves (Wichtl and Anton 1999). Spiraeoside plus rutin (in leaves) and vitexin are present in small amounts.

The *flavanolic* profile includes *catechins* and *proanthocyanidins*. The main monomer is (−) epicatechin. (+); catechin has been reported to a lesser extent (Thompson et al. 1972).

Hawthorn *proanthocyanidins* contain the more active oligomeric forms (two to six catechin subunits ; Haslam 1996) and higher polymers. Ammon and Kaul (1994a) reported amounts of 2.9–3.8% total proanthocyanidins comprising 1.9–2.7% oligomers in leaves.

Oligomers have mainly a C_4–C_8 binding as the B_1–B_2–B_3–B_4 dimers, widespread in the plant kingdom. The B_2 dimer [(–) epicatechin (4β–8)-(–) epicatechin] is abundant: 0.1–0.5% in leaves or unripe fruits (Thompson et al. 1972) together with B_1 dimer [(–) epicatechin-(4–8)-(+) catechin] (up to 0.1% in the fruit). Together with the C_1 trimer [(–) epicatechin (4β–8)-(–) epicatechin (4β–8)-(–) epicatechin], a B_5 dimer of (–) epicatechin (C_4–C_6) is also present in noticeable yields.

1.2.4 Other Components

Hawthorn also contains chlorogenic acid (caffeyl-3-*O*-quinic) and amines (tyramine, β-phenylethylamine, *O*-methoxyphenylethylamine, trimethylamine, dimethylamine, ethanolamine, some of them contributing to cardiovascular effects (Wagner and Grevel 1982; Occhiuto et al. 1986).

Triterpenoids and triterpenic acids (ursolic, crataegic and oleanolic) were identified (Occhiuto et al. 1986). Recent works on a triterpenic fraction of *C. monogyna* containing cycloartenol as the main compound showed anti-inflammatory effects after oral administration in rats or mice (Ahumada et al. 1997), while an anti-tumor action was also observed in extracts containing triterpenoid from *Crataegus* and *Viscum* (Saenz et al. 1997).

2 In Vitro Approaches

2.1 Review of Phenolic Production in Tissue Culture

Studies dealing with this subject are summarized in Table 1.

Schrall and Becker (1977) analyzed secondary metabolites of *Crataegus* and *Ginkgo* cultures, especially proanthocyanidins. Callus and suspension cultures initiated from aseptic plantlets of *C. monogyna* and *C. oxyacantha* (= *laevigata*) were maintained on MS medium. They observed a far better production of total phenolics (250 mg/100 g fresh weight) in light-exposed callus cultures than callus cultured in the dark over a 10-week culture period. Besides hydroxycinnamic and hydroxybenzoic acids (p-coumaric, caffeic, ferulic, anisic, vanillic, syringic and gentisic acids), catechins (mainly (–)epicatechin), and oligomeric proanthocyanidins were found in callus of *C. oxyacantha* and in suspension cultures of both *Crataegus* species. *C. oxyacantha* suspensions produced smaller amounts of condensed procyanidins than the plant body. The beginning of tracheal differentiation was observed in the better

Table 1. Review: phenolics in *Crataegus* tissue cultures

References, species	Medium/culture	Observations
Schrall et al. (1977)	MS/Callus; CSC	Acids = chlorogenic, p-coumaric, caffeic, ferulic, anisic, vanillic, syringic, gentisic.
Mainly *C. monogyna* (also *C. oxyacantha*)		Catechins = (–) epicatechin, (+) catechin. Procyanidins = dimeric and oligomeric.
	MS/callus	Light enhances growth and total phenols production.
	MS/callus	ANA induces better synthesis of total phenols, catechins, procyanidins.
	MS/callus; CSC	Less procyanidin polymers than plant leaf.
Kartnig et al. (1993) *C. monogyna*	MS/callus; CSC	Vitexin-2″-O-rhamnoside, rutin, hyperoside (major), in both types vitexin measurable only in CSC. Traces of quercetin, luteolin. Procyanidins. Higher ratio oligomeric/polymeric procyanidins than plant. Red color observed in calluses at end of subculture. More flavonoids and procyanidins in calluses than in CSC. General decrease with time.
Bahorun et al. (1994a) *C. monogyna*	Gamborg B_5 modified/Callus	Chlorogenic acid, hyperoside (main flavonoid), isoquercitrin (traces), (–)-epicatechin, procyanidins (global, oligomers, polymers), dimeric B_2 proanthocyanidins (major), anthocyanins. B_2 dimer content similar to plant flower bud. Long term cultures (2 years). Anthocyanins, nearly absent in flower bud. Calluses reddish-colored.

growing and higher procyanidin-producing light-exposed cultures. Changes in phytohormone composition showed a better influence of naphtaleneacetic acid (NAA) than 2,4-D on total phenols, catechins and procyanidin production. Reduced growth, however, resulted from increased concentrations. Kinetin was more favorable to both growth and phenolic synthesis than gibberellic acid.

Kartnig et al. (1993) studied flavonoid and procyanidin content of three cultures originating from *C. monogyna*: callus and cell suspensions (CSC) initiated from shoot tips of a cultivated I and a wild II tree (Kartnig et al. 1993). The medium used was that of Murashige and Skoog, either supplemented with indole-acetic acid/kinetin in the case of callus or with NAA/kinetin for both callus and cell suspensions.

The cultures were maintained in continuous light, one callus line originating from cultivated tree I showed a 900% maximal growth rate. The authors noted at the end of the 3- to 4-week culture, in addition to the commonly observed brownish appearance, the presence of violet pigmented areas. This observation seemed consistent with our reports on selected reddish anthocyanin-rich calluses initiated from *C. monogyna* flower buds (Bahorun et al. 1994a).

Various amounts of flavonoid glycosides were reported in the three cultures, with maximum values from 5.9 mg/100 g dry weight (DW) in cell suspension culture (CSC) up to 19.3 mg/100 g DW in the callus line, both originating from wild tree II. Hyperoside (from 2.7 to 12.5 mg/100 mg DW) was the major component, followed by vitexin-2″-O-rhamnoside, and rutin. Measurable amounts of vitexin were only detected in CSC.

Aglycones like quercetin were found in traces, accompanied by apigenin in the callus strain obtained from the cultivated tree or by luteolin in cell strains initiated from the wild tree. Variations in the flavonoid content were more marked in callus than in cell suspensions over a 6-month culture period. With time, a decline in the values was observed. The proanthocyanidin yields were always lower than in the parent leaves (more than 2600 mg/100 g DW). Calluses contained 13–17 mg/100 g DW and cell suspensions 50 mg, all these values decreasing after 6 months of culture.

2.2 Phenolics in Tissue Cultures from *C. monogyna* Flower Buds

2.2.1 Phenols in Callus Cultures

Callus cultures established from floral buds of *C. monogyna* have been maintained since 1991. The early steps were performed in the dark, then in a 16/8 light–dark period. A previous experience with *Fagopyrum esculentum* calluses (Moumou et al. 1987, 1992) led us to submit them to continuous light exposure, when red areas appeared after nine subcultures (27 weeks) in a 16/8 light–dark period. A low phenolic yield was initially observed (weeks 0–27), then a progressive enhancement occurred (stabilized after 63 weeks), especially following the appearance of the reddish zones.

Stabilized red calluses produced (Table 2) phenolic acids (mainly chlorogenic acid), catechins: (–)-epicatechin), total (GPC), oligomeric (OPC), and polymeric proanthocyanidins (PPC), flavonoids (mainly hyperoside with traces of isoquercitrin) and anthocyanins. HPLC data confirmed the presence of individual compounds (Rt, min): (1) chlorogenic acid (10.05), (2) proanthocyanidin dimer B_2 (13.19), (3) (–) epicatechin (16.22), (4) hyperoside (35.58), (5) isoquercitrin (38.10).

Production of total phenols and proanthocyanidins in relation to growth during 40 days is depicted in Fig. 2. Optimum increase in callus growth occurred from days 24–32 with a dry weight increase of 177–179 mg while maximal production of total phenolics (5.890 g/100 g DW, day 28) was observed. An intense phenol synthesis took place during the initial growth period (ca. days 4–16), but with a low biomass.

Table 2. Comparative phenolic production in hawthorn flower buds (from the original parent tree), callus cultures and cell suspensions. Values in g phenols/100 g dry weight. Day values = days of maximal production during 40-day subcultures

	Phenols		
	Flower buds Production	Callus cultures Maximal production	Cell suspensions Maximal production
Total phenols	9.97	5.89 (day 28)	7.26 (day 32)
Proanthocyanidins			
(GPC)	4.03	2.96 (day 32)	3.29 (day 32)
(PPC)	2.48	1.73 (day 32)	1.32 (day 32)
(OPC)	1.49	1.23 (day 28)	1.98 (day 32)
B_2 dimer	0.34	0.32 (day 24)	0.52 (day 32)
Epicatechin	1.31	0.84 (day 28)	1.64 (day 24)
Flavonoids			
Total	1.03	nd	0
Hyperoside	0.73	0.14 (day 36)	0
Vitexin-2″-O-rhamnoside	0.25	0	0
Isoquercitrin	0.02	0.02 (day 36)	0
Chlorogenic acid	1.62	0.77 (day 24)	0.78 (day 32)
Anthocyanins	0	0.288 (day 28)	0.153 (day 36)

A similar production pattern is observed for total proanthocyanidins (GPC: maximal synthesis at day 32 with 2.950 g/100 g DW (Fig. 2), oligomeric proanthocyanidins (OPC: maximum synthesis at day 28 with 1.227 g/100 g DW) and polymeric proanthocyanidins (PPC: maximum synthesis at day 32 with 1.727 g/100 g DW).

Proanthocyanidin B_2 variations seem analogous (Fig. 3) to those of GPC and OPC fractions with a rapid increase from day 4 to 16 followed by a maximum yield at day 24 (0.321 g/100 g DW).The major catechin produced, (–) epicatechin, also increased from day 4 to 16 and a maximum yield was observed at day 28 (0.838 g/100 g DW; see Fig. 3). A relation seems to exist between (–) epicatechin and the B_2 dimer as both have the same production profiles with maximum amounts between days 24 and 28. In addition, in both cases, a decrease is recorded on day 20 [(–) epicatechin: 542 mg/100 g DW and B_2 dimer: 60 mg/100 g DW]. Literature data show that epicatechin-rich plant organs generally contain proanthocyanidin B_2 as the major dimer (Thompson et al. 1972).

Concerning flavonoids, hyperoside (quercetin-3-galactoside) was the main component and showed little variation, the amount remaining within the range 0.061–0.143 g/100 g DW (Fig. 4). Isoquercitrin (quercetin-3-glucoside) was detected in trace amounts (0.020 g/100 g DW, day 36).

Anthocyanin levels, relatively constant from days 12–24, reached a maximum at day 28 with 0.288 g/100 g DW (Fig. 4). Synthesis of the main

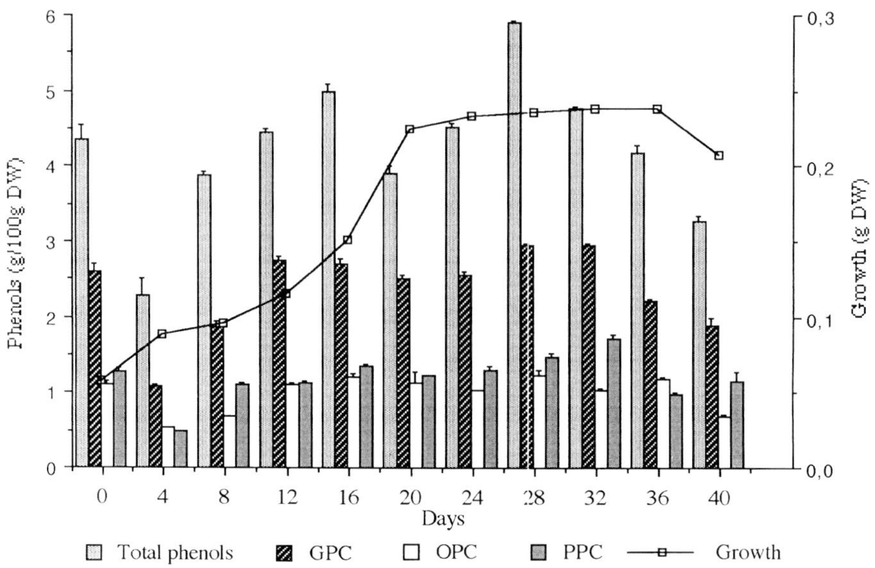

Fig. 2. Total phenols, proanthocyanidins (g/100 g DW), and growth of callus cultures (g DW)

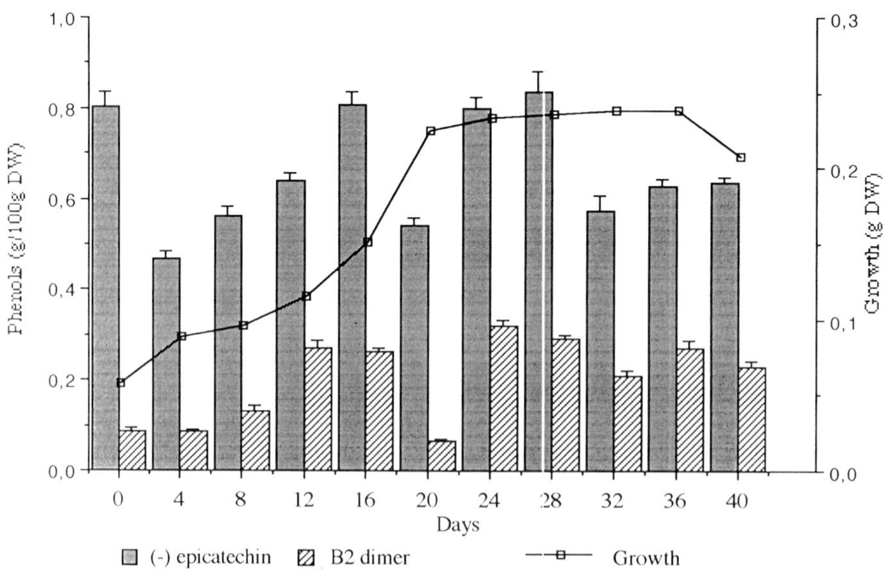

Fig. 3. Dimeric proanthocyanidin B_2, (–) epicatechin (g/100 g DW) and growth (g DW) of callus cultures

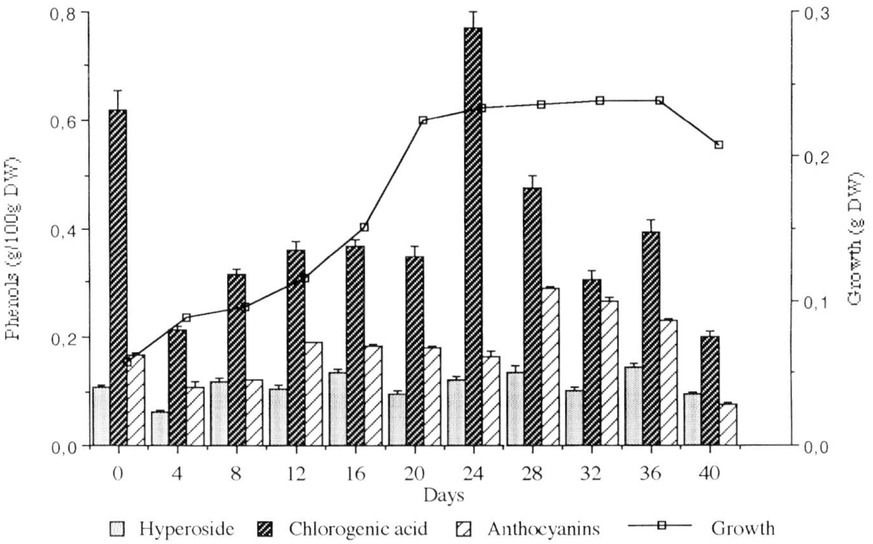

Fig. 4. Hyperoside, chlorogenic acid, anthocyanin content (g/100g DW) and growth (g DW) of callus cultures

phenolic acid detected, chlorogenic acid, had a peak production at day 24 (769 mg/100g DW; Fig. 4).

2.2.2 Phenols in Suspension Cultures

Cell suspension cultures established after 2-year callus culture showed a much higher biomass production between days 24 and 32 with a 12-fold increase. The phenolic profile was similar to that in calluses except for flavonoid derivatives which were not detected (Table 2).

Total phenolic contents were higher in cell suspensions than in calluses (7.26g/100g DW at day 32; Fig. 5). Higher levels of total proanthocyanidins (3.29g/100g DW against 2.347g/100g) and *oligomers* (1.98g/100g DW at day 32 against 1.23g/100g) were produced in cell suspension cultures (Fig. 5). A lower yield of polymeric proanthocyanidins (0.687g/100g DW) was observed.

Higher amounts were noted for proanthocyanidin B_2 content (0.52g/100g DW) and (–) epicatechin (1.64g/100g DW at day 32; Fig. 6). Chlorogenic acid (0.78g/100g DW) was more abundant, while anthocyanin content was much lower in cell suspension (0.77g/100g DW) than in calluses (Fig. 7).

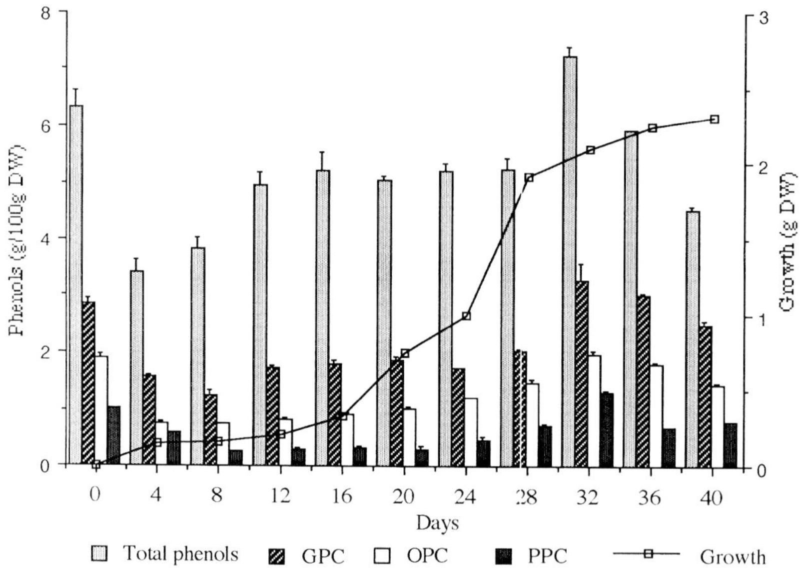

Fig. 5. Total phenols, proanthocyanidin contents (g/100 g DW), and growth of cell suspension (g DW)

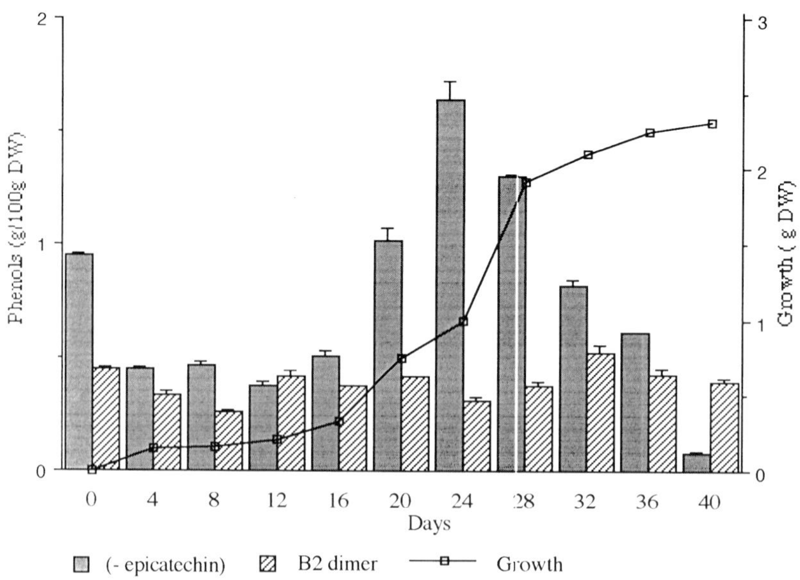

Fig. 6. Epicatechin, dimeric procyanidin B_2 contents (g/100 g DW) and growth (g DW) of cell suspension

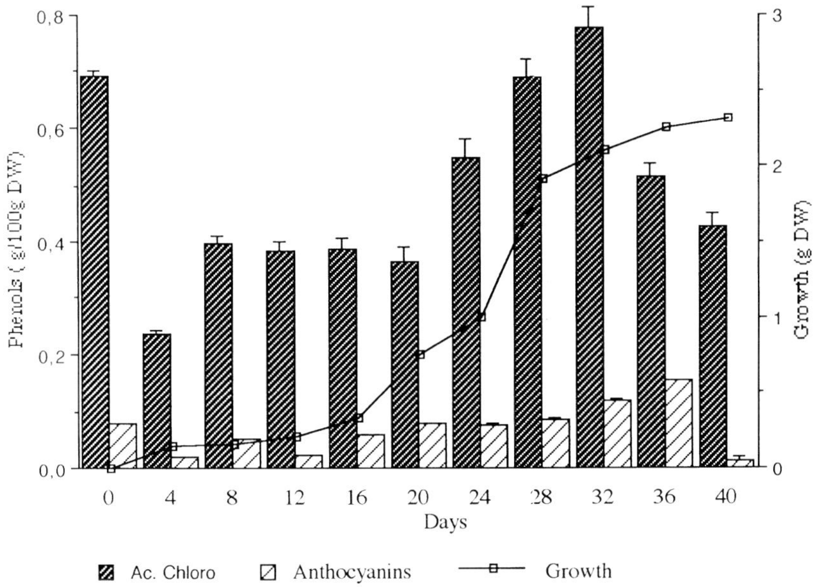

Fig. 7. Chlorogenic acid, anthocyanin contents (g/100 g dry weight) and growth (g DW) of cell suspension

2.2.3 Comparison of in Vivo/in Vitro Polyphenolic Contents

A comparison between callus, cell suspension and fresh flower bud – the parent plant organ (Table 2) – showed that flavonol production was generally higher in cell suspensions at the end of culture (day 32). Levels of oligomeric proanthocyanidins (1.98% against 1.49), B_2 dimer (0.52% against 0.34) and (–) epicatechin were much higher in cell suspension cultures than in the parent plant organ. These substances have already been reported for their high pharmacological potency (Ricardo Da Silva et al. 1991; Meunier et al. 1994). Moreover, it was previously shown (Bahorun et al. 1994b) that the floral bud from the fresh plant contained the highest amounts of phenolics from almost all categories tested when compared to leaves, flowers and fruits at various stages. Thompson et al. (1972) obtained 0.1–05% B_2 dimer in leaves or fruits of *Crataegus*. The plant bud produced more total phenols than cell cultures. This seems to be clearly influenced by the high total proanthocyanidin content, and by important levels of polymeric proanthocyanidins. The latter were shown to be less active in anti-lipid peroxidation assays on plant fresh organs (Bahorun et al. 1994b). The bud is also characterized by a flavonoid production which was absent in cell suspensions. In addition, anthocyanins were mainly obtained in calluses. These red pigments seem to play a role as visual marker for the synthesis of other polyphenolic classes as observed after the 27th week of callus cultures and as previously described in *F. esculentum* callus cultures (Moumou et al. 1992).

2.3 Influence of Precursors in Callus Cultures

Literature data abound in methods to enhance secondary metabolite production in cell cultures. Precursor incorporation is one of the most common, alone or combined with others (Banthorpe 1994). We tried to incorporate phenylalanine, shikimic acid, and gallic acid. Phenylalanine was shown to enhance the content in rosmarinic acid of *Coleus blumei* cell suspensions (Zenk et al. 1977). Shikimic acid is an important precursor in the formation of gallic acid and its derivatives (Dewick and Haslam 1969). In a previous work, incorporation of gallic acid by callus cultures of *Fagopyrum esculentum* (Moumou et al. 1992) yielded high levels of procyanidin B_2-3′-*O*-gallate, a potent antioxidant (Hatano et al. 1989). Saijo (1983) obtained epigallocatechin by incorporation of shikimic acid in *Camellia sinensis* cell culture. Our aim was, therefore, to obtain an enhanced production of dimeric procyanidins and eventually gallic esters.

2.3.1 Phenylalanine and Gallic Acid

Incorporation of phenylalanine (0–10–50–100 mg/l) did not modify the production of total phenols, flavonoids and proanthocyanidins (GPC). B_2 dimer production was slightly increased (1.12 times that of the control) with 10 mg/l phenylalanine. Gallic acid addition clearly depressed growth (except for 10 and 50 mg/l) and the content of the majority of phenols was lowered, despite sporadic increases in epicatechin, procyanidin B_2 and chlorogenic acid in tissues affected by necrosis. Moreover, no gallate derivative of epicatechin or dimer was detected. The fact that these derivatives are absent from the hawthorn plant may therefore provide an explanation for their absence in our in vitro cultures. It should be noted that the incorporation of gallic acid enhanced the in vitro formation of galloyl esters already present in *Fagopyrum* (Moumou et al. 1992).

2.3.2 Shikimic Acid

2.3.2.1 Effects on Growth

A stimulation of growth was observed from the 12th day of culture for all concentrations and became optimal in the presence of 50 mg/l shikimic acid at days 16 and 20. Higher concentrations (100 and 250 mg/l) reduced growth.

2.3.2.2 Effects on Phenolic Production

Optimum production always occurred either after 12 or 20 days of culture. In this respect, only data relating to the 12th and 20th day of culture have been presented in Table 3.

Table 3. Phenolic (g/100 g DW) content of callus cultures and oxygen species scavenging (IC_{50}) activities of their extracts at different periods of a 40-day subculture. IC_{50} values as mg cell culture DW

	Shikimic acid (S.A.)	Day	Phenol content		
			Maximum with S.A.	Control	Increase (×)
Total phenols	250	12	7.22	4.61	1.57
Proanthocyanidins (GPC)	250	12	3.62	3.05	1.19
B_2 dimer	10	20	0.31	0.05	6.00
Epicatechin	250	20	1.85	0.36	5.00
Hyperoside	100	12	0.26	0.12	2.27
Chlorogenic acid	250	12	0.67	0.29	2.29
Anthocyanins	250	20	0.30	0.18	1.66

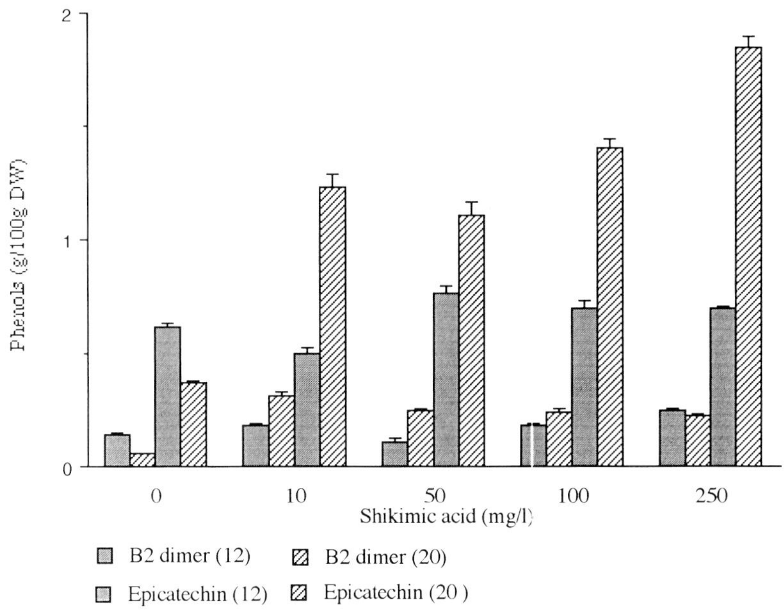

Fig. 8. B_2 dimer and epicatechin contents at days 12 and 20 of a 20-day subculture of calluses fed with different shikimic acid concentrations (mg/l)

Maximum yield of total phenols (7.22 g/100 g DW was obtained after 12 days in the presence of 250 mg/l, corresponding to a 1.57-fold time increase compared to control cultures (Table 3). After 20 days of culture, insignificant increases were observed in the presence of all concentrations used. For total proanthocyanidins, 250 mg/l shikimic acid induced, after 12 days, the highest production (3.619 g/100 g DW; 1.19-fold increase; Table 3).

The production of B_2 dimer varied after the 12th day with different shikimic acid concentrations (Fig. 8). Maximum production was in the presence of

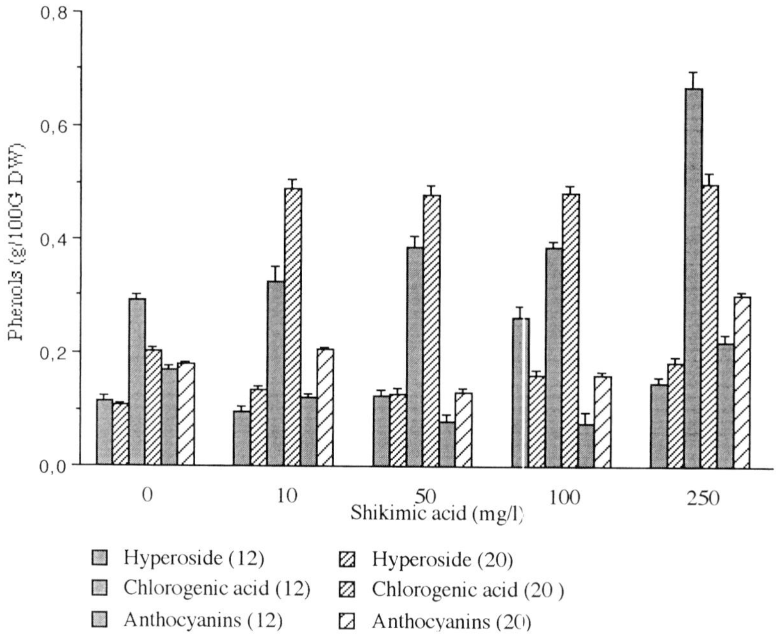

Fig. 9. Hyperoside, chlorogenic acid and anthocyanin contents at days 12 and 20 of a 20-day subculture of calluses in presence of different shikimic acid concentrations (mg/l)

10 mg/l after 20 days (0.309 g/100 g DW). This represents a sixfold increase with respect to control cultures (0.055 g/100 g DW) and it must be noted that this yield was obtained with the lowest shikimic acid concentration. (–) Epicatechin synthesis was also markedly influenced until day 12, whereas after 20 days the addition of 250 mg/l shikimic acid triggered a fivefold increase (1.846 g/100 g DW; Fig. 8).

The hyperoside content (Table 3, Fig. 9) obtained with the addition of 100 mg/l of shikimic acid was more than twice (0.263 g/100 g DW) the amount recorded in control calluses (0.116 g/100 g DW) after 12 days. During this assay, chlorogenic acid levels increased (0. 668 g/100 g DW) at day 12 under the influence of a 250 mg/l dose (Fig. 9). This was also the case for *anthocyanins* (0.320 g/100 g at day 12; Fig. 9).

In conclusion, production of all phenolics was enhanced in calluses. Total proanthocyanidin production (3.6 g/100 g DW) seemed to be very close to the synthesis observed in the floral bud while (–) epicatechin levels (1.84 g/100 g DW) were higher than the parent organ. Nevertheless, suspension cultures without incorporation remained generally more productive.

Table 4. H_2O_2, HOCl and O_2 scavenging of reference phenols corresponding to major components in hawthorn tissue cultures. IC_{50} values in mg/l and in µM/l

	Phenols	IC_{50}					
		H_2O_2		HOCl		O_2	
		(mg/l)	(µM/l)	(mg/l)	(µM/l)	(mg/l)	(µM/l)
Flavanols	Epicatechin	0.351	1.21	0.96	3.31	1.53	5.27
	B_2 dimer	0.58	1.00	1.52	2.62	2.48	4.28
Flavonoid	Hyperoside	0.54	1.17	1.42	3.08	2.36	5.13
Phenolic acid	Chlorogenic acid	1.13	3.19	2.82	7.96	4.97	13.75

3 Oxygen Species Scavenging by Callus Extracts

3.1 Reference Phenols

Table 4 compares IC_{50} values of references, expressed either as mg/l or also as µM/l, the latter preferably used by the authors to reflect the respective efficiency of individual compounds (Hatano et al. 1989). Weight- as well as molar-based values in the three systems show the antioxidant capacity of both flavanols and of the main flavonoid, hyperoside. Chlorogenic acid was the less potent scavenger. On a molar basis, the decreasing order of efficacy was: B_2 dimer > hyperoside ≈ epicatechin > chlorogenic acid. Expressed in weight values, the order was epicatechin > hyperoside ≈ B_2 > chlorogenic acid.

3.2 Callus Extract Activity

Scavenging potency remained low and more or less constant (Table 5) during the early growth period. After day 8 or 16, the 50% inhibitory concentration (IC_{50}) values decreased to reach minimal values in the three systems at day 28. This period corresponded to the maximal polyphenol production stage which related to the highest observed activity. The respective production kinetics of total phenols and proanthocyanidins (GPC, OPC, PPC) are compared with IC_{50} values in Fig. 10 and those of B_2 dimer, epicatechin, chlorogenic acid and anthocyanins in Fig. 11. At day 28, the extract was found to contain the highest content of total phenols, global procyanidins, epicatechin, and anthocyanins which could positively influence the observed activity. The yields of oligomeric proanthocyanidins, B_2 dimer and hyperoside were not far from maximum during the same period. By comparison, the flower bud extract was far more active than those of calluses with a respective IC_{50} in H_2O_2, HOCl and O_2 of 2.90, 8.49, and 12.62 mg/l, and the bud phenolic content was also much higher, as seen previously in Table 2.

Table 5. Polyphenolic composition (g/100 g DW) and scavenging activities of *Crataegus monogyna* callus extracts (IC_{50} values in mg DW) towards hydrogen peroxide (H_2O_2), hypochlorous acid (HOCl) and superoxide anion (O_2). Assays on extracts from days 0 (initiation), 8 and 16 (active phenolic production phase), day 24, day 28 (maximal polyphenolic production) and day 40 (end)

Days	Total Phe	Proanthocyanidins								IC_{50}		
		GPC	OPC	PPC	B_2	E	Hyp	CA	Antho	H_2O_2	HOCl	O_2^-
0	5.04	2.80	1.41	0.97	0.18	0.80	0.16	0.22	0.12	29.36	75.26	127.72
8	3.88	1.88	0.68	1.11	0.13	0.56	0.12	0.31	0.12	33.48	76.71	147.00
16	4.96	2.70	1.21	1.34	0.26	0.81	0.14	0.37	0.18	34.63	64.88	150.28
24	4.54	2.56	1.03	1.30	0.32	0.80	0.12	0.77	0.16	23.67	57.66	103.20
28	5.89	2.94	1.22	1.49	0.29	0.84	0.14	0.48	0.29	17.59	44.52	77.22
40	3.28	1.88	0.68	1.16	0.23	0.63	0.09	0.20	0.077	28.70	75.50	125.43

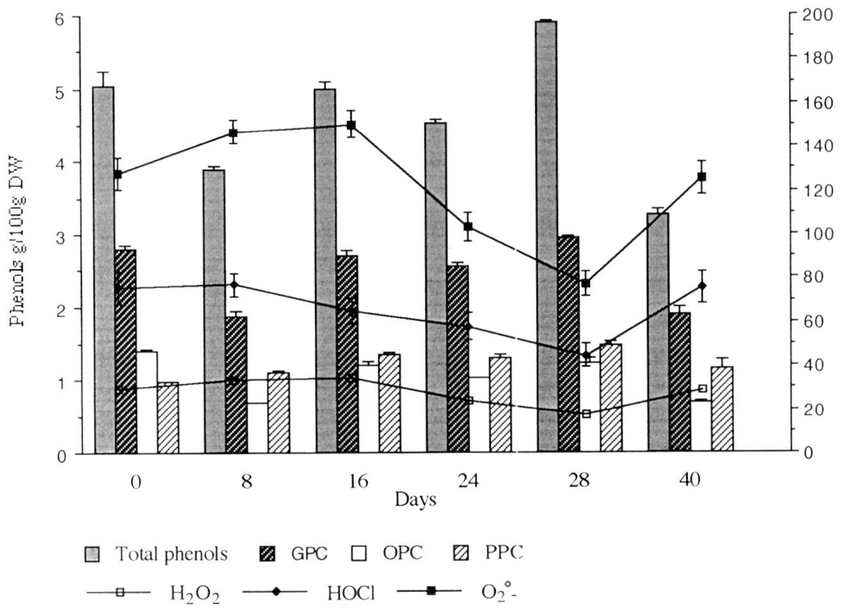

Fig. 10. Total phenols, proanthocyanidin contents of callus cultures and oxygen species scavenging IC_{50} of extracts during a 40-day subculture

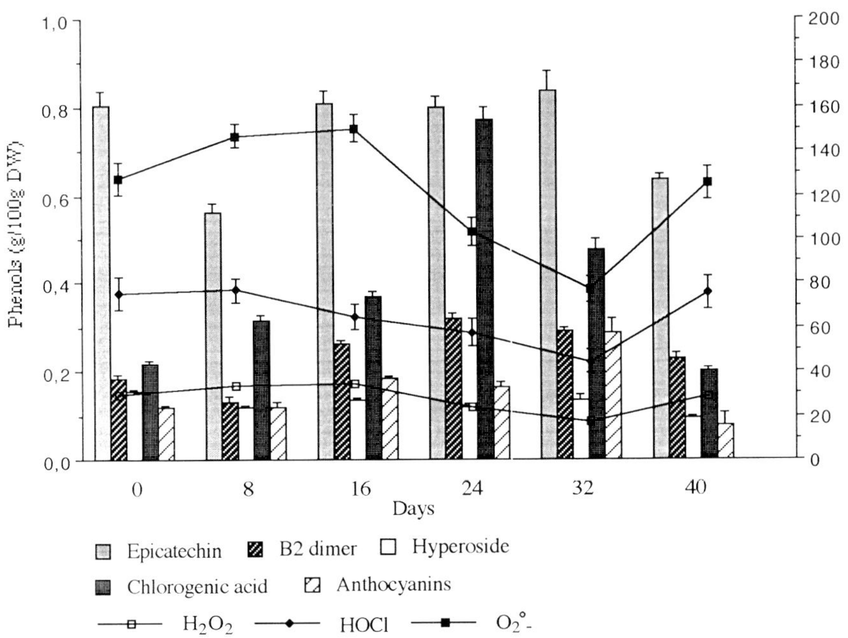

Fig. 11. B_2 dimer, epicatechin, chlorogenic acid, anthocyanin contents of callus cultures and oxygen species scavenging IC_{50} (mg DW) of extracts during a 40-day subculture

Nevertheless, a unique correlation in cell cultures to phenolic yield is not completely evident and could require complementary observations as can be seen in the case of the 24th-day extracts which have several phenolic contents similar to those of day 28, but exhibiting lower activities.

4 Summary and Conclusion

Hawthorn is one of the major modern plant sources for phytotherapeutics applied in mild heart disorders and neurotony. Its activities and those of standardized extracts are demonstrated in many modern ex vivo or in vivo experiments in animals and in a growing number of clinical studies in humans. The main actions have long been attributed to its high content in polyphenols, flavonoids and flavanols – mainly oligomeric proanthocyanidins which seem more efficient in myocardial and vascular effects. These polyphenols are also present in other pharmaceutical plants and in plant foods. The intake of polyphenol-rich foods such as onions, tea, grapes, and wine plays a role in the prevention of heart disease, atherosclerosis and some degenerative processes. Many pharmacological properties were shown with various intensities according to the kind of phenol: antioxidative properties (inhibition of lipid peroxidation, radical scavenging, protection of human low density lipoproteins against oxidation), enzyme inhibitions (elastase, collagenase, lipo- and cyclooxygenase) etc.

Studies on phenolics in hawthorn were made mainly on tissue cultures of the widespread *Crataegus monogyna*. As early as 1977, Schrall et al. aimed to obtain active phenols by means of vitrocultures and found phenolic acids together with proanthocyanidins, including unidentified dimers. The study of Kartnig et al. (1993) showed the possibility of not only obtaining flavonoids, but also proanthocyanidins, and like Schrall, they found a higher proportion of proanthocyanidin oligomers in tissue culture than in the plant. These oligomers have been demonstrated in several plants and hawthorn to be more active.

Our aim was to establish a long-term, high phenolic-producing tissue culture from hawthorn. The floral bud produced calluses which, during the first 63 weeks of their establishment, showed the formation of reddish areas. Due to prior experience in buckwheat tissue cultures, the presence of anthocyanins led us to expose the cultures to continuous light which induced higher phenolic synthesis. Calluses and suspension cultures showed similar phenolic profiles, especially for proanthocyanidins and catechins, except that cell suspensions produced very few anthocyanins and practically no flavonoids. Feeding with shikimic acid markedly enhanced the production of one of the most interesting substances, B_2 dimeric proanthocyanidin (also present in fruit and vegetables). Our cell suspension produced more B_2 dimer and epicatechin than the plant flower bud in these experiments. The callus extracts exerted important oxygen-scavenging activities, clearly bound to phenolics as seen by the comparison between IC_{50} of young (days 0–12) and older (20 days), phenolic-rich calluses. Despite the fact that the bud and other plant-part extracts are far better antioxidants, while containing much higher yields of phenols, an optimization and systematic study of the production in cell suspension of derivatives such as B_2 dimer could be interesting. Proanthocyanidin B_2 and other type B dimers or trimers are absent from the catalogues of fine chemicals and are needed for further pharmacological experiments, particularly on their metabolism in organisms.

5 Protocol

5.1 Tissue Culture

5.1.1 Callus Establishment and Culture

Floral buds deprived of sepals, petals, anthers, stigmas and pistils are sterilized 7 min in $Ca(OCl)_2$ 70 g/l, rinsed four times in sterile water for 5, 10, 15 and 20 min. Cultures are made in sterile Petri dishes (90-mm diameter) containing 25 ml nutrient medium solidified by 0.5% agar (Biokar type E). Callus culture medium comprised B_5 mineral solution elements (Gamborg et al. 1968) with additional (per liter): 30 g sucrose, 100 mg casein hydrolysate, 100 mg myoinositol, 0.5 mg pyridoxine, 0.5 mg nicotinic acid, 0.1 mg thiamine (Murashige and Skoog 1962), 2 mg glycine, 0.5 mg kinetin, 2 mg 2,4-D adjusted to pH 5.6 before sterilization. After 6-weeksculture in the dark at 22 °C, calluses are transferred to a 16/8 light — dark period at 22 °C. After formation of red-pigmented areas (nine subcultures in our case), calluses are transferred to and kept under permanent light (50 $\mu mol\, m^{-2} s^{-1}$) at 22 °C with subculturing every 21 days (four calluses per Petri dish, each 125–150 mg fresh weight). Experiments are performed for 40-day periods, except for precursor feeding.

For feeding tests, calluses are grown for 20 days with 4-day analysis intervals on basal solid medium supplemented with phenylalanine (5, 10, 50, 100 mg/l), or shikimic acid (10, 50, 100, 250 mg/l), or gallic acid (10, 50, 100, 250 mg/l).

5.1.2 Suspension Cell Cultures

Calluses with stable phenolic production (2 years old in our first attempt) are transferred into flasks containing 200 ml liquid medium (same composition as above) and the resulting suspension grown under permanent light at 22 °C on a rotary shaker (70 rpm) with 1-month subculture intervals.

5.2 Polyphenol Analysis

5.2.1 Extraction

Ten grams of crushed tissues are macerated at 4 °C for 24 h with 300 ml 70% methanol, then in 300 ml 70% acetone and finally in 300 ml absolute methanol. The gathered filtrates are evaporated at low pressure to obtain an aqueous extract which is divided into two parts. Part 1 (5 g fresh weight) is carefully evaporated to dryness and the residue dissolved in absolute methanol for storage at –20 °C (extract I). The part 2 extract is extracted with ethyl acetate (4 × 100 ml). The dehydrated ethyl acetate phase is evaporated to dryness and dissolved in methanol (extract II). The remaining water phase is evaporated and taken up in methanol (extract III).

5.2.2 Thin Layer Chromatography

TLC (Silicagel) in toluene/acetone/formic acid (3/3/1, v/v/v) is used for proanthocyanidins (Lea et al. 1979), or ethyl acetate/formic acid/water (8/1/1, v/v/v) for flavonoids (Lamaison and Carnat 1991). Proanthocyanidins/catechins are revealed (orange to red, visible) with anisaldehyde-sulfuric reagent at 105 °C, and flavonoids at 365 nm (yellow, orange, green) by 2-aminoethyldiphenylborate (1%, methanolic) reagent followed by 5% methanolic polyethyleneglycol 4000.

5.2.3 Colorimetric Analysis

5.2.3.1 Total Phenols

Total phenols are measured in extract I according to Singleton et al. (1965): 0.5 ml extract is added to 7 ml distilled water and 0.5 ml Folin-Ciocalteu's reagent (Merck), then after 3 min mixed with 2 ml 20% Na_2CO_3 After 1 min in a boiling water bath, absorption of the cooled solution is measured at 680 nm and compared to a gallic acid standard. Results are expressed in mg gallic acid/100 g DW.

5.2.3.2 Proanthocyanidins

According to Porter et al. (1986), proanthocyanidin content is measured by adding 0.5 ml methanolic extract to 6 ml n-butanol/conc. HCl (19:1v/v) and 0.2 ml 2% (w/v) $NH_4Fe(SO_4)_2.12H_2O$ in 2 M HCl. The tightly capped tubes are mixed, heated 40 min in a water bath at 95 ± 0.2 °C. Extinction of the cooled solution is recorded at 550 nm and the amount of proanthocyanidin is expressed in mg cyanidin chloride/100 g DW when compared to a cyanidin chloride (Extrasynthèse) standard in chlorhydric butanol.

Global proanthocyanidins (GPC) contents are obtained in total extracts (extracts I), while oligomeric proanthocyanidins (OPC) values are measured from ethyl acetate extracts (extracts II). The value of polymeric proanthocyanidin (PPC) is calculated according to the difference between the proanthocyanidin content of total and ethyl acetate extracts.

5.2.3.3 Flavonoids

Global flavonoids in plant flower buds are measured according to Lamaison and Carnat (1991): 1 ml extract I diluted to 1/5 in methanol is added to 1 ml of 2% methanolic $AlCl_3$ $6H_2O$. Extinction is measured after 10 min at 394 nm and compared to a hyperoside standard (Extrasynthèse) and results expressed as mg hyperoside/100 g DW. This method cannot be applied to measure extract I from calluses and cell suspension because of a much lower flavonoid production and interference in $AlCl_3$ of anthocyanins present in these tissue culture extracts. In this case, flavonoids are individually determined by HPLC.

5.2.3.4 Anthocyanins

Anthocyanin content, expressed in mg cyanidin chloride/100 g DW is given by the extinction of 0.5 ml methanolic extract added to 6 ml n-butanol-concentrated HCl (95:5 v/v) at 550 nm and compared to an n-BuOH-HCl solution of cyanidin chloride (Extrasynthèse).

5.2.4 HPLC Analysis

Methanolic extracts are filtered on Millipore 0.45 µm, 20 µl injected onto a Merck Ultrasphere RP18 column (5 µm; 4.6 × 150 mm) and analyzed with LKB HPLC apparatus (Controller 2152, Pump 2150, Detector 2151) and Shimadzu CR 5 A integrator. Elution (flow rate 0.7 m/min) is performed in the order: 0–10 min = 100% of A, 10–30 min = 0–15% of B in A, 30–50 min = 15% of B in A, 50–60 min = 15–25% of B in A, 60–90 min = 25–100% of B in A, 90–100 min = 100% of A (solvent A = acetonitrile/water, 1/9 v/v, pH 2.6; solvent B = acetonitrile, 1/1 v/v, pH 2.6). Column temperature is 25 °C and detection made at 280 nm for proanthocyanidins and phenolic acids, 360 nm for flavonoids. (–) epicatechin, chlorogenic acid, hyperoside and isoquercitrin are identified by comparison with commercial samples. Authentic B_2 dimer obtained from *F. esculentum* callus cultures (Moumou et al. 1992) is also used as a standard.

5.3 Pharmacological Methods

Reference substances as well as cell methanolic extracts evaporated to dryness are dissolved in phosphate buffer saline (PBS) 0.5 M, pH 7.4. For tissue or plant parts, the 50% inhibitory concentration (IC_{50}) values are uniformly expressed as corresponding dry matter weight. All values are the mean of three determinations.

5.3.1 *Hydrogen Peroxide Scavenging*

The hydrogen peroxide amount is measured according to Pick and Keisari (1980). Various concentrations of hawthorn extracts or standard solutions (100 µl), are added to 100 µl H_2O_2 (120 nmol/l), brought to 1 ml with PBS pH 7.4, left 15 min at 37 °C, then overlaid with 1 ml phenol-red solution (0.2 mg/ml) containing 17 U/ml horseradish peroxidase (HRP) type II (Gressier et al. 1993). Addition of 50 µl of 1 N NaOH is made and the H_2O_2 measured by absorbance at 610 nm, with the calculation of the concentration read from a standard curve of H_2O_2.

5.3.2 *Superoxide Anion Scavenging*

Inhibition of O_2 generated by hypoxanthine-xanthine oxidase (Aruoma et al. 1989) is determined by the addition to 50 µl EDTA (30 nmol/l) of 5 µl hypoxanthine (30 mmol/l) in 50 nmol/l KOH, plus 50 µl ferricytochrome C (0.5 mmol/l) and 150 µl hawthorn extract in a total volume of 1.5 ml buffered KH_2PO_4 (50 nmol/l) (Gressier et al. 1993). Adding 100 µl xanthine oxidase (1 U/ml) starts the reaction and the reduction rate of ferricytochrome C is measured at 550 nm. The O_2 amount generated is calculated with an extinction coefficient $E_{550} = 2.1 \times 10^{-2} \mu mol^{-1} cm^{-1}$, and percentage of inhibited O_2 production expressed in IC_{50}.

5.3.3 *Hypochlorous Acid Scavenging*

Chlorination of taurine is realized (Weiss et al. 1982) by adding 100 µl sodium hypochlorite (600 nmol/l) to 100 µl taurine (150 nmol/l) together with 100 µl hawthorn extract in a total volume of 1 ml PBS, pH 7.4 (Gressier et al. 1993). Potassium iodide (10 µl of 2 mol/l solution) is added and absorbance read at 350 nm. Measurements according to an extinction coefficient of $2.3 \times 10^{-2} \mu mol^{-1} cm^{-1}$ (E_{350}) are expressed in IC_{50} values.

References

Ahumada C, Saenz T, Garcia D, De La Puerta R, Fernandez A, Martinez E (1997) The effects of a triterpene fraction isolated from *Crataegus monogyna* Jacq. on different acute inflammation models in rats and mice. Leucocyte migration and phospholipase A_2 inhibition. J Pharm Pharmacol 49:329–331

Ammon HPT, Kaul R (1994a) *Crataegus*, Herz-Kreislauf-Wirkungen von Crataegusextrakten, Flavonoiden und Procyanidinen. Teil 1: Historisches und Wirkstoffe. Dtsch Apoth Ztg 134: 2433–2436

Ammon HPT, Kaul R (1994b) *Crataegus*, Herz-Kreislauf-Wirkungen von Crataegusextrakten, Flavonoiden und Procyanidinen. Teil 2: Historisches und Wirkstoffe. Dtsch Apoth Ztg 134: 2521–2535

Ammon HPT, Kaul R (1994c) *Crataegus*, Herz-Kreislauf-Wirkungen von Crataegusextrakten, Flavonoiden und Procyanidinen. Teil 3: Historisches und Wirkstoffe. Dtsch Apoth Ztg 134: 2631–2636

Aruoma OI, Halliwell B, Hoey BM, Butler J (1989) The antioxidant action of *N*-acetylcysteine: its reaction with hydrogen peroxide, hydroxyl radical, superoxide, and hypochlorous acid. Free Rad Biol Med 6:593–597

Bahorun T, Trotin F, Vasseur J (1994a) Comparative polyphenolic productions in *Crataegus monogyna* callus cultures. Phytochemistry 37:1273–1276

Bahorun T, Trotin F, Vasseur J, Pinkas M (1994b) Antioxidant activities of *Crataegus monogyna* extracts. Planta Med 60:323–328

Banthorpe DV (1994) Secondary metabolism in plant tissue culture: scope and limitations. Nat Prod Rep:303–328

Blesken R (1992) *Crataegus* in der Kardiologie. Fortschr Med 110:290–292

Brevoort P (1997) Der Heilpflanzenmarkt der USA – Ein überblick. Z Phytother 18:155–162

Brevoort P (1998) The booming U.S. botanical market. A new overview Herbal Gram Issue 44:33–48

Chatterjee SS, Koch E, Jaggy H, Krzeminski T (1997) In vitro und in vivo Untersuchungen zur kardioprotectiven Wirkung von oligomeren Procyanidinen in einem *Crataegus*-Extract aus Blättern mit Blüten. Arzneim Forsch (Drug Res) 47:821–825

Coste H (1937) Flore descriptive et illustrée de la France. Librairie Scientifique et Technique Albert Blanchard, Paris

Couplan F, Styner E (1994) Guide des plantes sauvages comestibles et toxiques. Delachaux et Niestlé, Lausanne

Courtney SP, Manzur MI (1985) Fruiting and fitness in *Crataegus monogyna*: the effects of frugivores and seed predators. Oikos 44:398–406

Dewick PM, Haslam E (1969) Phenol biosynthesis in higher plants. Gallic acid. Biochem J 113:537–541

Fournier P (1947) Le livre des plantes médicinales et vénéneuses de France. Lechevalier Ed, Paris

Gamborg UL, Miller A, Ojima K (1968) Nutrient requirements of suspension cultures of Soybean root cells. Exp Cell Res 50:151–158

Gressier B, Cabanis A, Lebègue S, Brunet C, Dine T, Luyckx M, Cazin M, Cazin JC (1993) Comparison of effect in vitro of two thiol containing drugs on human neutrophils hydrogen production. Methods Find Exp Clin Pharmacol 15:101–105

Grisvard P, Chaudun V, Chouard P, Guillaumin A (eds) (1964) Le Bon Jardinier, 152th edn. La Maison Rustique, Paris

Haslam E (1996) Natural polyphenols (vegetable tannins) as drugs: possible modes of action. J Nat Prod 59:205–215

Hatano T, Edamatsu R, Hiramatsu M, Moti A, Fujita Y, Yasuhara T, Yoshida T, Okuda T (1989) Effects of the interaction of tannins with co-existing substances. VI. Effects of tannins and

polyphenols on superoxide anion radical, and on 1,1-diphenyl-2-picrylhydrazyl radical. Chem Pharm Bull 37:2016–2021
Heide L (1996) Phytotherapy in Germany. Its role in self-medication and in medical prescribing. Nat Med 50:259–264
Kartnig T, Kögl G, Heydel B (1993) Production of flavonoids in cell cultures of *Crataegus monogyna*. Planta Med 59:537–538
Keller K (1996) Herbal medicinal products in Germany and Europe: experiences with national and European assessment. Drugs Inf J 30:933–948
Kurzmann M, Schimmer O (1996) Weissdorn. Dtsch Apoth Ztg 136:2759–2764
Lamaison JL, Carnat A (1990) Teneurs en principaux flavonoïdes des fleurs et des feuilles de *Crataegus monogyna* Jacq. et de *Crataegus laevigata* (Poiret) DC. Pharm Acta Helv 65:315–320
Lamaison JL, Carnat A (1991) Teneurs en principaux flavonoïdes des fleurs et des feuilles de *Crataegus monogyna* Jacq. et de *Crataegus laevigata* (Poiret) DC. en fonction de la végétation. Plant Med Phytother XXV:12–16
Lea AGH, Bridle P, Timberlake CF, Singleton VL (1979) The procyanidins of white grapes and wines. Am J Enol 30:289–300
Loew D (1997) Phytotherapy in heart failure. Phytomedicine 4:267–271
Mabberley J (1987) The plant-book. Cambridge University Press, Cambridge
Martindale (1996) Reynolds JEF (ed) The extra Pharmacopoeia, 31th edn. Royal Pharmaceutical Society, London
Meunier MT, Villie F, Bastide P (1994) Etude de l'interaction des oligomères proanthocyanidoliques de *Cupressus sempervirens* L sur l'élastase et les élastines, J Pharm Belg 49:453–461
Morris R (1997) Plants for a future (A resource for edible and other useful plants) *Crataegus* species – the hawthorns. http://www.scs.leeds.ac./pfaf/crataegs
Moumou Y, Trotin F, Pinkas M, Dubois J, Vasseur J (1987) Production de polyphénols par des colonies tissulaires de Sarrasin. Ann Pham Fr 45:255–260
Moumou Y, Trotin F, Vasseur J, Vermeersch G, Guyon R, Dubois J, Pinkas M (1992) Procyanidin production by *Fagopyrum esculentum* callus cultures. Planta Med 58:516–519
Murashige T, Skoog F (1962) A revised medium for rapid growth and bioassays with tobacco tissue cultures. Physiol Plant 15:473–497
Nikolov N, Seligmann O, Wagner H, Horowitz RM, Gentili B (1982) Neue Flavonoid glykoside aus *Crataegus monogyna* und *Crataegus pentagyna*. Planta Med 44:50–53
Oberdorfer E (1983) Pflanzensoziologische Exkursions Flora. Ulmer, Stuttgart
Occhiuto F, Costa R, Brigoglio F, Tommasini A, De Pasquale (1986) Etude comparée de l'activité cardiovasculaire des pousses de feuilles et des fleurs de *Crataegus oxyacantha* L. I. Activité électrique et tension artérielle chez le rat. Plant Med Phytother 20:37–51
Pick E, Keisari Y (1980): A simple method for the measurement of hydrogen peroxide produced by cells in culture. J Immunol Methods 938:161–170
Porter LJ, Hrstich LN, Chan BG (1986) The conversion of procyanidins and prodelphinidins to cyanidins and delphinidins. Phytochemistry 25:225–230
Rameau JC, Mansion D, Dume G (1989) Flore Forestière Franÿaise, Guide écologique illustré. Institut de Développement Forestier (Ed), Paris
Ricardo Da Silva JM, Darmon N, Fernandez Y, Mitjavila S (1991) Oxygen free radical scavenger capacity in aqueous models of different procyanidins from grape seeds. Agric Food Chem 39:1549–1552
Rice-Evans CA, Miller NJ, Bolwell PG, Bramley PM, Pridham JB (1995) The relative antioxidant activities of plant-derived polyphenolic flavonoids. Free Rad Res 22:375–383
Saenz MT, Ahumada MC, Garcia MD (1997) Extracts from *Viscum* and *Crataegus* are cytotoxic against larynx cancer cells. Z Naturforsch [C] 52:42–44
Saijo R (1983) Pathway of gallic acid biosynthesis and its esterification with catechins in young tea shoots. Agric Biol Chem 47:455–461
Schrall R, Becker H (1977) Produktion von Catechinen und oligomeren Proanthocyanidinen in Callus- und Suspensionskulturen von *Crataegus monogyna*, *C. oxyacantha* und *Ginkgo biloba*. Planta Med 32:297–307

Schüssler M, Hölzl J, Fricke U (1995) Myocardial effects of Flavonoids from *Crataegus* species. Arzneim Forsch (Drug Res) 45:842–845

Singleton VL, Rossi JA Jr (1965) Colorimetry of total phenolics with phosphomolybdic-phosphotungstic acid reagents. Am J Enol Vitic 16:144–153

Tauchert M, Ploch M, Hübner WD (1994) Wirksamkeit des Weissdorn-Extraktes LI132 im Vergleich mit Captopril. Münch Med Wschr 136:S27-S33

Thompson RS, Jacques D, Haslam E, Tanner RJN (1972) Plant proanthocyanidins. Part 1: introduction; the isolation structure and distribution in nature of plant procyanidins. J Chem Soc Perkin I:1387–1399

Tutin TG, Heywood VH, Burges NA, Moore DM, Valentine DH, Walters SM, Webb DA (1968) Flora Europaea. Cambridge University Press, Cambridge

Van Acker SABE, Van Den Berg DJ, Tromp M, Griffioen DH, Van Bennekom WP, Van Der Vijgh WJF, Balot A (1996) Structural aspects of antioxidant activity of flavonoids. Free Rad Biol Med 20:331–342

Wagner H, Grevel J (1982) Herzwirksame Drogen. IV: Kardiotone Amine and *Crataegus oxyacantha*. Planta Med 45:98–101

Weihmayr T, Ernst E (1996) Die therapeutische Wirksamkeit von *Crataegus*. Fortschr Med 114:27–29

Weiss SJ, Klein R, Slivka A, Wei M (1982) Chlorination of taurine by human neutrophils. J Clin Invest 70:598–607

Wichtl M, Anton R (1999) Plantes thérapeutiques. Technique et Documentation, Paris

Zenk MH, El-Shagi H, Arens H, Stockigt J, WeilerEW, Deus B (1977) Plant tissue culture and its bio-technological application In: Barz W, Reiharn E, Zenk MH (eds) Formation of the indole alkaloids serpentine and ajmalicine in cell suspension cultures of *Catharantus roseus* in plant tissue culture and its botanical application. Springer, Berlin Heidelberg New York, pp 27–43

4 *Dionaea muscipula* Ellis (Venus Flytrap): In Vitro Cultures and in Vitro Production of Secondary Metabolites

K. KUKUŁCZANKA[1] and J. BUDZIANOWSKI[2]

1 Introduction

In 1768, William Young, the royal botanist, imported living plants of the Venus fly-trap to England. They were shown to John Ellis, a member of the Royal Society, who recognised the Venus as a carnivorous plant. He wrote a letter and sent it with a dried plant to the Swedish scientist, Carl von Linné. Among others Ellis wrote: "Nature may have some views towards its nourishment in forming the upper joint of its leaf like a machine to catch food: upon the middle of this lies the bait for the unhappy insect that becomes its prey . . . the two lobes rise up, grasp it fast, lock the rows of spines together, and squeeze it to death . . . the small erect spines are fixed near the middle of each lobe, over the glands, that effectually put an end to all its struggles". Linné gave this species the name *Dionaea muscipula* Ellis (Slack 1985; Juniper et al. 1989). This name comes from the Greek word *Dionaia*, the goddess of love.

The very restricted natural occurrence of this unique species led to the investigation of the methods of its cultivation and propagation. Moreover, extracts of *D. muscipula* are used against malignant diseases.

1.1 Distribution and General Morphology

The carnivorous plant *Dionaea muscipula* Ellis (the Venus flytrap) is a monotypic genus belonging to the Droseraceae family. This native and endemic plant species is restricted to the coast of south-east North Carolina and the coastal corner of South Carolina in the USA, where it favours damp soil which is predominantly sand with a small proportion of peat (Slack 1985).

Dionaea muscipula Ellis is a rhizomatous and perennial plant. The vegetative shoot of the Venus flytrap is a rosette of specialised leaves radiating from an elongated and fresh rhizome. The leaves have broad, winged petioles ending in a blade which is a trap (Lloyd 1942). The modified blade traps insects. It is kidney-shaped to circular, the inner part with reddish granular dots, variously

[1]Botanical Garden, University of Wrocław, ul. Sienkiewicza 23, 50-335 Wrocław Poland
[2]Department of Pharmaceutical Botany, Karol Marcinkowski University of Medical Sciences in Poznań, ul. św. Marii Magdaleny 14, 61-861 Poznań, Poland

Biotechnology in Agriculture and Forestry, Vol. 51
Medicinal and Aromatic Plants XII (ed. by T. Nagata and Y. Ebizuka)
© Springer-Verlag Berlin Heidelberg 2002

coloured and patterned in shades of green, yellow and red, is hinged along the centre and fringed with up to 8-mm marginal bristles (Fig. 1C); each half of the blade bears three sensitive hairs on the upper surface which, when touched in succession by an insect, cause the two halves to snap rapidly together, thus trapping the prey.

The flowers are radially symmetric and they form a cluster on leafless stalks 15–45 cm in height. Five sepals, five petals, white with green veins, spatulate 1.1–1.3 cm, stamens 15–20 cm. Ovary superior, one-celled; one style. Fruit an ovoid capsule 3–4 mm, which splits irregularly. Seeds are numerous, black (Hamlet 1995).

D. muscipula has one of the most sophisticated systems for catching and digesting insects, and hence it is probably the most fascinating carnivorous plant. The leaf-lobes shut rapidly like a pair of jaws at a rate similar to that of a mammalian (Juniper et al. 1989; Shahinpoor and Thomson 1995). Two successive bends of the trigger hairs in a time interval shorter than 20 s are sufficient for a quick closure of the trap within 0.5 s. On the physiological and biochemical level, the stimuli of the trigger hairs evoke action potentials, which spread over the trap tissues with a velocity of ca 10 cm/s, whereas changes in the concentrations of cytosolic-free Ca^{2+} ions are considered to play a key role in this mechanism (Hodick and Sievers 1986; Trębacz et al. 1996; Trębacz and Sievers 1998). However, in spite of intensive studies, the trapping mechanism of *Dionaea* is still not fully understood (Trębacz, pers. comm.). Then, the prey is digested by enzymes (proteases) present in the fluid excreted by the numerous digestive glands located on the inner faces of the lobes. It seems that naphthoquinone - plumbagin also plays a role in the digestion process due to the inhibition of bacterial growth, thus eliminating competition from bacteria, as well as predigestion by oxidative modification of proteins of the prey (Juniper et al. 1989; Galek et al. 1990).

1.2 Importance

Dionaea muscipula, a legally protected carnivorous species, arouses increasing interest. This plant has been cultivated for 200 years in collections of various botanical gardens, where it is conventionally propagated. Recently, the Venus flytrap has attracted the attention of numerous hobby gardeners, and professional gardeners are absorbed in its commercial production. Also, many researchers have carried out comprehensive physiological and biochemical studies of this species since it produces digestive enzymes.

Because of the above-mentioned trends, the possibility of a quick and large-scale propagation of this plant is considered necessary.

Dionaea muscipula also has medicinal importance since it is considered to be an anticancer drug with oncolytic, anti-proliferative and immunomodulatory effects. The concept of such an application was raised in the 1970s by Keller (1987) who based it on his own theory that carnivorous plants, which

Fig. 1. *Dionaea muscipula*. **A** An in vitro culture on 1/2 MS medium (December 1998); **B** plants transferred to moss peat, covered with moss (*Sphagnum*) and kept in a glass jar to ensure high humidity. **C** Plants in pots. *1* Two-lobed lamina forming the trap, *2* petiole

can break down primitive epithelia, might be able to reduce malignant tissue (Hauser 1988). The commercial preparation containing the expressed juice of the fresh plants cultivated in a greenhouse, named Carnivora, has been produced on a large scale in Germany since 1981 (Hauser 1988; Rote Liste 2000). Repeated clinical trials in a number of patients showed limited or no remission (Dietzel et al. 1985) and in consequence, the efficacy of the *Dionaea* preparation in the treatment of human cancer was not considered to be proved scientifically (Hauser 1988). In vitro investigations showed moderate antiproliferative activity of Carnivora on five tumour cell lines (sensitive- and drug-resistant) at high concentrations of 100 μg/ml and higher under continuous long-term exposure (Todorov et al. 1995a,b, 1996a,b, 1998, Todorov and Ilarianova 1996). Currently, the preparation is indicated for unconventional anticancer treatment (Rote Liste 2000). Extract of *Dionaea* was found to be cytotoxic at concentrations of 1–0.1 mg/ml (Wagner et al. 1986) and more recent investigations have also indicated such effects of *Dionaea* extracts at the high dose level of 100 μg/ml (Drees et al. 1996). This activity was related to the major 1,4-naphthoquinones present in the species, plumbagin (Zenk et al. 1969) and hydroplumbagin 4-*O*-glucoside (Kreher et al. 1990; Fig. 6). The former compound, otherwise well known from its occurrence in several plant families like Droseraceae (Culham and Gornall 1994), Nepenthaceae, Plumbaginaceae, Dioncophyllaceae and Ebenaceae (Hegnauer 1989, 1990), as well as a broad pharmacological potency (Gujar 1990; Raggazzi et al. 1993), showed in vitro immunosuppressive cytotoxicity at concentrations 1–0.01 mg/ml and a reversed effect (immunostimulating) at very low concentrations (Wagner et al. 1986). The dual effect of plumbagin was observed on human granulocytes in in vitro studies, where the compound exerted cytotoxic effects at high concentrations (100 μg–100 ng/ml) and stimulation of phagocytosis at low concentrations (100 pg–10 fg/ml) and no effect at dosages of 2.5 ng–25 pg/ml. Similar dose-dependent activity was observed in a parallel study for several well-known cytotoxic or cytostatic agents (e.g. vincristine; Wagner et al. 1988). It should be noted that plumbagin was found to be toxic in mice; LD_{50} = 28 mg/kg i.p. (Wurm et al. 1984). Furthermore, hydroplumbagin glucoside showed immunostimulating activity as it exerted a high phagocytosis coefficient K = 1.6 in an in vivo carbon-clearance assay in mice at a dose of 0.1 mg/kg i.p. (Kreher et al. 1988a), as well as enhanced in vitro proliferation of T-lymphocytes in very low concentrations of 100 pg–1 pg/ml (Kreher et al. 1988b). Minor naphthoquinones of the species, droserone and chloroplumbagin (Kreher et al. 1990), were inactive (Kreher et al. 1988a). Since the effect of active naphthoquinones was observed with unusually small doses, it was, therefore, concluded that the antitumour effect of the drug Carnivora depended on the induction of the immune system and was due to the low content of plumbagin and hydroplumbagin glucoside in the daily dosage of the drug (i.e. 7.7 and 67.4 mg per 2 ml, respectively) (Kreher et al. 1988a).

1.3 Conventional Practices for *Dionaea* Propagation

Conventional cultivation of the Venus flytrap is difficult. Encke (1958) and Slack (1985) described its propagation from seeds and leaf cuttings which can be slow and not very effective. The seeds sown before winter germinate in spring. Seedlings initially grow very slowly. They take from 5 to 7 years to reach maturity. In late spring, fully expanded leaves of the Venus flytrap are excised in order to obtain leaf cuttings. After a few weeks, at the base of each petiole, a bud is formed which develops into a shoot and plantlet. Enke (1958) also propagated the Venus flytrap from shoots with pieces of rhizomes. *Dionaea* is cultivated and propagated on medium consisting of sphagnum peat moss and sand (fine gravel) under glass or in a cold greenhouse.

2 In Vitro Approaches

2.1 Establishment of Tissue Cultures and Plant Regeneration

The in vitro cultures of *Dionaea* allow a large-scale propagation of this plant (Beebe 1980; Parliman et al. 1982a,b; Hutchinson 1984; Kukułczanka et al. 1989).

Rapid and mass in vitro propagation of *Dionaea* is possible owing to its vigorous development and growth as well as a considerable inborn regeneration potential of its leaves.

2.2 Propagation from Seeds

The seeds which are to be propagated in vitro have to be sterilised, because a surface sterilisation of the plant tissue with alcohol or Clorox frequently kills it. Even if the tissue remains alive, it turns brown (Beebe 1980). Teng (1999), however, successfully used flower stalks to initiate *Dionaea* plantlets under in vitro condition.

Beebe (1980) stored seeds at 4 °C prior to sowing them. Seeds germinated over a 10–55 day period (Beebe 1980; Hutchinson 1984; Kukułczanka and Cząstka 1987). Over 70% of the seeds germinated between the 10th and 35th day.

The seeds were sterilised for 15–20 min in 5% calcium hypochlorite (Hutchinson 1984), or in 3% sodium hypochlorite (Kukułczanka and Cząstka 1987), or in Clorox (Beebe 1980). Just prior to sterilisation, the seeds were dipped in 70% ethanol and/or in 0.1% Tween 20 for 1 min and finally rinsed three times with sterile water. They were subsequently transferred to tubes on agar medium (8–10 g/l) with 20–30 g/l sucrose (Beebe 1980; Parliman et al. 1982b; Hutchinson 1984; Kukułczanka and Cząstka 1987).

MS medium (Murashige and Skoog 1962) is most frequently used (Beebe 1980; Parliman et al. 1982a,b), but Hutchinson (1984) applied LS medium (Linsmaier and Skoog 1965). Kukułczanka and Cząstka (1987) in the laboratory of the Wrocław University Botanical Garden tested and compared the germination of *Dionaea* seeds taking place on various media for orchids. All those media were comprised of MS medium. The investigations proved that seeds germinated most effectively and grew most vigorously on RM medium according to Reinert and Mohr (1967). The media for orchids are slightly acid, and initially the pH of RM medium is 5.5.

In order to stimulate germination of the *Dionaea* seeds and growth of the plantlets, Beebe (1980) supplemented MS medium with such growth substances as auxin and cytokinin in doses which were multiplied ten-fold: auxin (NAA) from 1×10^{-8} to 1×10^{-5} M and cytokinin (BA) from 1×10^{-7} to 10^{-5} M. Low doses of growth regulators stimulated the growth of seedlings, while NAA in the concentration of 10^{-5} M was lethal for seedlings, and BA in the concentration of 10^{-5} M resulted in a growth modification which manifested itself in the rapid growth of callus, but also in the inhibition of root growth. Kukułczanka and Cząstka (1987) proved that the seedlings cultured on RM medium without exogenous auxin and cytokinin developed vigorously and after a few months, they produced leaves with blade traps.

2.3 Growth and Development of Plants

Early growth of the *Dionaea* seedlings was very slow (Kukułczanka 1991). After a few months (4–7) the seedlings produced rosettes and leaves with the extended blade traps. In the following months, rhizomes were produced and their lateral buds developed into new shoots and rooted plantlets. In order to stimulate the growth of plants, researchers supplemented the medium with small doses of auxin and cytokinin.

The *Dionaea* cultures were kept in a growth chamber at 20–26 °C (Parliman et al. 1982a; Hutchinson 1984; Kukułczanka and Cząstka 1987; Kukułczanka 1991). Lower temperatures, i.e. 18–23 °C, stimulated an effective development of the blade traps, and higher temperatures, i.e. 25–26 °C, induced the development of new shoots (Kukułczanka and Cząstka 1991).

The Venus flytrap is a photophilic plant. Hutchinson (1984) reported maximum growth and regeneration of this plant at $100\,nE\,m^{-2}s^{-1}$. He also recorded encouraging results at a photon flux density of 75 or $125\,nE\,m^{-2}s^{-1}$. On the other hand, Teng (1999) has reported that the regeneration rate increases when dark re-culture of the *Dionaea* plantlets is applied before taking explants. A photoperiod of 16–18 h is recommended. Kukułczanka and Cząstka (1987) and Kukułczanka et al. (1989) successfully applied daylight conditions supplemented with fluorescent light.

Dionaea is a boggy plant and grows on acid soils. Hutchinson (1984), who compared shoot growth of plantlets grown on media with various pH values

(3.7, 4.7, 5.7 and 6.7), reported that the plants responded best to pH 5.7. In the first tests, Parliman et al. (1982a,b) applied the media with pH 5.8, and in the next, 4.9. In the experiments carried out by Beebe (1980), as well as by Kukułczanka and Cząstka (1987) and Kukułczanka et al. (1989), the medium was adjusted to a pH ranging from 5.5 to 5.6 before autoclaving.

2.4 Propagation from Shoots and Rhizomes

Growth and branching out of the *Dionaea* plantlets cultured in vitro on various media were vigorous and rapid. Parliman et al. (1982b) described the formation of new plantlets. They distinguished the plantlets derived immediately from the parental plant tissue as adventitious bud-derived plants (Fig. 2A,B) and the plantlets derived from lateral buds (Fig. 2B).

For propagation purposes, various explants and various media were used (Parliman et al. 1982a,b; Hutchinson 1984; Kukułczanka 1991). Hutchinson

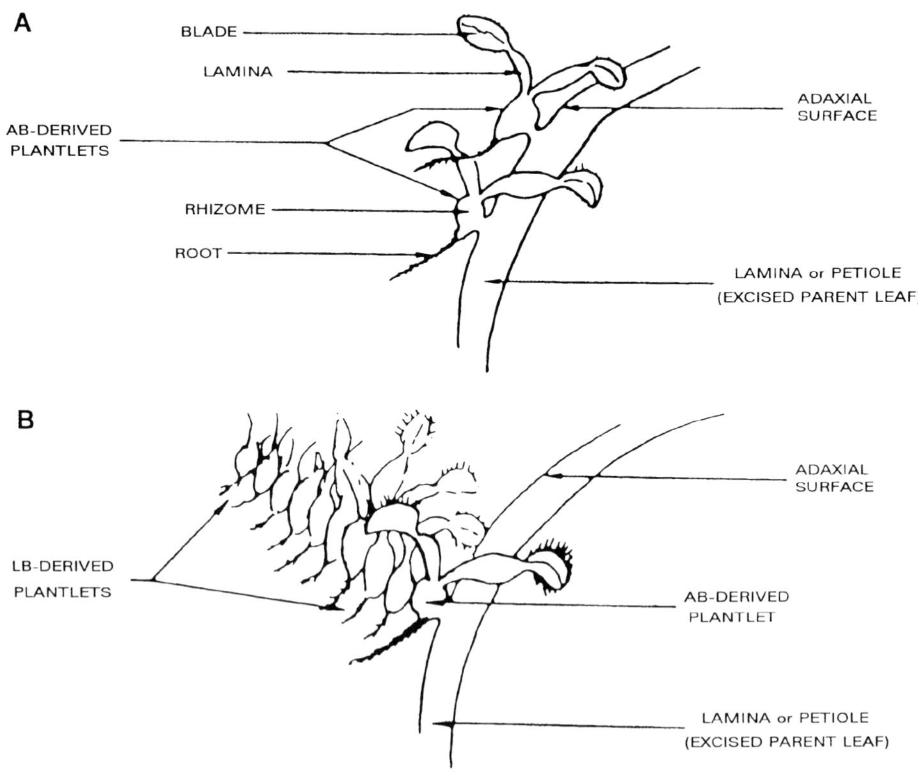

Fig. 2. **A** Adventitious bud-derived *Dionaea* plantlets (*AB*) arising directly from the adaxial surface of a leaf; **B** long chain of lateral bud-derived *Dionaea* plantlets (*LB*). (Parliman et al. 1982b)

(1984) used the explants from shoot tips 3–5 mm in length from the seedlings transferred to LS medium. For rapid asexual propagation, Parliman et al. (1982a) used the LB explants assumed to have arisen as lateral shoots from nodes within parental rhizome structures which ranged in size from 10 to 20 mm in length. The AB explants, only 5–10 mm in length, were taken from single leaves cultured in vitro. Parliman et al. (1982a) used MS and ZB media (Zimmerman and Broome 1980). Kukułczanka and Cząstka (1991) excised single shoot explants with a few leaves and a piece of rhizome each and cultured them on RM medium.

Parliman et al. (1982a) stated that a half-strength MS medium supplemented with 1.9 mg/l α-naphthaleneacetic acid (NAA) and 0.2 mg/l 6-benzylaminopurine (BA) produced 14 or more rooted plantlets in 40–60 days of culture. The same medium produced optimum plantlet size and quality. Supplementing the basal MS medium with 0.3 or 1.0 mg/l gibberellic acid (GA_3) decreased the number of explants and increased the size of plantlets prior to acclimatisation. The media containing higher or lower concentrations of IAA (indole-3-acetic acid), NAA, 2,4-D (2,4-dichlorophenoxyacetic acid), BA, 2iP [i.e. 6-(-γ,γ-dimethylallylamino) purine] produced fewer plantlets while increasing deleterious effects.

Hutchinson (1984) reported optimum proliferation on LS medium with 10 nM kinetin and 0.5 nM NAA. In 8 weeks, a single explant formed ca. ten new shoots. Kinetin stimulated better proliferation than BAP [6-(benzylamino)-9-(2-tetrahydropyranyl)-9H-purine], and IBA (indole-3-butyric acid) stimulated better rooting than NAA. Also, Hutchinson (1984) noted that emission of nitrogen from the LS medium reduced propagation, but this was partly restored by addition of casein hydrolysate.

Kukułczanka and Cząstka (1991) used slightly larger shoot explants grown on RM medium without exogenous auxin and cytokinin. In 3 months, they observed the formation of 20–30 side shoots from lateral buds.

2.5 Propagation from Leaves

Excised *Dionaea* leaves cultured in vitro are characterised by a great regeneration potential. Adventitious bud-derived plantlets arise directly from an adaxial surface (Parliman et al. 1982b). Formation of adventitious buds is mostly observed at the basal part of the petiole and lamina as well as at the base and the apex of the trap part (Fig. 3). Exogenous cytokinins were found to stimulate the formation of adventitious buds along the whole petiole axis (Kukułczanka et al. 1989). Cytokinins, particularly BA, stimulated production of meristematic tissue (callus), in which numerous growth centres originated and numerous adventitious buds were formed. A higher percent of excised leaves with fully expanded blade traps survived in vitro and produced more adventitious buds compared to immature leaves (Kukułczanka et al. 1989; Kukułczanka and Sarowska, unpubl.).

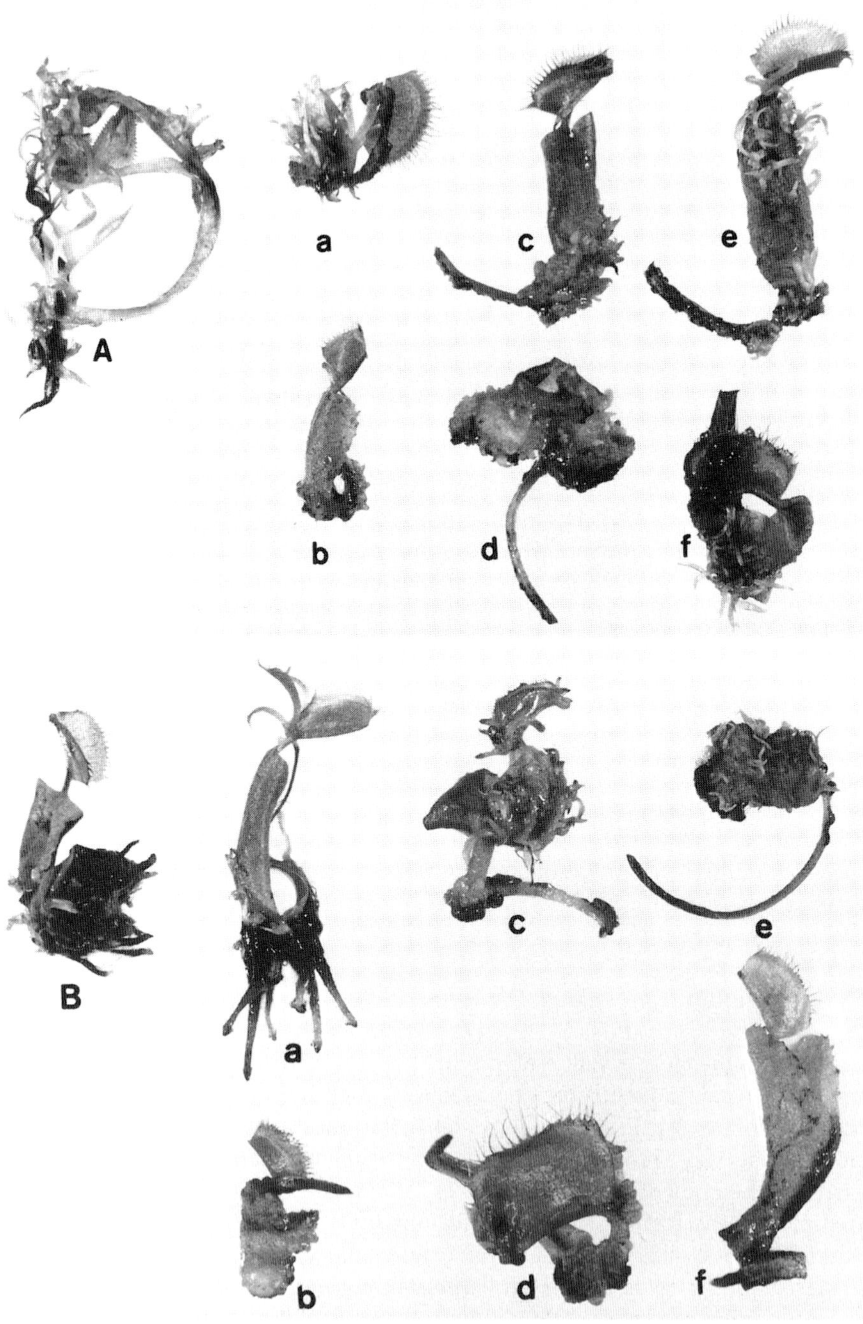

Fig. 3. Regeneration and organogenesis from excised leaves of *Dionaea muscipula* after 2-month culture: **A** RM medium, **B** RM medium + 0.2 mg/l NAA; cytokinin addition; 0.2 mg/l (*a*) and 1.0 mg/l (*b*) kin; 0.2 mg/l (*c*) and 1.0 mg/l (*d*) BA; 0.2 mg/l (*e*) and 1.0 mg/l (*f*) 2iP

When dividing the excised, etiolated leaves, Teng (1999) observed that the percentage of adventitious buds derived from petioles was higher than that of the same buds regenerated from blade traps. He also found that adventitious buds were initiated mainly on the adaxial surface of petiole and trap, and only occasionally on their abaxial side.

Parliman et al. (1982b) reported that whole excised leaves of *Dionaea* produced up to 15 AB and LB plantlets in 60–90 days, cultured on 0.5 MS medium supplemented with 1.9 mg/l NAA and 0.2 mg/l 2iP. In the experiments of Kukułczanka et al. (1989), a single leaf grown on RM medium supplemented with both kinetin and NAA at 0.2 or 1.0 mg/l produced over 20 plantlets.

In further experiments (Kukułczanka and Sarowska, unpubl.), the effects of various cytokinins on the regeneration ability of excised leaves were compared (Table 1, Fig. 3). Bud formation was most effectively stimulated by BA, but this cytokinin retarded shoot and plantlet development and inhibited root formation. The effect of kinetin and 2iP on bud and shoot formation proved to be smaller; however, they only slightly retarded the plantlet rooting. Cytokinins at 0.2 mg/l stimulated the plantlet development better. Also, the percentage of explants that survived was higher than that at 1 mg/l. The plantlets responded best to basal RM medium and RM medium supplemented with 0.2 mg/l 2iP or kinetin plus 0.2 mg/l NAA (Fig. 5).

After 8 weeks the leaf explant mainly formed adventitious buds and shoots (Fig. 3) and after 4 months, chains of lateral bud-derived plantlets were also formed (Fig. 4). A single leaf produced over 20 plantlets.

Parliman et al. (1982b) applied an interesting method. The excised leaves were dipped in various, concentrated solutions of BA or 2iP and NAA or IBA for times ranging from 10 s to 24 h before placing them on the MS medium. This treatment improved the production of plantlets compared with that involving the undipped leaves. The leaves dipped in 2.1 mg/l 2iP for 24 h produced the largest average number of AB- and LB-derived plantlets.

2.6 Long-Term Storage

Long-term in vitro cultures appear to be excellent plant material for further studies and can be considered as a germ plasm of an endangered species.

Kukułczanka (1991) investigated the effect of temperature drop on the subcultures of *Dionaea muscipula*. The plantlets were grown at 20–24 °C for 2 months and thereafter the cultures were cooled down to 16, 8–10, 5–6 or 0–2 °C. One hundred percent of the plantlets survived a rapid temperature drop and a continued temporary cooling.

After 6 months the plantlet size decreased with the decrease in storage temperature. At 0–2 °C the fresh weight of plantlets was seven times lower and their growth was totally inhibited. With the drop in temperature, the content of sugars, proteins and lipids considerably increased (Kukułczanka et al. 1991).

Table 1. Regeneration of *Dionaea muscipula* excised leaves (per 1 explant)

RM medium + growth regulators (mg/l)	After 2 months				After 4 months				
	Explant rejection (%)	Plantlets no.	Buds no.	Roots no.	Explant necroses (%)	Plantlets	Buds no.	Roots no.	Fresh weight (g)
Control	80	7.3	10.1	5.2	15	17.9	0.33	15.3	8.32
0.2 kin	57	9.1	13.2	0.3	25	20.6	0.52	15.9	6.83
1.0 kin	53	0.8	16.0	0.1	45	21.3	0.64	10.4	5.70
0.2 BA	60	8.3	21.0	0.0	30	28.7	8.3	5.2	6.09
1.0 BA	50	0.4	28.6	0.0	35	18.4	10.6	1.3	4.80
0.2 2iP	73	8.6	15.8	1.5	20	21.7	0.9	15.8	9.12
1.0 2iP	60	0.8	13.4	0.3	30	16.1	3.2	8.4	6.40
0.2 NAA	60	2.3	6.7	9.6	40	17.3	1.8	24.3	7.20
0.2 kin, 0.2 NAA	67	4.1	9.3	7.1	10	20.1	0.9	25.2	8.45
1.0 kin, 0.2 NAA	50	0.4	10.1	0.2	36	18.5	3.3	12.1	5.90
0.2 BA, 0.2 NAA	37	8.3	19.2	0.0	47	24.3	8.6	8.4	4.32
1.0 BA, 0.2 NAA	30	0.6	15.3	0.0	56	16.8	10.3	1.7	2.94
0.2 2iP, 0.2 NAA	57	5.4	15.1	0.6	15	21.9	1.1	20.4	10.80
1.0 2iP, 0.2 NAA	43	0.8	10.4	0.3	25	17.3	2.4	11.2	7.65

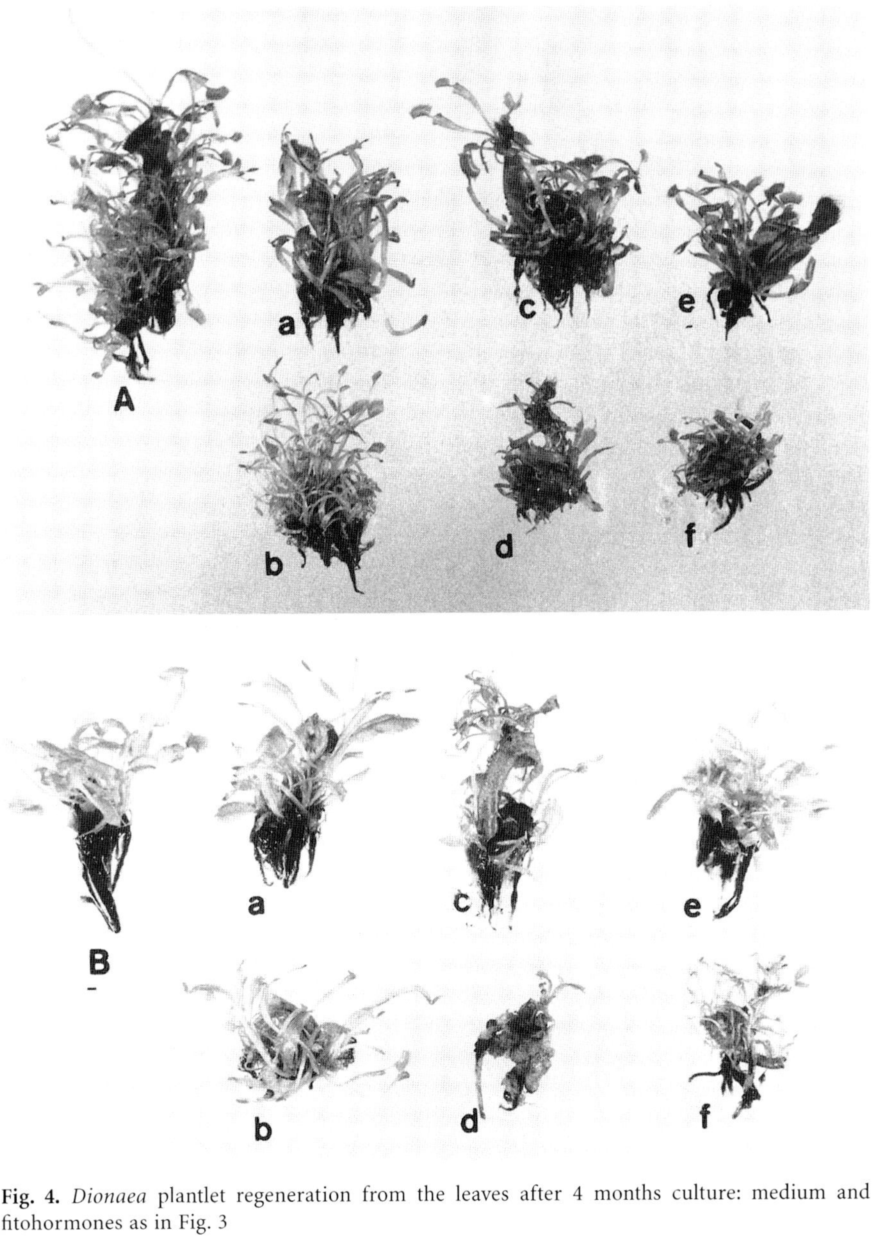

Fig. 4. *Dionaea* plantlet regeneration from the leaves after 4 months culture: medium and fitohormones as in Fig. 3

Plantlets kept continuously at 20–24 °C attained their maximum size after 3–4 months. After 7–8 months, the cultures were senescent.

Dionaea plantlets which, after 5–6 months of cold, were transferred to fresh RM medium and cultured at 20–24 °C, renewed their growth after a few days. The plantlets kept previously at 5–6 °C grew and rooted most rapidly

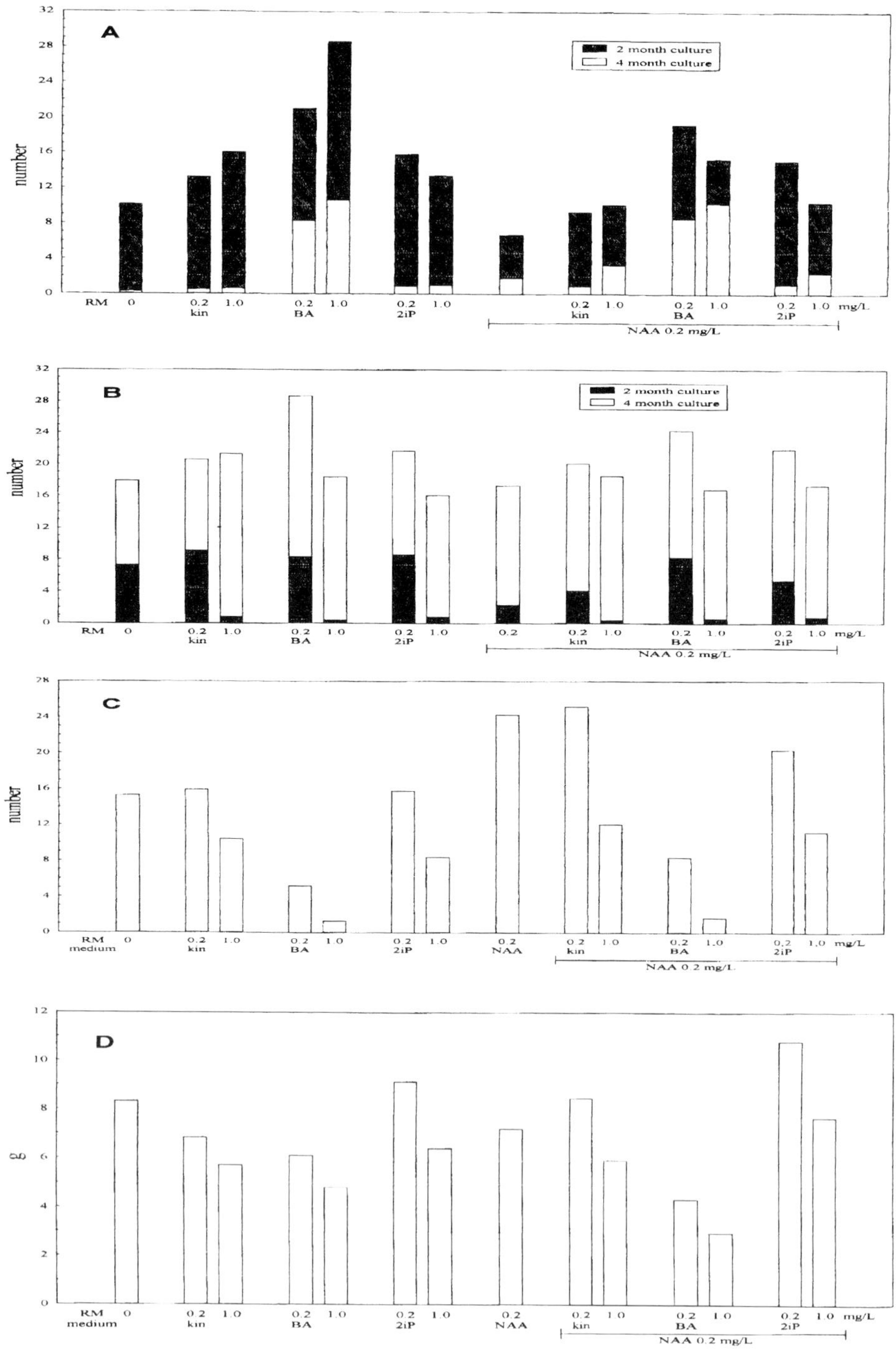

Fig. 5. *Dionaea* regeneration from excised leaves after **A** 2 and **B** 4 months or **C** and **D** 4 months of culture. **A** Bud initiation, **B** plantlet regeneration, **C** root formation, **D** fresh weight

(Parliman 1982a, Kukułczanka 1991). No variations were observed in *Dionaea* plants stored for many months at low temperatures.

2.7 Ex Vitro Culture

Vigorous plantlets of the Venus flytrap adapted themselves best to glasshouse conditions. Addition of IBA, an auxin, to medium before weaning favoured the rooting (Hutchinson 1984). Cooling off the plantlets cultured in vitro to 5 °C for 2–4 weeks before weaning improved their adaptation and rooting (Kukułczanka and Cząstka 1991).

The growth of plantlets potted up in a 1 : 1 ratio of sphagnum and a mixture of sphagnum or highmoor peat and broken-up pine bark was satisfactory. A temperature of 15–20 °C and increased relative humidity favoured plant development. Watering was done by infiltration or irrigation (Kukułczanka 1991).

After a few weeks/months the plants flowered and fruited. Then they either became dormant or died. Plants were transferred to a cool glasshouse at 8–12 °C. Those plants, which had survived the winter, produced developing buds after a few weeks/months (Kukułczanka 1991).

3 Secondary Metabolites

3.1 Compounds of Plants from Natural Sources or Greenhouse Cultivation

D. muscipula from natural sources was found to contain anthocyanin – cyanidin 3-*O*-glucoside (Di Gregorio and Di Palma 1966), flavonoid aglycones – quercetin and myricetin (Jay and Lebreton 1972) and 1,4-naphthoquinones, which are considered to be very characteristic of the family Droseraceae (Culham and Gornall 1994) like plumbagin as a dominant compound (Zenk et al. 1969; Kreher et al. 1990), chloroplumbagin, droserone, and a bound form of plumbagin – hydroplumbagin 4-*O*-glucoside (Kreher et al. 1990). In the latter paper, it has also been proved that chloroplumbagin is a natural, not artefactual, product, and the structure of hydroplumbagin glucoside was revised. The quantitative content of those compounds determined by HPLC methods was: plumbagin 205 mg, chloroplumbagin 9 mg, droserone 2.4 mg, hydroplumbagin glucoside 640 mg per 100 g fresh wt. (Kreher et al. 1988a). There were also some compounds found which are related to plumbagin, like diomuscinone and diomuscipulone (Miyoshi et al. 1984; Table 2, Fig. 6). In the latter paper the source of plant material was not given. Sterols, like β-sitosterol, stigmasterol and cholesterol, were also detected (Gershbein et al. 1995).

Naphthoquinones and related compounds

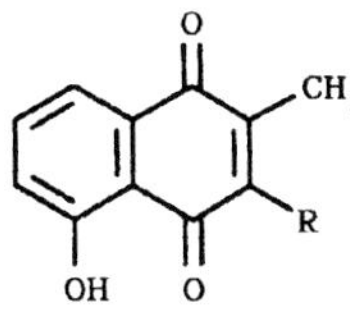

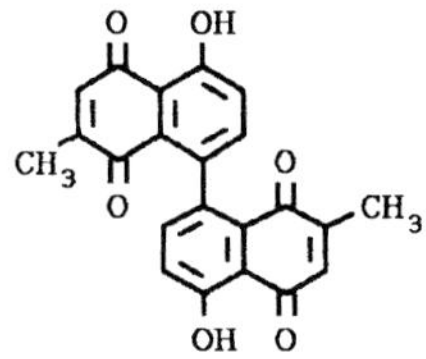

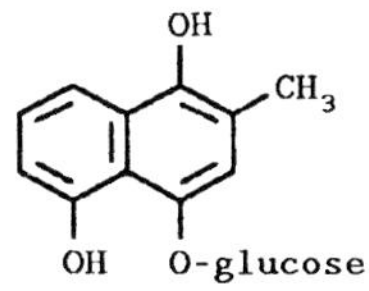

R	
H	plumbagin
Cl	chloroplumbagin
OH	droserone

8,8'-biplumbagin (maritinone)

hydroplumbagin glucoside

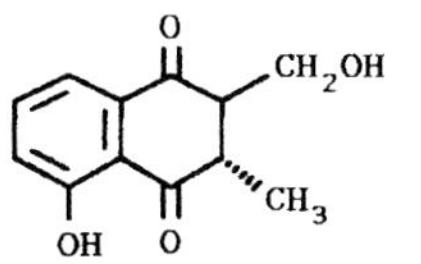

diomuscinone

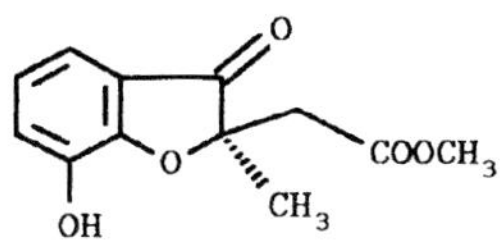

diomuscipulone

Phenolic acids derivatives

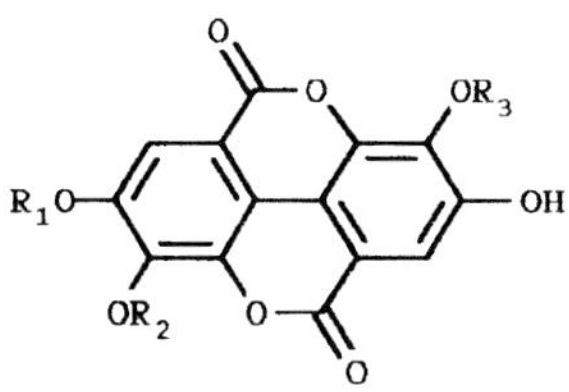

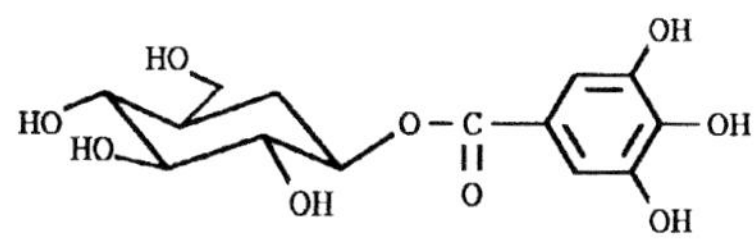

1-O-galloyl-β-glucoside

R_1	R_2	R_3	R_4	
H	H	H	H	ellagic acid
H	CH_3	H	H	3-O-methylellagic acid
H	CH_3	CH_3	H	3,3'-di-O-methylellagic acid
glc	CH_3	CH_3	H	3,3'-di-O-methylellagic acid 4-O-glucoside
glc	CH_3	CH_3	glc	3,3'-di-O-methylellagic acid 4,4'-di-O-glucoside

Flavonoids

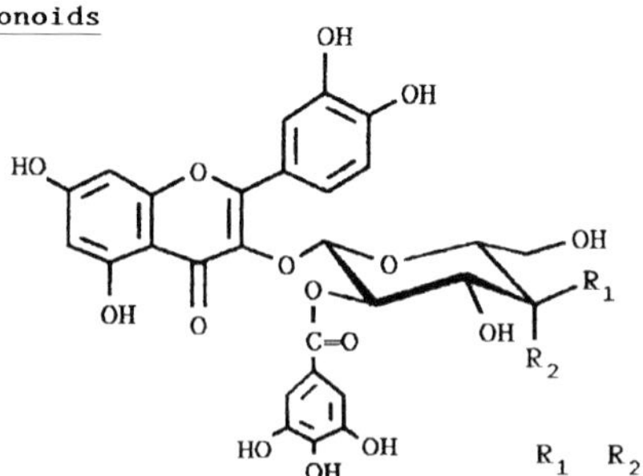

R_1	R_2	
OH	H	quercetin 3-O-(2"-O-galloyl)glucoside
H	OH	quercetin 3-O-(2"-O-galloyl)galactoside

Fig. 6. Structures of phenolic compounds from *Dionaea muscipula*

Table 2. Phenolic compounds in *Dionaea muscipula*

Compound	Reference
Naphthoquinones	
Plumbagin	Zenk et al. (1969) Miyoshi et al. (1984) Kreher et al. (1990) Fujii et al. (1992) Pakulski and Budzianowski (1996a)
3-Chloroplumbagin	Kreher et al. (1990) Pakulski and Budzianowski (1996a)
Droserone	Kreher et al. (1990) Budzianowski (unpubl., 1999)
Hydroplumbagin 4-*O*-glucoside	Kreher et al. (1990) Budzianowski (unpubl., 1999)
88′-Biplumbagin	Pakulski and Budzianowski (1996a)
Diomuscinone	Miyoshi et al. (1984)
Diomuscipulone	
Flavonoids	
Cyanidin 3-*O*-glucoside (anthocyanin)	Di Gregorio and Di Palma (1968)
Delphinidin 3-*O*-glucoside (anthocyanin)	Ichiishi et al. (1999)
Quercetin	Jay and Lebreton (1972)
Myricetin	
Kaempferol 3-*O*-galactoside	Pakulski and Budzianowski (1996b)
Kaempferol 3-*O*-glucoside	
Quercetin 3-*O*-glucoside	
Quercetin 3-*O*-(2″-*O*-galloylglucoside)	
Quercetin 3-*O*-galactoside	
Quercetin 3-*O*-(2″-*O*-galloyl)galactoside	
Phenolic acids derivatives	
Ellagic acid	Pakulski and Budzianowski (1996a)
3-*O*-methylellagic acid	
3,3-di-*O*-methylellagic acid	
3,3-di-*O*-methylellagic acid 4-*O*-glucoside	
3,3-di-*O*-methylellagic acid 4,4′-di-*O*-glucoside	
1-*O*-galloyl-β-glucose	

3.2 Secondary Metabolites in Plants From in Vitro Culture

Dionaea obtained on MS medium was utilised as a source of free plumbagin for investigation of some of its biological properties (Fujii et al. 1992). It should be noted that other sources of this compound are species of the genus *Drosera* (Sundew) of the same family (Crouch et al. 1990; Finnie and van Staden 1993) or it is commercially available. Our phytochemical studies on the whole fresh plants obtained on Reinert-Mohr medium (Kukułczanka 1991) led to the isolation of compounds belonging to the three classes of phenols: naphthoquinones, phenolic acid derivatives and flavonoids (Table 2; Fig. 6). They were naphthoquinones: plumbagin (as a major compound), chloroplumbagin, 8,8′-biplumbagin; phenolic acid derivatives: ellagic acid, 3-*O*-methylellagic acid, 3,3′-di-*O*-methylellagic acid, 3,3′-di-*O*-methylellagic 4-*O*-glucoside,

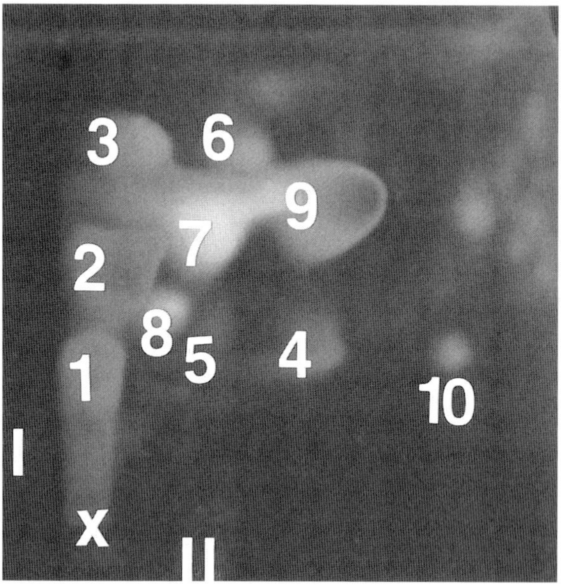

Fig. 7. 2D-TLC on cellulose of butanol fraction of *Dionaea* from in vitro culture (1/2 MS medium). *X* Origin. Mobile phases: *I* 1-butanol-acetic acid-water 4:1:5, *II* acetic acid-water 3:17. Detection: UV 365 nm light after spraying with 1% $AlCl_3$. Compounds: *1* ellagic acid, *2* 3-*O*-methylellagic acid, *3* 3,3′-di-*O*-methylellagic acid, *4* 3,3′-di-*O*-methylellagic acid 4-*O*-glucoside, *5* 3,3′-di-*O*-methylellagic acid 4,4′-di-*O*-glucoside; *6* kaempferol 3-*O*-glucoside and galactoside, *7* quercetin 3-*O*-glucoside and galactoside, *8* 2″-*O*-galloyl esters of quercetin 3-*O*-glucoside and galactoside; *9* hydroplumbagin 4-*O*-glucoside, *10* 1-*O*-galloylglucose. Assignments of spots of compounds according to Pakulski (1994)

4,4′-di-*O*-glucoside and 1-*O*-galloylglucose; flavonoids: kaempferol and quercetin 3-*O*-glucosides and 3-*O*-galactosides, 3-*O*-(2″-*O*-galloyl)glucoside and 3-*O*-(2″-*O*-galloyl)galactoside (Pakulski 1994, Pakulski and Budzianowski 1996a,b). Moreover, the nonphenolic compounds – α- and β-carotene – have also been isolated (Pakulski 1994). However, in the above-mentioned studies we could not isolate hydroplumbagin glucoside which was previously isolated by Kreher et al. (1990) from the plants for the production of the drug Carnivora and much earlier also from *Drosera rotundifolia* L. grown in vivo (Vinkeborg et al. 1969). The presence of this glucoside in our material was suspected from the steady emergence of free plumbagin during preparative chromatography of water fractions of the extract, as well as the presence of spots of this compound on a two-dimensional thin layer chromatogram (2D-TLC; Fig. 7) compared with that of *Drosera intermedia* Hayne, from which it was isolated later (Budzianowski 1996). This glucoside is very unstable and releases free plumbagin by spontaneous hydrolysis and oxidation as observed on the sample from *Drosera intermedia* (Budzianowski 1996).

In our work, this compound was successfully obtained by methods previously applied for the isolation of naphthohydroquinone glucosides from *Drosera* species (Budzianowski 1995, 1996). Thus, the fresh plants of *Dionaea* from in vitro culture on MS medium maintained at our department in Poznań (Fig. 1A) were extracted by a two-fold maceration with methanol within 4 days, and the extract processed to obtain a distillate of the water extracted from tissues and chloroform, 1-butanol and water-soluble fractions. The butanol fraction, upon subsequent column chromatography over alumina and Sephadex LH-20, yielded pure hydroplumbagin 4-*O*-glucoside (Budzianowski, unpubl.), ^{1}H- and ^{13}C-NMR (Fig. 8) and UV spectra of which exactly matched

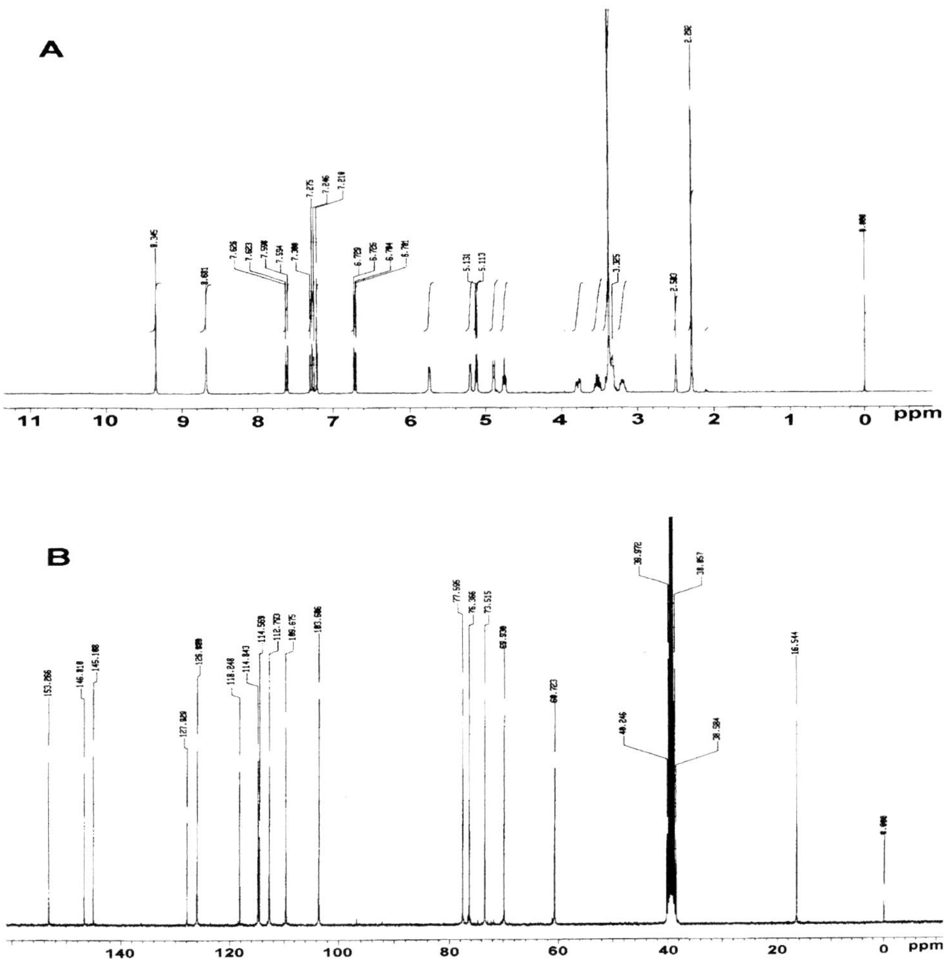

Fig. 8. NMR spectra (in DMSO-d_6) of hydroplumbagin glucoside isolated from *Dionaea muscipula* from in vitro culture (1/2 MS medium). **A** ^{1}H-NMR (300 MHz), **B** ^{13}C-NMR (75 MHz)

those reported for this compound (Kreher et al. 1990). The water distillate contained plumbagin, while the chloroform fraction showed additionally the presence of previously reported naphthoquinones (Pakulski 1994; Pakulski and Budzianowski 1996a) and also that of droserone, reported by Kreher et al. (1990); all by co-chromatography on silica gel plates against authentic samples (droserone was from *Drosera gigantea* Lindley; Budzianowski et al. 1998; Budzianowski 2000).

The butanol fraction by 2D TLC on cellulose (Fig. 7) showed the presence of all compounds previously found in the material from Reinert-Mohr medium. (Pakulski 1994; Pakulski and Budzianowski 1996a,b). It should be noted that the composition of phenolics in *Dionaea muscipula* seems to be very similar to that of species of the genus *Drosera* of the same family (Budzianowski et al. 1993a,b). However, the noticeable, small differences are: (1) the complete absence of 7-methyljuglone and its bound form, rossoliside (7-methyl*hydro*juglone 4-*O*-glucoside), in *Dionaea* (Pakulski and Budzianowski 1996a; Budzianowski 1996, 1997), which are frequently present in the genus *Drosera* L. (Vinkeborg et al. 1970; Sampara-Rumantir 1971; Culham and Gornall 1994; Budzianowski et al. 1998; Samaj et al.. 1999); (2) the apparent presence of 3-*O*-methylellagic acid in *Dionaea* (Fig. 6), which could only be detected in trace amounts in *Drosera* species and was isolated from *Drosera aliciae* Hamet (Budzianowski et al. 1996), whereas 1-*O*-galloylglucose was isolated from *D. spathulata* Labill. (Budzianowski et al. 1995). Amongst the flavonoids, the only difference between these two genera is the site of galloylation which is at the C-2″ position in *Dionaea* (Pakulski and Budzianowski 1996b; Fig. 6) but at the C-6″ in the case of *D. aliciae* (Budzianowski et al. 1996).

Although no quantitative determination was made for the above compounds in *Dionaea* from in vitro culture, we can estimate from the isolation yields reported in the above-cited works that the major constituents are plumbagin, hydroplumbagin glucoside and ellagic acid methyl ethers.

Interestingly, the red pigmentation of the leaves due to the anthocyanins – cyanidin and delphinidin 3-*O*-glucosides – appeared in *Dionaea* cultured on 1/2 MS agar medium with sufficiently diluted nitrogen components (NH_4NO_3 and KNO_3) and increased sucrose content (1.5%) in the medium (Ichiishi et al. 1999).

4 Conclusions and Prospects

Dionaea muscipula Ellis is a carnivorous plant, the natural occurrence of which is restricted to a small area in North America. The plant catches its prey in snap traps by a rapid movement at a rate similar to that of animals. Hence, it is considered as a model for a biomimetic material with built-in sensors and actuations (Shahinpoor and Thomson 1995).

Dionaea muscipula can effectively be mass-propagated in vitro. Seeds disinfected by surface sterilisation germinated in vitro produce numerous seedlings, which after a few months developed into plantlets. For mass propagation, excised shoots with pieces of rhizomes or excised leaves are used. The researchers have been unanimous in their opinion that the propagation rate is high enough to allow the production of numerous plants in a short time.

One shoot produces 20–30 lateral shoots in 3 months, producing ca. 380,000 plants per annum. One excised leaf, which undergoes two multiplication cycles per annum, can produce 4000 shoots and plantlets per annum (Kukułczanka 1991).

Propagation of plants on a half-strength MS medium and RM medium with or without such growth regulators as cytokinins (2iP and kinetin) and auxins (NAA and IBA) at low concentrations gives satisfactory results. The optimum conditions are: temperature 20–26 °C, light 100 nE $m^{-2} s^{-1}$, photoperiod 16 h.

Because no plant variations have been reported, it is possible to develop a potential procedure of revitalising continuously grown in vitro cultures of *Dionaea muscipula*. A drop in temperature to 5 °C allows the long-term storage of cultures and stimulates rooting.

The plant is considered to have anticancer properties and the expressed juice of the fresh plants provides the commercially available drug designed for unconventional treatment (Rote Liste 2000). Both extracts of *Dionaea*, as well as its constituent, plumbagin, were shown to have dual in vitro cytotoxic and immunostimulating activity, like some known anticancer agents. It is interesting that another carnivorous plant with anticancer properties, *Sarracenia flava* (golden trumpet, Sarraceniaceae), containes no naphthoquinones and the active principles appeared to be terpenes (Bhattacharyya et al. 1976; Miles et al. 1976; Juniper et al. 1989).

The plants obtained by micropropagation on either Reinert-Mohr or diluted MS medium produce such phenolic compounds like naphthoquinones, phenolic (ellagic and gallic) acids derivatives and flavonoids. Many of them are present in *Drosera* species obtained by in vitro micropropagation. It should be emphasised that ellagic acid is well known for its chemopreventive, anticancer properties (e.g. Josephy and Snieckus 1992; Ahn et al. 1996). Pharmacologists conclude that further pharmacological and clinical investigations to evaluate the usefulness of preparations from *Dionaea muscipula* in cancer therapy are desirable (Todorov et al. 1996c, 1998). Our suggestion is that further possibilities in this area should be searched for within ellagic acid derivatives and maybe other, so far unknown, constituents of *Dionaea* like non-phenolic compounds, in analogy with the case of *Sarracenia* (see above).

5 Protocol

5.1 Tissue Culture

Material: seed, seedlings, in vitro plantlets
Explants: excised shoots or leaves
Media: 0.5 MS or RM, pH 4.9–5.7
Growth regulators: 0.2 mg/l 2iP or 0.2–1.0 mg/l kinetin, 0.2–1.0 mg/l NAA, or IBA
Optimum conditions: temperature 20–26 °C, long-term subcultures 5–16 °C
Photon flux density: 75–125 (optimum 100) $nE\,m^{-2}\,s^{-1}$
Photoperiod: 16 h

5.2 Phytochemical Analysis

Analytical procedures (Budzianowski 1995, 1996): Fresh plants are plunged into boiling MeOH for a few minutes, left for maceration not longer than 7 days, evaporated at 40 °C in vacuo to dryness, collection of the distillate of water separately (Budzianowski 1995). Distillate is extracted with toluene. The dry extract is dissolved in water and extracted (three times) with $CHCl_3$ and 1-BuOH, saturated with water. Toluene, $CHCl_3$ and 1-BuOH fractions are concentrated to dryness as above.

Free naphthoquinones: toluene and $CHCl_3$ fractions, and reference plumbagin, dissolved in $CHCl_3$, are applied to a silica gel plate and developed in mobile phase: toluene-HCOOH 99:1 (Kreher et al. 1990; Budzianowski 1995). Detection: daylight – yellow bands.

Polar phenols: BuOH fraction dissolved in MeOH is applied to 2D TLC cellulose plate and developed in 1-BuOH-HOAc-H_2O 4:1:5, upper phase – first direction and HOAc-H_2O 3:17 – second direction. Detection: UV light 365 nm after spraying with 1% $AlCl_3$ solution in ethanol followed by heating at 105 °C. Hydroplumbagin glucoside – red (also red in daylight); phenolic acids derivatives – mauve and blue; flavonoids – yellow.

Acknowledgements. The authors are grateful to Dr. Barbara Thiem, in our department at Poznań, for preparing some photographs; Prof. Kazimierz Trębacz, Maria Curie-Skłodowska University, Lublin, Poland, for comments on *Dionaea* physiology; Carnivora-Forschungs GmbH, Nordhalben, Germany, for providing papers on pharmacological investigations.

References

Ahn D, Putt D, Kresty L, Stoner GD, Fromm D, Hollenberg PF (1996) The effects of dietary ellagic acid on rat hepatic and esophageal mucosal cytochrome P450 and phase II enzymes. Carcinogenesis 17:821–828

Bhattacharayya J, Kokpol U, Miles DH (1976) The isolation from *Sarracenia flava* and partial synthesis of betulinaldehyde. Phytochemistry 15:432–433

Beebe JD (1980) Morphogenetic responses of seedlings and adventitious buds of the carnivorous plant *Dionaea muscipula* in aseptic culture. Bot Gaz 141:396–400

Budzianowski J (1995) Naphthoquinones of *Drosera spathulata* from in vitro cultures. Phytochemistry 40:1145–1148

Budzianowski J (1996) Naphthohydroquinone glucosides of *Drosera rotundifolia* and *D. intermedia* from in vitro cultures. Phytochemistry 42:1145–1147

Budzianowski J (1997) 2-Methyl naphthazarin 5-*O*-glucoside from the methanolic extracts of in vitro cultures of *Drosera* species. Phytochemistry 44:75–77

Budzianowski J (2000) Naphthoquinone glucosides of *Drosera gigantea* from in vitro cultures. Planta Med 66:667–669

Budzianowski J, Pakulski G, Skrzypczak L (1993a) Ellagic acid derivatives and further naphthoquinones from *Dionaea muscipula* and four species of the genus *Drosera* from in vitro cultures. 41th Annu Congr Med Plant Res, 31 August–4 September 1993, Düsseldorf, Abstr, p 79. Planta Med 59 (Suppl A 654)

Budzianowski J, Skrzypczak L, Kukułczanka K (1993b) Phenolic compounds of *Drosera intermedia* and *D. spathulata* from in vitro cultures. Acta Hortic 330:277–280

Budzianowski J, Pakulski G, Skrzypczak L (1995) Phenolic metabolites of Droseraceae species from in vitro cultures. 43th Annu Congr Med Plant Res, Halle (Saale), Abstr, p 34

Budzianowski J, Ratajczak L, Skrzypczak L (1996) Flavonoids of Droseraceae species from in vitro cultures. 44th Annu Congr Med Plant Res, Prague, Abstr, p 140

Budzianowski J, Ratajczak L, Kromer K (1998) Phenolic compounds in further species of the genus *Drosera* L. from in vitro cultures. 46th Annu Congr Med Plant Res, 30 August–4 September 1998, Vienna, Abstr, p C19

Crouch IJ, Finnie JF, van Staden J (1990) Studies on the isolation of plumbagin from in vitro and in vivo grown *Drosera* species. Plant Cell Tissue Organ Cult 21:79–82

Culham A, Gornall RJ (1994) The taxonomic significance of naphthoquinones in the Droseraceae. Biochem Syst Ecol 22:507–515

Di Gregorio GJ, Di Palma JR (1966) Anthocyanin in *Dionaea muscipula* Ellis (Venus's flytrap) Nature (Lond) 212:1264–1265

Dietzel U, Sauer R, Reichardt U (1985) Erfahrungen mit Carnivora. Fortschr Med 103:760–781

Drees M, Berger DP, Dengler WA, Fiebig GH (1996) Direct cytotoxic effects of preparations used as unconventional methods in the clonogenic assay and in nude mice. Contrib Oncol 51:115–122; Chem Abstr 1996 125:75628u

Encke F (1958) *Dionaea* L. In: Encke F (ed) Pareys Blumengärtnerei, vol 1, 2nd edn. Paul Parey, Berlin, pp 713–714

Finnie JF, van Staden J (1993) *Drosera* spp. (Sundew): micropropagation and the in vitro production of plumbagin. In: Bajaj YPS (ed) Biotechnology in agriculture and forestry, vol 24. Medicinal and aromatic plants V. Springer, Berlin Heidelberg New York, pp 164–177

Fujii N, Yamashita Y, Arima Y, Nagashima M, Nakano H (1992) Induction of topoisomerase II-mediated DNA cleavage by the plant naphthoquinones plumbagin and shikonin. Antimicrob Agents Chemother 36:2589–2594

Galek H, Osswald WF, Elstner EF (1990) Oxidative protein modification as predigestive mechanism of the carnivorous plant *Dionaea muscipula*: an hypothesis based on in vitro experiments. Free Radical Biol Med 9:427–434

Gershbein LL, Brown MF, Hoepsch JP, Young DC (1995) Sterol composition and distribution in carnivorous plants, *Sarracenia flava*, *Sarracenia purpurea* and *Dionaea muscipula*. Trans Ill Acad Sci 88:13–20; Chem Abstr 1995 123:52269x

Gujar GT (1990) Plumbagin, a naturally occurring naphtoquinone. Its pharmacological and pesticidal activity. Fitoterapia 61:387–394

Hamlet EH (1995) *Dionaea*. In: Cullen J, Alexander JCM, Brady A (eds) The European garden flora, vol 4, part 2. Dicotyledons. Cambridge Univ Press, Cambridge, p 88

Hauser SP (1988) Carnivora – Phytotherapeuticum zur Behandlung maligner Erkrankungen. Schweiz Rundsch Med (PRAXIS) 77:283–287

Hegnauer R (1989) Chemotaxonomie der Pflanzen, vol 8. Birkhäuser, Basel

Hegnauer R (1990) Chemotaxonomie der Pflanzen, vol 9. Birkhäuser, Basel

Hodick D, Sievers A (1986) The influence of Ca^{2+} on the action potentials in mesophyll cells of *Dionaea muscipula* Ellis. Protoplasma 133:83–84

Hodick D, Sievers A (1989) Action potentials of *Dionaea muscipula* Ellis. Planta 179:32–34

Hutchinson JF (1984) In vitro propagation of *Dionaea muscipula* Ellis (Venus fly trap). Sci Hortic 22:189–194

Ichiishi S, Nagamitsu T, Kondo Y, Iwashina T, Kondo K, Tagashira N (1999) Effects of macrocomponents and sucrose in the medium on in vitro red-color pigmentation in *Dionaea muscipula* Ellis and *Drosera spathulata* Labill. Plant Biotechnol (Tokyo) 16 (3):235–238

Jay M, Lebreton P (1972) Chemotaxonomic research on vascular plants 26. The flavonoids of Sarraceniaceae, Nepenthaceae, Droseraceae and Cephalotaceae, a critical study of the order Sarraceniales. Nat Canadienne 99:607–613

Josephy PD, Snieckus VA (1992) Ellagic acid and synthetic analogs as inhibitors of mutagenesis and carcinogenesis. Cell Mol Targets Chemoprev 1992:147–160

Juniper BE, Robins RJ, Joel DM (1989) The carnivorous plants. Academic Press, London

Keller H (1987) Pharmaceutical containing the digestive juices of carnivorous plants for treatment of malignant and chronic diseases. Ger Offen DE 1987 3:619; Chem Abstr 1988 109:79697h

Kreher B, Neszmelyi A, Polos K, Wagner H (1988a) Structure elucidation of plumbagin-analogues from *Dionaea muscipula* and their in vitro immuno-modulating activities in vitro and in vivo. Proc Int Symp Hungarian Chem Soc, Sopron (Hungary) 24-27.06.88, pp 1–16

Kreher B, Neszmelyi A, Polos K, Wagner H (1988b) Structure elucidation of plumbagin analogues from *Dionaea muscipula* and their in vitro immunological activity on human granulocytes and lymphocytes. 36th Annu Congr Med Plant Res, Freiburg, Abstr, p 45

Kreher B, Neszmelyi A, Wagner H (1990) Naphthoquinones from *Dionaea muscipula*. Phytochemistry 29:605–606

Kukułczanka K (1991) Micropropagation and in vitro germplasm storage of Droseraceae. Bot Gard Micropropagation News 1:38–42

Kukułczanka K, Cząstka B (1987) Propagation of *Drosera* L. in vitro conditions, Wiad Bot Biul Ogrodów Bot 31:61–64

Kukułczanka K, Cząstka B (1991) Vegetative propagation of selected species of Droseraceae taxa and in vitro gene bank establishment. Prace Ogr Bot PAN 1:55–61

Kukułczanka K, Cząstka B, Arczewska A (1989) Regeneration from leaves of *Dionaea muscipula* Ellis cultured in vitro. Acta Hortic 251:155–160

Kukułczanka K, Kromer K, Klonowska B, Cząstka B (1991) The temperature effect on the preservation of the plant cultures of Droseraceae in vitro. Prace Ogr Bot PAN 1:63–68

Linsmaier EM, Skoog F (1965) Organic growth factor requirements of tobacco tissue cultures. Physiol Plant 18:100–127

Lloyd FE (1942) The carnivorous plants. Chronica Botanica. Waltham, MA, USA, pp 177–194

Miles DH, Kokpol U, Bhattacharayya J, Attwood JL, Stone KE, Bryson TA, Wilson C (1976) Structure of sarracenin. An unusual enol diacetal monoterpene from the insectivorous plant *Sarracenia flava*. J Am Chem Soc 98:1569–1573

Miyoshi E, Shizuri Y, Yamamura S (1984) Isolation and structures of diomuscinone and diomuscipulone from *Dionaea muscipula*. Phytochemistry 23:2385–2387

Murashige T, Skoog F (1962) A revised medium for rapid growth and bio-assays with tobacco cultures. Physiol Plant 15:473–497

Pakulski G (1994) Związki fenolowe z gatunku *Dionaea muscipula* Ellis pochodzącego z kultur in vitro. (Phenolic compounds from the species *Dionaea muscipula* Ellis of in vitro culture origin) Dissertation, Faculty of Pharmacy, K.Marcinkowski University of Medical Sciences, Poznań

Pakulski G, Budzianowski J (1996a) Ellagic acid derivatives and naphthoquinones of *Dionaea muscipula* from in vitro cultures. Phytochemistry 41:775–778

Pakulski G, Budzianowski J (1996b) Flavonoids of *Dionaea muscipula* from in vitro cultures. Planta Med 62:95–96

Parliman BJ, Evans PT, Rupert EA (1982a) Tissue culture of single rhizome explants of *Dionaea muscipula* Ellis ex L., the Venus fly-trap, for rapid asexual propagation. J Am Soc Hortic Sci 107 (2):305–310

Parliman RJ, Evans PT, Mazur AR (1982b) Adventitious bud differentiation and development in leaf cuttings of *Dionaea muscipula* Ellis ex L. (Venus fly-trap) cultured in vitro. J Am Soc Hortic Sci 107 (2):310–316

Ragazzi E, de Biasi M, Pandolfo L, Chinellato A, Caparrotta L (1993) In vitro effects of naphthoquinones isolated from *Drosera* species. Pharmacol Res 27(Suppl 1):87–88

Reinert RA, Mohr HC (1967) Propagation of *Cattleya* by tissue culture of lateral bud meristems. Proc Am Soc Hortic 91:664–671

ROTE LISTE (2000) Verzeichnis von Fertigarzneimitteln der Mitglieder des Bundesverbandes der Pharmazeutischen Industrie. Cantor, Aulendorf

Šamaj J, Blehová A, Repčák M, Ovečka M, Bobák M (1999) *Drosera* species (Sundew): in vitro culture and the production of plumbagin and other secondary metabolites. In: Bajaj YPS (ed) Biotechnology in agriculture and forestry, vol 43. Medicinal and aromatic plants XI. Springer, Berlin Heidelberg New York, pp 105–135

Sampara-Rumantir N (1971) Rossoliside. Pharm Weekbl 106:653–664

Shanhinpoor M, Thompson MS (1995) The Venus flytrap as a model for a biomimetic material with built-in sensor and actuations. Mater Sci Eng C C2:229–233; Chem Abstr 1996 124:6160r

Slack A (1985) Karnivoren: Biologie und Kultur der Insektenfangenden Pflanzen. Ulmer, Stuttgart, pp 243–245 (15–19), 167–173

Teng WL (1999) Source, etiolation and orientation of explants affect in vitro regeneration of venus fly-trap (*Dionaea muscipula*). Plant Cell Rep 18:363–368

Todorov DK, Ilarianova MV(1996) Antitumor activity of the *Dionaea muscipula* Ellis. Preparation Carnivora New in vitro and in vivo on several animal and human tumors, sensitive and resistant to antitumor drugs. Carnivora-Forschungs GmbH Nordhalben Firm's Information

Todorov DK, Ilarianova MV, Maneva K (1995a) Antiproliferative activity of Carnivora on two human glioma cell lines. Carnivora-Forschungs GmbH Nordhalben Firm's Information

Todorov DK, Ilarianova MV, Maneva K (1995b) Antiproliferative activity of Carnivora on multidrug-resistant sarcoma 180/dox constant tumor cell line. Carnivora-Forschungs GmbH Nordhalben Firm's Information

Todorov DK, Ilarianova MV, Maneva K (1996a) Experimental studies on the antiproliferative activity of Carnivora on human leukemic CEM cell line – multidrug-resistant (MDR). Carnivora-Forschungs GmbH Nordhalben Firm's Information

Todorov DK, Ilarianova MV, Maneva K (1996b) Experimental studies on the antiproliferative activity of Carnivora on human leukemic CEM cell line – multidrug-resistant (MDR). Carnivora-Forschungs GmbH Nordhalben Firm's Information

Todorov DK, Ilarionowa MV, Timcheva KB, Pajeva IK (1998) Antitumor activity of *Dionaea muscipula* E. preparation Carnivora in animal and human tumors sensitive and resistant to antitumor drugs. Biotechnol Biotechnol Equip 1998 (2):61–66

Trębacz K, Sievers A (1998) Action potentials evoked by light in traps of *Dionaea muscipula* Ellis. Plant Cell Physiol 39:369–372

Trębacz K, Bush MB, Hejnowicz Z, Sievers A (1996) Cydopiazonic acid disturbs the regulation of cytosolic calcium when repetitive action potentials are evoked in *Dionaea* traps. Planta 198:623–626

Vinkeborg J, Sampara-Rumantir N, Uffielie OF (1969) Presence of hydroplumbagin glucoside in *Drosera rotundifolia*. Pharm Weekbl 104:45–49

Vinkeborg J, Sampara-Rumantir N, Uffelie OF (1970) Rossoliside, a glucoside from *Drosera rotundifolia* L. Pharm Weekbl 105:414

Wagner H, Kreher B, Jurcic K (1986) Immunological investigations of naphthoquinone containing plant extracts, isolated quinones and other cytostatic compounds in cellular immunosystems. 34th Annu Congr on Med Plant Res, Hamburg, Germany, Abstr, pp 69–70

Wagner H, Kreher B, Jurcic K (1988) In vitro stimulation of human granulocytes and lymphocytes by pico and femtogram quantities of cytostatic agents. Arzneim-Forsch Drug Res 38:273–275

Wurm G, Grimm H, Geres U, Schmidt H (1984) Plumbagin: Reaktivität, toxicität und mikrobielle aktivität des in *Drosera* und *Plumbago* arten vorkommenden. Dtsch Apoth Ztg 124:2128–2132

Zenk MH, Fürbringer M, Steglich W (1969) Occurence and distribution of 7-methyljuglone and plumbagin in the Droseraceae. Phytochemistry 8:2199–2200

Zimmerman RH, Broome OC (1980) Blueberry micropropagation. Proc Conf Nursey Culture – Applications and Feasibility, 21–22 April 1980, Beltsville, USA, pp 44–47

5 In Vitro Culture and the Production of Secondary Metabolites in *Hyoscyamus reticulatus L.*

I. IONKOVA[1]

1 Introduction

Tropane alkaloids constitute one of the distinctive groups of secondary metabolites of the Solanaceae and many plants containing them have long been utilized for their medicinal, hallucinogenic, and poisonous properties (Evans 1979). *Hyoscyamus* plants are a natural source for the isolation of hyoscyamine (atropine) and scopolamine, 6–7 epoxide of hyoscyamine. Both alkaloids are of medicinal importance because of their suppressive activity on the parasympathetic nervous system. In addition, scopolamine is also applied to suppress the central nervous system, whereas hyoscyamine excites it. Ratios of hyoscyamine content to scopolamine content vary markedly between plant species. These differences result in a higher commercial demand for scopolamine than for hyoscyamine (and its racemic form atropine). Both appear in the USA in the list of the ten most used compounds of plant origin (Straus 1989). Because many tropane alkaloid-producing species accumulate hyoscyamine as the major alkaloid and scopolamine in minor quantities, it is of commercial importance to increase scopolamine content in these species (Hashimoto and Yamada 1992). Moreover, these plants also synthesize the calystegines, a pseudotropine-derived group of alkaloids, found in considerable amounts in *Atropa* and *Hyoscyamus* species, exerting a strong inhibitory activity on β-glycosidases, which can be of great importance for cell recognition (Portsteffen et al. 1994). The latter can possibly result in treatment of some types of cancers.

This chapter mainly deals with *H. reticulatus* and describes the optimization of culture conditions for growth and productivity of tropane alkaloids from transformed *H. reticulatus* roots. Methodologies to enhance secondary metabolite levels are also discussed, together with a new technique of permeabilization of *H. reticulatus*-hairy roots for the release of stored tropane alkaloids. These recent developments are largely based on our published and unpublished experimental results.

[1]Department of Pharmacognosy, Faculty of Pharmacy, Medical University, Sofia, 2 Dunav Str., 1000 Sofia, Bulgaria

Biotechnology in Agriculture and Forestry, Vol. 51
Medicinal and Aromatic Plants XII (ed. by T. Nagata and Y. Ebizuka)
© Springer-Verlag Berlin Heidelberg 2002

1.1 Distribution and Importance of the Plant

The four genera of subtribe Hyoscyaminae (Scopolia, Physochlaina, Prezewalskia, Hyoscyamus) produce hyoscyamine as the usual major alkaloid with some hyoscine, but some *Hyoscyamus* species may contain hyoscine as the principal alkaloid.

Hyoscyamus plants are herbaceous, differing essentially from other solanaceous genera by having a fruit composed of a thornless seed capsule with a cover. The genus consists of 14 species: *H. albus* L. (including var. *desertorum*, var. *canariensis*), *H. arenarius* Dun., *H. aureus* L., *H. ceratophyllus* Fisch., *H. eminens* Kunce, *H. falezlez* Coss., *H. grandiflorus* Franch., *H. muticus* L., *H. niger* L. (including var. *agrestis*, var. *bohemicus*, var. *pallidus*), *H. physaloides* L., *H. pseudophysaloides* Roth., *H. pussilus* L., *H. reticulatus* L. and *H. senecionis* Willd (Jackson 1977).

Hyoscyamus reticulatus, henbane, is an annual or biennial plant, native to arid and semi-arid regions of Egypt, south-west Asia, Iran, and Turkey. The plant consists of an erect, simple ore branched stem, covered with basal and cauline leaves petiolate, upper leaves sessile, pinnatipartite, lobate or entire. Calyx is 10–20 mm, fruit 20–30 mm, not constricted at middle, teeth cuspidate, pungent, patent to deflexed. The corolla is pale yellowish, soon becoming purplish-violet with darker reticulate veins, 20–30 mm, almost campanulate with yellow anthers. It is found on cereal fields, waste places, roadsides, vineyards from 10 to 1700 m (Deivis 1978).

1.2 *H. reticulatus* as a Source of Tropane Alkaloids

Most pharmacological effects of *Hyoscyamus* drugs can be attributed to tropane alkaloids – hyoscyamine and scopolamine. They have a wide range of pharmaceutical applications, including uses as mydriatica, spasmolytica, analgetica, sedativa, antiperspirants, antidotes to intoxications, or for their beneficial effects in cases of asthma, Parkinson's disease, ulcers, and motion sickness (Cordell 1978). These species also contain various other tropane alkaloids: apohyoscyamine, norhyoscyamine, littorine, tropine, cuscohygrine, tigloidine, tigloyloxytropane (Evans 1979). Recently, the importance of the henbane drug has decreased, being often replaced by the pure tropane alkaloids, which allow a more controlled and specific application (Strauss 1989). The *Hyoscyamus* genera produce hyoscyamine as the major alkaloid. Scopolamine is more active than hyoscyamine, and is formed from hyoscyamine via 6-hydroxyhyoscyamine. The extent to which this conversion occurs often depends on the age of the plant, particular variety or species, and geographical source, and is an important factor in determining the ultimate use and pharmacological properties of the plant (Evans 1979). Other secondary metabolites detected in *Hyoscyamus* plants are: flavonoids (rutin), chlorogenic acid, coumarins, tannins (Hoppe 1975).

Table 1. Content of hyoscyamine and scopolamine during different growth periods in *Hyoscyamus reticulatus* L. plant

Growth period	Roots		Leaves	
	Hyoscyamine %/DW	Scopolamine %/DW	Hyoscyamine %/DW	Scopolamine %/DW
1. Plant with ten leaves	0.093	0.036	0.201	0.164
2. Plant before flowering	0.163	0.061	0.631	0.320
3. Plant during seed formation	0.226	0.049	0.231	0.154

Pelt et al. (1967) reported that the main tropane alkaloid of the *H. reticulatus* plant is hyoscyamine in the range from 0.033 to 0.056% dry weight (DW), followed by less scopolamine from 0.011 to 0.015% DW, depending on the strain and the age of the plant and growth conditions. In our investigation (Ionkova 1992b) with the *H. reticulatus* plant, we have determined the content of hyoscyamine and scopolamine during different growth periods – the plant with ten leaves, plant before flowering and plant during seed formation (Table 1). The content of both alkaloids in the leaves and roots is maximal before the flowering stage, and maximal flower scopolamine content is before flowering. The hyoscyamine and scopolamine contents in the leaves are 0.631 and 0.320% DW, respectively (plant before flowering), whereas the production in the intact roots at the same time is even four times less. The changes in the content of both main alkaloids in *H. reticulatus* were greater in the leaves than in the roots during different growth periods (Table 1). In contrast to the data of Pelt et al. (1967), the maximum hyoscyamine and scopolamine yields (11 and 20 times higher, respectively) were found in the leaves before flowering. In addition, apoatropine and littorine were detected. The roots produced trace amounts of hygrine, tropine and cuscohygrine (Ionkova 1992b).

2 In Vitro Approaches

Hyoscyamus species are amenable to different in vitro techniques which offer possibilities for plant improvement and tropane alkaloid production. Numerous laboratories have directed a great deal of time and effort in attempts to produce scopolamine and hyoscyamine with plant cell cultures. Basic plant cell culture techniques described in several books can also be applied to cell and tissue culture of *Hyoscyamus* species (Strauss 1989). However, the alkaloid levels in disorganized cells are very low. For many years, unsuccessful attempts were made to produce tropane alkaloids by means of cell suspension cultures (Verpoorte et al. 1991). In the last 10 years, the application of differentiated organ cultures has led to significant advances in the in vitro production of secondary products (Mano et al. 1989; Robins et al. 1990; Ionkova and Alfermann

1994; Ionkova 1996, 1999). Transformed roots induced by infection of plants with *Agrobacterium rhizogenes* have been investigated for several years as a mean of producing plant secondary metabolites in vitro. As the biosynthesis of tropane alkaloids is obviously restricted to an organized state of the tissue, root cultures are a more promising system for alkaloid production.

Infection of *Hyoscyamus* plant cells by *Agrobacterium rhizogenes* results in "hairy root" disease at the sites of infection (Ionkova et al. 1989, 1994). This morphogenic event is due to the transfer of genetic information by Ri plasmids from bacteria to the plant cells. The integration and expression of the transferred DNA (T-DNA) of Ri plasmids cause metabolic changes mainly determined by genes involved in auxin syntheses. Two regions, TL- and TR-DNA, were found to be integrated and stably conserved in the plant genome.

Solanaceous hairy root cultures are the most intensively studied because of their ability to synthesize relatively high and stable amounts of tropane alkaloids (Petri and Bajaj 1988). The transformed roots grow rapidly and branch profusely on hormone-free medium. They are able to synthesize specific metabolites, opines, which are catabolized by free-living Agrobacteria. Transformation of tropane-alkaloid-producing plants readily yields the so-called hairy roots, which can be cultured easily in vitro and produce alkaloid levels comparable to the parent plant. In hairy roots of various *Hyoscyamus* species, alkaloid content ranged from 0.1 to 0.7% DW (Verpoorte et al. 1993). The production of tropane alkaloids by root cultures of a variety of solanaceous species has been investigated in the last few years by a number of groups (Yamada and Hashimoto 1988; Christen and Roberts 1993; Ionkova et al. 1994; Boitel-Conti et al. 1995; Ionkova 1995b, 1996; Sevon et al. 1998), attracted by the considerably higher productivity of such cultures in comparison to the equivalent cell suspension cultures. In view of the apparently predictable relationship between the biosynthetic properties of transformed roots and those of parent plants, it should prove possible to understand the variations seen in root cultures in terms of the properties of the different genera, species and individuals. Screening plant populations for individuals showing particularly favorable characteristics – such as unusually high alkaloid productivity, or the production of a particular alkaloid without significant accumulation of other unwanted alkaloids – should thus be an important step in the development of commercially and scientifically useful transformed cultures (Parr et al. 1990).

2.1 Establishment and Growth of Hairy Root Clones of *H. reticulatus*

The studies on in vitro cultures of *H. reticulatus* began with the report on transformed roots and its production of secondary metabolites (Ionkova 1992b). In our current investigation on the constituents and distribution of tropane alkaloids in normal and transformed root cultures of solanaceous plants (Ionkova et al. 1989, 1990, Ionkova 1992b), we established various hairy roots of *H. reticulatus* (Fig. 1). Four different bacterial strains were used in our experiments – TR 105, R 1601, LBA 9402 and ATCC 15834 – to produce hairy root clones. Young leaves or stems of sterile *H. reticulatus* were used for the induction of hairy roots. The susceptibility of *H. reticulatus* to infection with different bacterial strains was variable. The difference in virulence could be explained by the plasmids harbored by bacterial strains (Cardelli et al. 1985). Our experiments showed that transformation by TR 105 and R 1601 occurs at a low frequency compared to results observed using strains ATCC 15834 and LBA 9402 (Fig. 2). Similar results were found working with *Astragalus* species (Ionkova 1999). The different *Agrobacterium* strains also influenced the growth, morphology and alkaloid production of hairy roots.

Agrobacterium rhizogenes – ATCC 15834 – showed the best virulence characteristics. ATCC 15834 belongs to the agropine-type strains eliciting hairy roots containing agropine, manopine, manopinic acid and agropinic acid (Petit et al. 1983). The opine production of different transformed root clones was examined by paper electrophoresis to confirm whether transformation with Ri plasmid had occurred. Only agropine was detected in the freshly isolated hairy root extract, indicating that they were transformed. Opines could not be found in the non-transformed root extracts of the same species. The variability of the opine content in hairy roots transformed by *Agrobacterium rhizogenes* ATCC 15834 has been previously reported (Christen et al. 1989).

Bacteria were eliminated on agar-solidified MS medium (Murashige and Skoog 1962), supplemented with 0.5 mg/ml Claforan. After 3 weeks of culture, the tips of newly formed roots were excised and placed on fresh medium of the same composition. The roots were cultured at 25 °C in the dark, at 80 ppm and subcultured at 4-week intervals. The 36 clones transformed by ATCC 15834 were cultured in liquid medium, and their growth rates and alkaloid contents were determined. Marked differences were found in the growth rates of these 36 clones. The most rapidly growing clones were Hr-3, Hr-13, Hr-15, Hr-27 and Hr-31. Clones Hr-7 and Hr-21 grew much more slowly. The growth index for clone Hr-3 was ten times that of clone Hr-7.

From the 24 clones transformed with LBA 9402, clone Hr-L12, producing high levels of tropane alkaloids (1.44% DW), was selected. The root cultures transformed with the two different strains of *A. rhizogenes* showed different characteristics. We presume that the Ri-plasmids induced various clones with different genotypes inducing variation of the alkaloid productivity.

2.2 Tropane Alkaloid Production of Transformed Roots of *H. reticulatus*

Various reliable procedures for the laboratory extraction of tropane alkaloids from *Hyoscyamus* and *Datura* drugs were worked out and discussed (Straus 1989). Numerous methods were used for its detection and determination, including titration, colorimetry, UV-spectrophotometry, TLC, GC/MS, HPLC, immunoassays, biological assays (mydriatic response, neuroreceptor binding).

Fig. 1. *Hyoscyamus reticulatus* L. in vitro plant and hairy root cultures

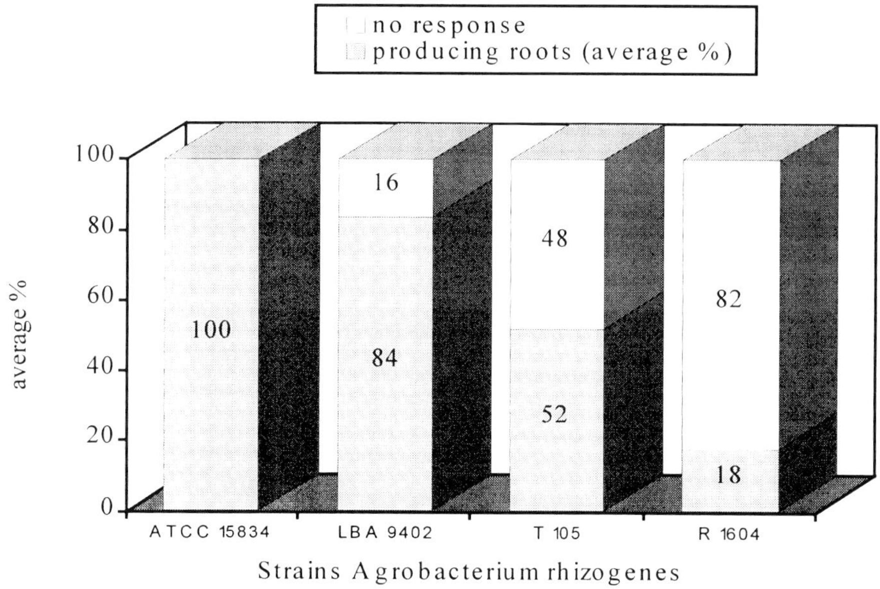

Fig. 2. Response of *Hyoscyamus reticulatus* L. to different bacterial strains

In order to analyze the variation in the content of tropane alkaloids, the content of scopolamine and hyoscyamine, the major alkaloids of *H. reticulatus* were determined after 28 days of cultivation, because the alkaloid content reached its maximum value at this day of culture. Some differences were found in the contents of these alkaloids among the clones. In contrast to the parent plant, many of the established clones tended to synthesize scopolamine preferentially rather than hyoscyamine. Both alkaloids were present in the culture media of many of the hairy root clones. There was no clear relationship between growth and alkaloid production.

The selection of a large number of hairy roots is a prerequisite for establishing hairy root clones that produce large amounts of alkaloids. Our screening led to the selection of the high scopolamine-producing clone Hr-3 (from the transformant with ATCC 15834) and the high alkaloid-producing clone Hr-L12 (transformant with LBA 9402).

Clone Hr-3 produced the largest amount of tropane alkaloids among the ATCC-clones. The spectrum is dominated by scopolamine (46%), hyoscyamine (35%), acetyltropine (9%), littorine (3%), 6-hydroxyhyoscyamine (1.3%) with a further five compounds. The ratio of scopolamine to hyoscyamine in normal roots was 0.9. In contrast to the parent plant, the production of scopolamine in this clone was 1.3 times greater than that of hyoscyamine. The scopolamine content was low during the early stage of culture, but increased markedly during the stationary phase of growth. This difference in tropane alkaloid production between normal and transformed roots could be due to the different state of development in the transformed roots, to the absence of long-distance transport and/or to a high hyoscyamine-6-β-hydroxylase activity. Only in clone Hr-21 was no scopolamine detected. The differences in scopolamine levels of *H. reticulatus* plant and hairy roots, however, showed that the role of aerial tissue in epoxidation may be minor. Our results indicate that existing substantial genetic variation is indeed carried over to the in vitro transformed root system, as shown by the characteristic differences in the patterns of tropane alkaloids found in intact plant (four alkaloids) and transformed roots (ten alkaloids) (Table 1). The ratio of scopolamine/hyoscyamine and alkaloid spectrum in the transformed roots of *H. reticulatus* differed from that in the parent plant, which seems to indicates that Ri-T-DNA does have effects on qualitative and quantitative changes in the alkaloid composition.

2.2.1 *Towards Improved Productivity in* H. reticulatus-*Hairy Roots*

To date, several experiments have been carried out which attempt to alter levels of secondary metabolites specifically in hairy roots. The fact that we found a high amount of these secondary metabolites motivated us to study different factors which might influence the alkaloid production. It has already been noted (Zehra et al. 1999) that site-specific insertion of Ri T-DNA strongly regulates the growth and secondary metabolite production and the insertion sites vary within each and every transformation event. A number of other factors such as the strain specificity of the T-DNA and genotype interaction of the T-DNA with the host plant cell may determine the ultimate productivity expression of hairy roots induced through T-DNA integration (Zehra et al. 1999).

2.3 Doubly Transformed Hairy Root Clones

We have investigated the possibility of increasing the number of T-DNA copies in the plant genome by the transformation of *Hyoscyamus reticulatus* plant cells, which are already transformed. Why? The TR-DNA is not present in all transformed root clones, moreover, if present, it might be integrated at a different locus distant from the TL-DNA region (Jaziri et al. 1993) and must be present in different copy numbers. Theoretically, a large number of T-DNA copies can be integrated into plant genomes, but the maximum number of the integrated T-DNA is still obscure. Byrne et al. (1983) have shown that hairy roots transformed by *Agrobacterium rhizogenes* strain 8196 have multiple copies of Ri T-DNA fragments of various sizes. Vilaine and Casse-Delbart (1987) reported that the transfer of TL-DNA alone, as well as the transfer of TR-DNA alone, leads to root induction in plant tissues transformed by agropine-type Ri plasmids such as pRiA4 and pRiHR1. Jouanin et al. (1987) have shown that in pRiA4 plasmid-transformed plants, the size of the TR-DNA (including the region homologous to *tms*) varies from about 5–28 kb, although

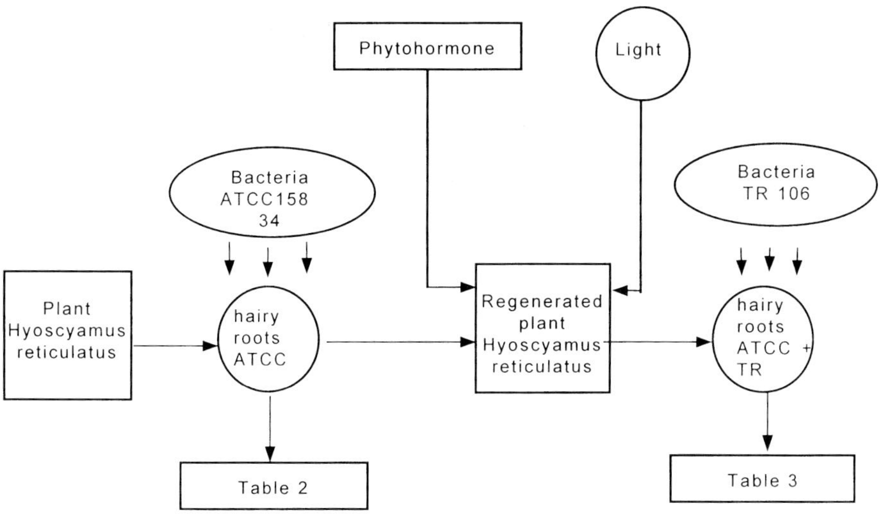

Fig. 3. The way to double transformation in the *Hyoscyamus reticulatus* plant

the size of the TL-DNA (19–20 kb) seems to be relatively constant. They also showed that the copy number varies from one to four per plant genome and that TL- and TR-DNAs are not always present in the same copy numbers: in some cases, the regions are linked together. Ambros et al. (1986) found that T-DNA is present in different chromosomal locations in different transformed lines. On the basis of this evidence, there may be considerable variation in the alkaloid production of *H. reticulatus* hairy roots as a result of the copy number, size and chromosomal location of the Ri T-DNA fragments integrated into the plant genome. For this reason we have investigated how the double transformation may affect the secondary metabolism in the *Hyoscyamus reticulatus* plant. In order to answer this question, we produced doubly transformed hairy roots with different *Agrobacterium* strains (Fig. 3).

2.3.1 Induction and Culture of Transgenic Plantlets

Hairy root segments (ca. 1 cm) from clone Hr-3 of *Hyoscyamus reticulatus* were placed onto MS (1/2 concentration) solid medium with 1.2 mg/l 6-benzylaminopurin (BAP), 0.1 mg/l indole-3-acetic acid (IAA), 2 g/l casein and 2% sucrose and incubated at 25 °C in the light. The roots turned green, developing calluses and started to differentiate. Transgenic plantlets were obtained by culturing the regenerated shoots (with callus phase) on solid S medium (Ionkova et al. 1990) under light. It is known that transgenic plants induced by Ri plasmid show abnormal features named "hairy root syndrome" (Tepfer 1984), such as shortened internode and abundant lateral root formation. The morphological traits of transgenic plantlets were compared to those of non-transformed sterile plants and showed remarkable differences in leaf morphology. The transgenic plantlets of *Hyoscyamus reticulatus* demonstrated wrinkled leaves and many axillary shoots.

2.3.2 Culture of Doubly Transformed Hairy Roots

In the presence of high levels of exogenous auxins, hairy root cultures of *Hyoscyamus reticulatus* formed callus spontaneously, from which suspension cultures were obtained. Hairy roots of *Hyoscyamus reticulatus* were generated from individual protoplasts prepared from suspension cultures, which had been initiated from hairy root tissue of clone Hr-3. This selection, applied at the single-cell-derived colony stage, can potentially generate clonal roots of elevated stable synthetic capacity.

Transgenic *Hyoscyamus reticulatus* plants were regenerated from hairy roots clone Hr-3. A second infection was achieved using the co-culture method of transgenic *Hyoscyamus reticulatus* leaf segments and *Agrobacterium rhizogenes* strain TR 106. A profusion of the roots appeared within 3–5 weeks. The doubly transformed clone Hr-D3 was subcultured, after elimination of the *Agrobacterium* with Claforan (Roth). Doubly transformed hairy roots (Fig. 3) could be maintained in MS liquid medium without phytohormone, though their growth was inferior to that of mono-transformed clone Hr-3. The definitive proof for this event has not been given, but confirmation of the double transformation transformant nature of the hairy root Hr-D3 was verified by the observation of three phenotypic features which distinguished them easily from ordinary roots in in vitro culture, i.e. active proliferation without phytohormone, lack of geotropism and abundant lateral branching. The double transformation could have an influence, not only on the growth rate, but on the alkaloid production of these cultured clones.

2.3.3 Alkaloid Production of Doubly Transformed Hairy Roots

Using a selected high productive strain Hr-D3, it was possible to increase the alkaloid production capacity. Tropane alkaloid contents in the doubly transformed roots were analyzed by HPLC and GC/MS (Table 3). There are great differences between singly and doubly transformed hairy root clones. Clone Hr-D3 showed the highest level of alkaloid production of all clones transformed with ATCC 15834. In contrast to clone Hr-3, the doubly transformed roots produced several new alkaloids which were not identified in Hr-3 (Table 2). Most interesting are: nicotine, acetyltropine, 2-methyl-butyryltropine, tigloyltropine, benzoyltropine, tropacocaine, 3-phenylacetyl-tropine, aposcopolamine, 6-hydroxyapoatropine, norscopolamine, 7-hydroxy-hyoscyamine. MS fragmentations of some alkaloids are shown in Fig. 4.

Table 2. Tropane alkaloid formation in *Hyoscyamus reticulatus* plant roots and transformed clone Hr-3 (the results were obtained using GCL)

	Compounds	Ret. times (DB 17)	Content (% DW)	
			Clone Hr-3	Normal roots
1	Hygrine	7.632	0.0011	–
2	Tropine	13.168	0.0064	–
3	α-Acetyltropine	18.890	0.030	–
4	β-Acetyltropine	20.182	0.005	–
5	Cuscohygrine	36.615	0.0012	–
6	Apoatropine	40.044	0.0023	0.110
7	Littorine	42.485	0.0104	0.060
8	Hyoscyamine	42.721	0.115	0.439
9	Scopolamine	45.784	0.151	0.425
10	6-β-Hydroxyhyoscyamine	47.049	0.0043	–

Table 3. Biochemical traits of doubly transformed hairy roots of *Hyoscyamus reticulatus* clone Hr-D3

Compounds	M+	[100%] Base peak
Hygrine	141	84
Tropinone[a]	139	82
3-α-Tropine	141	82
3-β-Tropine[a]	141	82
3-α-Acetoxytropane	183	124
Nicotine[a]	162	84
3-β-Acetoxytropane	162	84
3-α-Propioniloxytropane[a]	197	124
3-β-Propioniloxytropane[a]	197	124
Isobutyriloxytropane[a]	211	124
3-α-Tigloyloxytropane[a]	223	124
Cuscohygrine	224	84
Senecioyltropine[a]	223	124
3-β-Tigloyloxytropane (Tigloidine)[a]	223	124
Benzoyltropine[a]	245	124
Tropacocaine[a]	245	124
Phenylacetoxytropane[a]	259	124
Apoatropine	271	124
Aposcopolamine[a]	285	94
Hyoscyamine/Littorine	289	124
6-Hydroxyapoatropine[a]	287	94
Norscopolamine[a]	289	94
Scopolamine	303	94
7-Hydroxyhyosciamine[a]	305	94
6-Hydroxyhyoscyamine	305	94

[a] New for hairy root cultures of *H. reticulatus*.

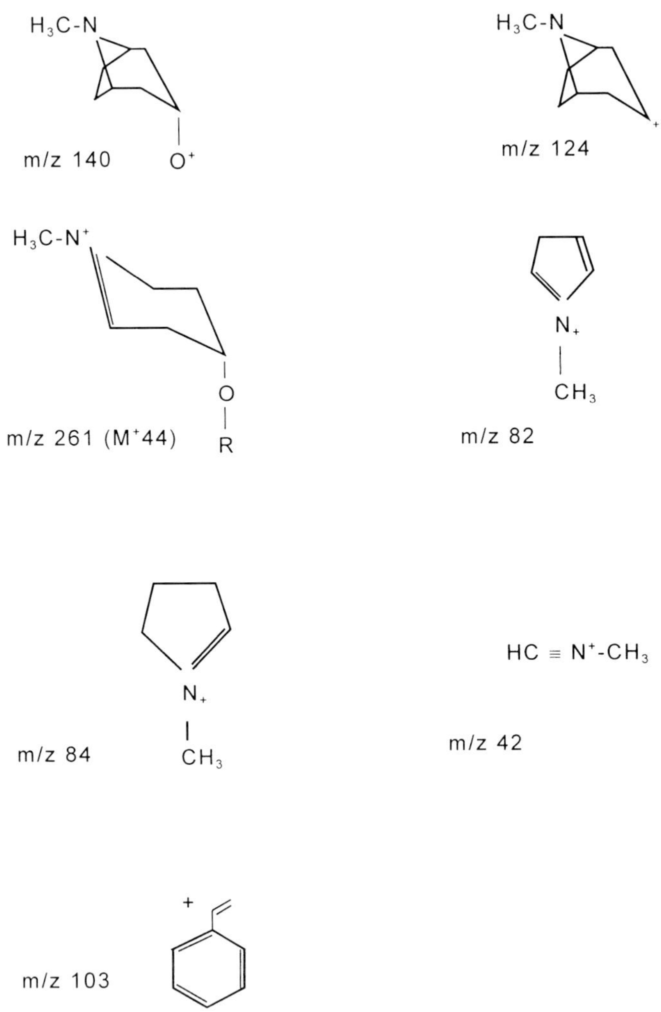

Fig. 4. MS fragmentation of tropane alkaloids in transformed root cultures of *Hyoscyamus reticulatus* L. clone HR-D3

Both singly and doubly transformed hairy roots contained scopolamine as a main alkaloid. Better productivity of scopolamine was found in clone Hr-D3. This alkaloid increased from 0.151% DW (Hr-3) to 0.212% DW (Hr-D3). In contrast to the singly transformed clone Hr-3, the doubly transformed clone produced a low amount of littorine (GC), 0.0082% DW. The total alkaloid level is higher in Hr-3D than in clone Hr-3. Moreover, our results showed that double transformation with different *Agrobacterium* strains might have an effect, not only on total alkaloid content, but also on the ratio of major (higher level of scopolamine) and minor (low amount littorine) alkaloids.

Working with *Hyoscyamus reticulatus* , we have shown that the quantitative balance of alkaloids produced can be markedly influenced. In contrast to the situation in many systems, in this case of tropane alkaloid production, the *Agrobacterium*-mediated transformation may be favorable for the full expression of potential biosynthetic capacity. Moreover, this effect might occur because the initial root clone consists of heterogeneous cells, even though this root line was established from a single root tip and thus the variation increases within the clones (Sevon et al. 1998). Thus, it may be more correct to state that it is not certain whether an increase in copy number would correlate with the increase in the production of alkaloids in plant cells. This might be due to a gene amplification or different patterns in enzyme activities.

2.4 Effect of Colchicine on Growth and Alkaloid Content of *H. reticulatus* Hairy Roots

An interesting method for selection of highly productive plants is polyploidization. The disadvantages associated with loss of fertility or an increase in susceptibility to plant pests, however, do not affect cultures in vitro (Ionkova et al. 1996). To conduct these investigations, a transformed clone Hr-3 of *H. reticulatus* (ATCC) was employed. Growth and production of tropane alkaloids were investigated using various culture media, supplemented with different concentrations of colchicine. The concentration of colchicine for our investigation were chosen according to our previous experiment with *Datura stramonium, var. bernhardii* (Ionkova et al. 1996). Colchicine, at concentrations of 0.04, 0.1, and 0.2%, was added after sterilization of the medium with a sterile filter (Millipore).

The intact plant had about 0.22% alkaloids in the roots and 0.95% in the leaves before flowering. The content of total alkaloids in hairy root clone Hr-3, cultivated in medium without colchicine, is about 0.32% of the dry material.

The treatment with colchicine was conducted in three concentrations, 0.04, 0.1 and 0.2%, for 14 days each. While 0.2% colchicine killed most of the roots, cultures grew well with 0.04 and 0.1% colchicine. These cultures were transferred to colchicine-free medium and subcultured every 28 days for the further investigations.

The growth characteristics of colchicine-treated hairy roots were not similar to those of the control (without colchicine). Within 14 days, the fresh weight doubled, compared to a four- to fivefold increase in the controls (without colchicine). After 28 days, the fresh weight of the 0.04% colchicine-culture reached 62% of the weight of the control and 47% of cultures with 0.1% colchicine in the medium. The morphology of treated hairy roots clearly differs from that of the control. These roots were thick and brownish.

The colchicine-treated root cultures produced substantial amounts of alkaloids (Fig. 5), as seen on HPLC analysis. In both cases (with 0.04

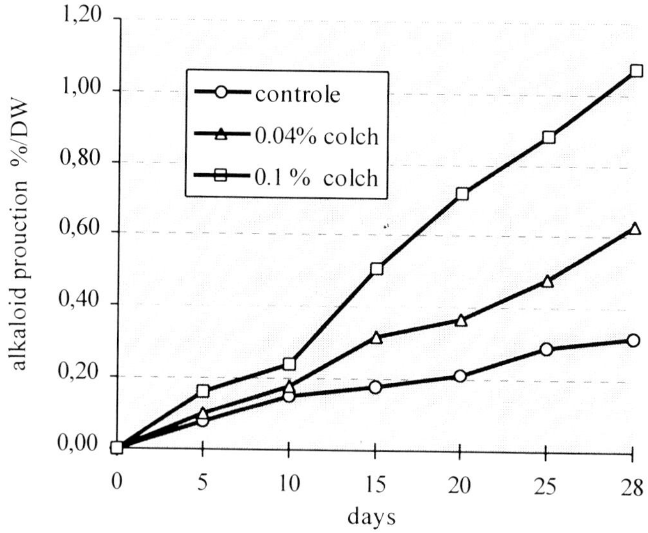

Fig. 5. Alkaloid production during different stages of growth, after colchicine treatment of *Hyoscyamus reticulatus* L. hairy roots, clone Hr-3

and 0.1%colchicine), hairy root cultures yielded higher levels of alkaloids than the non-treated hairy roots. Unexpectedly, the alkaloid profile showed only a few alkaloids –3-α-tropine, 3-α-acetoxytropin, cuscohygrine, apoatropin, hyoscyamine and scopolamine. Nearly 20% of the total alkaloids was acetoxytropine. It appears that in this culture, tropine is being left as the free alcohol rather than as acetylated. From the results of GC-MS analysis and HPLC determination of total alkaloids produced in each clone, it can be concluded that colchicine changes the relative yields of the alkaloids (increasing total alkaloid production to 1.0% DW), but does not increase the composition of tropane alkaloids in hairy roots of *Hyoscyamus reticulatus*.

2.5 Effect of Saponins on the Release of Stored Tropane Alkaloids of *Hyoscyamus reticulatus* Hairy Roots

There are three factors affecting accumulation: (1) de novo biosynthesis of the product, (2) compartmentalized accumulation of the product and (3) breakdown of the product (Verpoorte et al. 1993). An important factor to be considered in connection with tropane alkaloid production is compartmentalization, which seems to play a role in the regulation of alkaloid biosynthesis. In transformed root lines, the better compartmentalization may thus be one of the reasons for the improved alkaloid biosynthesis and accumulation to high levels.

Many different techniques of permeabilization of tropane alkaloids in cell cultures have already been described, but none of them have given a completely positive result (Boitel-Conti et al. 1995). The best producing clone Hr-L12 from clones transformed by bacterial strain LBA 9402 was chosen for our experiments, because of its ability to synthesize relatively high (1.44%) and stable (for more than 14 subcultures) amounts of tropane alkaloids.

Saponins have solubilizing properties and their influence on the water solubility of different compounds has been under discussion for a long time (Bader et al. 1996). The effect of a natural compound – glycyrrhizin as a permeabilizing agent on tropane alkaloids from *Hyoscyamus reticulatis* hairy roots – has been investigated. The release of the two major alkaloids (hyoscyamine and scopolamine) has been examined according to the saponin concentration in the medium and the treatment duration.

2.5.1 *Permeabilization Assay*

Transformed roots from clone Hr-L12 were separated in equal fractions and each of them was suspended in fresh medium with different concentrations of glycyrrhizin – 0.04, 0.06, 0.08, 0.1, 0.5 and 1.0% – to promote product release into the medium.

Both alkaloids were excreted in the culture medium. In the best case, the alkaloid concentration increased sevenfold in the culture medium compared to the control, depending on the concentration of glycyrrhizin (Fig. 6). After 28 days, the alkaloid concentration in the roots decreased to 0.26% DW (in the medium with the maximal saponin concentration, without loss of

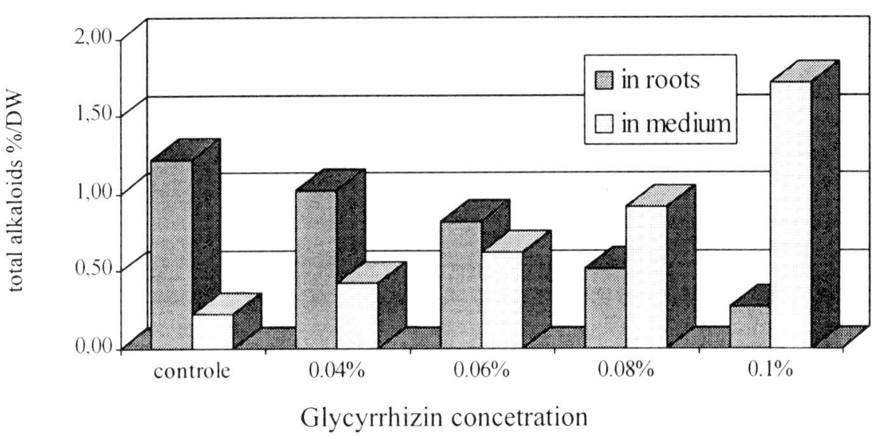

Fig. 6. Alkaloid excretion in the culture medium after 28 days of cultivation, of *Hyoscyamus reticulatis* L. hairy roots, clone Hr-L12 after treatment with different concentrations of glycyrrhizin

viability), and the alkaloids increased in the medium to 1.72%. This led to an increase in at least 53% of the total alkaloid content per saponin-treated flask in comparison with the control. The calculation of the alkaloid ratio demonstrates a difference between hyoscyamine and scopolamine excretion. For the 0.1% concentration of the permeabilizing agent, the maximum scopolamine excretion was obtained (0.72%). Furthermore, the increase in the total alkaloid content suggests a possible biosynthesis stimulation induced by the treatment.

2.5.2 Viability After Treatment

In order to evaluate the effects of permeabilization on viability, the fresh and dry weights of the roots were measured after 28 days of culture. The biomass increase was lowered with the increased glycyrrhizin concentration. After 28 days the fresh weight of the control was 8.82 g, and 1.58 g for the hairy roots with 0.1% concentration of the glycyrrhizin. A higher concentration leads to inhibition of growth and to a browning of the culture, and may be due to a possible cytotoxicity of the triterpenoid saponins (Fig. 7).

Our results demonstrate that saponins are very interesting natural agents for the permeabilization of hairy roots and tropane alkaloid accumulation in the medium, and as an external storage site for alkaloids. It was also shown that an important release of the alkaloids could be obtained without loss of viability. Furthermore, it was shown that saponins induced an increase in the total alkaloid content, as in the case of an elicitor treatment. The high levels of tropane alkaloids found in medium support the idea that in transformed roots of *H. reticulatus* saponins influenced transport of the alkaloids through cell

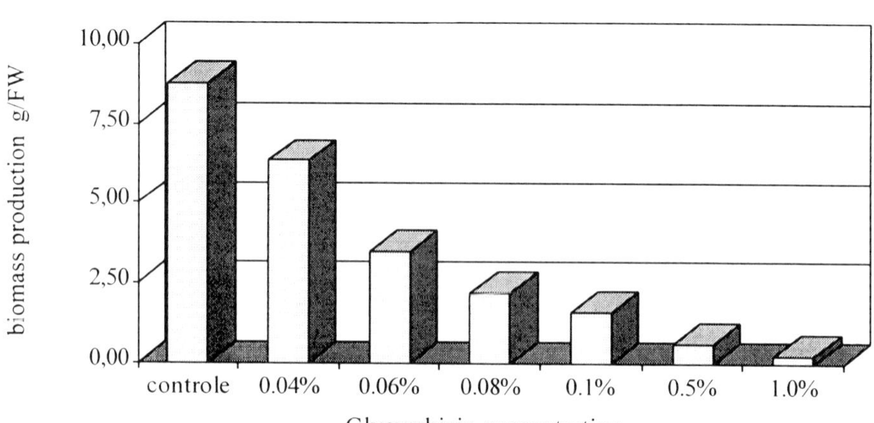

Fig. 7. Growth of transformed root clone Hr-L12 of *Hyoscyamus reticulatus* L. in culture medium with different concentrations of glycyrrhizin for 28 days of cultivation

membranes and may thus be one of the reasons that alkaloids are produced at high levels there.

3 Conclusion

We discussed the approaches applied to establish high producing lines of *H. reticulatus*-transformed roots that might meet the productivity requirements for a commercial process. In our studies with *H. reticulatus,* it was shown that in tropane alkaloid production, the *Agrobacterium*-mediated transformation may be favorable for the full expression of potential biosynthetic capacity. In a screening program, a number of cell lines are set up and the production of the compounds of interest was determined. The best producing lines were selected. Transformation with different *Agrobacterium* strains might have an effect, not only on total alkaloid content, but also on the ratio of the alkaloids. The effects of two natural agents (colchicine and glycyrrhizin) in the regulation of tropane alkaloid biosynthesis were demonstrated. The influence of saponins on the release of tropane alkaloids in culture medium may be applicable to a continuous culture system for alkaloid production. Lines capable of exporting large amounts of alkaloids into the medium offer the best operational system for large-scale production.

4 Protocol

4.1 Infection of Leaf Explants/Stem Tissues

1. Surface sterilize a leaf or stem tissue with a 10% dilution of commercial bleach (e.g., Domestos) for 10–15 min and rinse thoroughly in sterile water three to four times. Remove damaged tissues and trim so that the explant will fit inside an appropriate culture pot.
2. Using a sterile scalpel, make a number of small cuts and incisions in the major veins or in the stem tissue. Inoculate one to two drops of fresh (2-day-old) *A. rhizogenes* in YMB culture medium into the wounded portion of tissue.
3. Incubate the tissue explant for 2–3 days at 25 °C in darkness on MS medium with 2% sucrose.
4. Transfer explants to MS medium containing 500 mg/ml cefotaxime (e.g., Claforan, Hoechst AG). Root differentiation should be apparent within 2–4 weeks and will usually originate mainly from vascular tissue on leaf and stem tissue explants. Normally, we culture roots in 25 ml medium on an orbital shaker operated at 80–90 rpm, and select a single healthy growing root tip from the edge of the culture to establish a clonal hairy root culture.

Cefotaxime should be maintained at 500 mg/ml for two to three subcultures to prevent agrobacterial contamination.

Extraction of Alkaloids. In our investigation the plant material and hairy roots were lyophilized and ground. They were extracted as described earlier (Ionkova 1992a), with 0.2 M sulfuric acid for 2 h. Cell debris was removed by filtration under vacuum. After centrifugation, the residue was washed with 25 ml sulfuric acid. The combined acid filtrates were made alkaline (pH 12) by addition of 1 M NaOH and were extracted with $CHCl_3$. The chloroform-extracts were completely dried under vacuum. A similar procedure with the omission of the initial step (H_2SO_4) was used to extract the alkaloids from the medium. The alkaloid content was analyzed from the dry root materials and media, and expressed as a sum of the cells and medium. Leaves and roots of normal, pot-grown plants were analyzed by the same procedure.

HPLC Analysis. The alkaloid extracts were analyzed using an HPLC method on a C6 column. The solvent system used was acetonitrile/5% H_3PO_4. Flow rate was 1 ml/min. The alkaloids were detected by a UV detector at 210 nm. Standard scopolamine hydrochloride and hyoscyamine hydrobromide were used to calibrate the column.

TLC Densitometry. Silica gel 60 F254 plates (10/20 cm) were used and developed with chloroform-acetone-MeOH-ammonia (25%) (75:10:15:2). After development and drying, the plates were sprayed with Dragendorff reagent. The spots were measured with a TLC-Scanner (Shimadzu CS) at 520 nm. The mean values were calculated from the integration of nine spots corresponding to three different concentrations of hyoscyamine and scopolamine (standards).

GC-MS Analysis. The identification of alkaloids was confirmed by GLC or GC-mass spectrometry and compared to the value for authentic alkaloids or with data obtained from literature. Helium was used as carrier gas.

References

Ambros PF, Matzke AJM, Matzke MA (1986) Localization of *Agrobacterium rhizogenes* T-DNA in plant chromosomes by in situ hybridization. EMBO J 5:2073–2077

Bader G, Plohmann B, Hiller K, Franz G (1996) Cytotoxicity of triterpenoid saponins. Pharmazie 51:414–417

Boitel-Conti M, Gontier E, Laberche JC, Ducrowc C, Sangwan-Norreel BS (1995) Permeabilization of *Datura innoxia* hairy roots for release of stored tropane alkaloids. Planta Med 61:287–290

Byrne MC, Koplow J, David C, Tempe J, Chilton M (1983) Structure of T-DNA in roots transformed by *Agrobacterium rhizogenes*. J Mol Appl Genet 2:201–209

Cardarelli M, Spanol L, De Paolis A, Mauro ML, Nitali G, Constantino P (1985) Identification of the genetic locus responsible for non-polar root induction by *Agrobacterium rhizogenes* 1855. Plant Mol Biol 5:385–391

Christen P, Roberts MF (1993) Transformation in *Datura* species. In: Bajaj YPS(ed) Biotechnology in agriculture and forestry, vol 22. Plant protoplasts and genetic engineering III. Springer, Berlin Heidelberg New York, pp 157–171

Christen P, Roberts MF, Philipson JD, Evans WC (1989) High-yield production of tropane alkaloids by hairy root cultures of a *Datura candida* hybrid. Plant Cell Rep 8:75–77

Cordell GA (1978) Alkaloids. In: Mark HF, Othmer DF, Overberger CG, Seaborg GT (eds) Kirk-Othmer encyclopedia of chemical technology, vol 1. Wiley, New York, pp 883–943

Deivis TH (ed) (1978) Flora of Turkey, vol 6. University Press, Edinburg, pp 454–455

Evans WC (1979) Tropan alkaloids of the solanaceae. In: Hawkes JG, Leister RN, Skelding AD (eds) The biology and taxonomy of the Solanaceae. Linnean Society Symposium Series No 7. Academic Press, London, pp 241–255

Hashimoto T, Yamada Y (1992) Biosynthesis of scopolamine and an application for genetic engineering of medicinal plants. In: Plant tissue culture and gene manipulation for breeding and formation of phytochemicals. Proceedings of the German-Japanese Joint Meeting on Plant tissue culture, 19–21 February, pp 255–259

Hoppe HA (1975) Drogenkunde, Band 1, Angiospermen, 8 Auflage. Walter de Gruyter Verlag, Berlin, pp 596–598

Ionkova I (1992a) Potential of transformed root cultures from *Hyoscyamus aureus*, for biosynthesis of tropane alkaloids. Biotechnol BioE 6(3):41–43

Ionkova I (1992b) Alkaloid production of *Hyoscyamus reticulatus* plant and transformed root culture clone. Biotechnology BioE 6(2):50–52

Ionkova I (1995a) In vitro cultures and formation of secondary metabolites in Astragalus. In: Bajaj YPS (ed) Biotechnology in agriculture and forestry. Medicinal and aromatic plants VIII. Springer, Berlin Heidelberg New York, pp 97–138

Ionkova I (1995b) Increasing the effectiveness of obtaining tropane alkaloids in transformed root cultures of *Datura innoxia*. Pharm Sofia 1:17–20

Ionkova I (1996) Tropan alkaloids in doubly transformed root cultures of *Hyoscyamus reticulatus*. Pharm Sofia 5:18–23

Ionkova I (1999) Genetic transformation in *Astragalus* spp. In: Bajaj YPS (ed) Biotechnology in agriculture and forestry 45. Transgenic medicinal plants. Springer, Berlin Heidelberg New York, pp 55–72

Ionkova I, Alfermann AW (1994) *Althaea officinalis* L. (Marsh mallow): In vitro cultures and production of biological active compounds. In: Bajaj YPS (ed) Biotechnology in agriculture and forestry. Medicinal and aromatic plants VII. Springer, Berlin Heidelberg New York, 13–42

Ionkova I, Witte L, Alfermann AW (1989) Production of alkaloids in transformed root cultures of *Datura innoxia*. Planta Med 55:229

Ionkova I, Zvetanova V, Alfermann AW (1990) Production of tropane alkaloids in hairy root cultures of *Hyoscyamus albus*, Solanaceae, cultivated in vitro. Pharm Sofia 5:13–17

Ionkova I, WitteL, Alfermann AW (1994) Spectrum of tropane alkaloids in transformed roots of *Datura innoxia* and *Hyoscyamus gyoerffyi*, cultivated in vitro. Planta Med Short Com 60 (4):382–384

Ionkova I, Witte L, Alfermann AW (1996), Influence of colchicine on growth and alkaloid production in hairy roots of *Datura stramonium*, var. *bernhardii*. Principles regulating biosynthesis and storage of secondary products, Symposium, Halle, September 26–28, p 27

Jackson BD (ed) (1977) Index Kewensis, vol 1. Koeltz, Koenigstein, pp 1189–1190

Jaziri M, Yoshimatsu K, Homes J, Shimomura K (1993) Primary and secondary metabolism of plants and plant cell cultures. Leiden, Netherlands, Abstract L24

Jouanin E, Guerche D, Pamboukdjian N, Tourneur C, Casse-Delbart F, Tourneur J (1987) Structure of T-DNA in plants regenerated from roots transformed by *Agrobacterium rhyzogenes* strain A4. Mol Gen Genet 206:387–389

Mano Y, Ohkawa H, Yamada Y (1989) Production of tropane alkaloids by hairy root cultures of *Duboisia leichhardtii* transformed by *Agrobacterium rhizogenes*. Plant Sci 59:191–201

Parr AJ, Payne J, Eagles J, Chapman BT, Robins RJ, Rhodes MJC (1990) Variation in tropane alkaloid accumulation within the Solanaceae and strategies for its exploitation. Phytochemistry 29(8):2545–2550

Pelt JM, Younos C, Hayon JC (1967) Sur la constitution alcaloidique de quelques Solanacees d'Afghanistan: II. Comparaison de la constitution des feuilles et des graines de plusieurs espèces de Datura et Jasquiames. Ann Pharm Fr 25:101–106

Petri G, Bajaj YPS (1988) *Datura* spp: In vitro regeneration and the production of tropanes. In: Bajaj YPS (ed) Biotechnology in agriculture and forestry, vol 7. Medicinal and aromatic plants II. Springer, Berlin Heidelberg New York, pp 135–161

Portsteffen A, Dräger B, Narstedt A (1994), The reduction of tropinone in *Datura stramonium* root cultures by two specific reductases. Phytochemistry 37(2):391–400

Robins RJ, Parr AJ, Payne J, Walton NJ, Rhodes MJC (1990) Factors regulating tropane-alkaloid production in a transformed root culture of a *Datura candida* ◊ *D. aurea* hybrid. Planta 181:414–442

Sevon N, Hiltunen R, Oksman-Caldentey KM (1998), Somaclonal variation in transformed roots and protoplast-derived hairy root clones of *Hyoscyamus muticus*. Planta Med 64:37–41

Straus A (1989) *Hyoscyamus* spp: In vitro culture and production of tropane alkaloids. In: Bajaj YPS (ed) Biotechnology in agriculture and forestry, vol 7. Medicinal and aromatic plants. Springer, Berlin Heidelberg New York, pp 286–314

Tepfer D (1984), Transformation of several species of higher plants by *Agrobacterium rhizogenes* – sexual transformation of the transformed genotype and phenotype. Cell 37:957–967.

Verpoorte R, van der Heijden R, van Gulik WM, ten Hoopen HJG (1991) Economic feasibility of a plant cell biotechnological production In: Brossi A (ed) The alkaloids. vol 40. Academic Press, San Diego, p 1 and references cited therein

Verpoorte R, van der Heijden R, Schripsema J, Hoge JHC, ten Hoopen HJG (1993) Plant cell biotechnology for the production of alkaloids: Present status and prospects. J Nat Prod 56(2):186–207

Vilaine F, Casse-Delbart F (1987), Independent induction of transformed roots by the TL and TR regions of the Ri plasmid of agropine type *Agrobacterium rhizogenes*. Mol Gen Genet 206:17–23

Yamada Y, Hashimoto T (1988) Biosynthesis of tropane alkaloids. In: Bock G, Marsh J (eds) Application of cell and tissue culture. Wiley, New York, pp 199–212

Zehra M, Banerjee S, Sharma S, Kumar S (1999) Influence of *Agrobecterium rhyzogenes* strains on biomass and alkaloid productivity in hairy root lines of *Hyoscyamus muticus* and *H. albus*. Planta Med 65:60–63

6 Micropropagation of *Hypericum canariense* L. for the Production of Hypericin

SEBASTIANA MEDEROS-MOLINA[1]

1 Introduction

Shrubs or herbs of *Hypericum* L. species are distributed throughout the world. They are found in the Mediterranean region, Portugal, Spain, Canary Islands [Spain], Africa, Turkey, Australia, Tasmania, New Zealand, New Caledonia, Papua New Guinea, parts of South-East Asia, Sri Lanka, and from North to South America (Sampson and Parker 1930; Milne-Redhead 1953; Robson 1958, 1968, 1981, 1990, 1993; Bramwell and Bramwell 1974; Gunatilanka et al. 1982; Campbell and Delfose 1984; Campbell 1985; Kikuchi et al. 1985; Crompton et al. 1988; Sakar et al. 1988; Willis et al. 1997). Etymologically, the name *Hypericum* L. was first used by Linnaeus. It comes from *hyper* (over) and *eikon* (image), on account of the image that appears on the petals (Linnaeus 1753). According to other botanists, the name comes from *hypo* and *ereikn* or *erikn* meaning "plant that grows under heathers"; it could also come from *hyper* and *eikon* meaning "plant resembling a ghost's image or plant with an air of mystery" (Calvert 1930; Milne-Redhead 1953; Robson 1958).

1.1 Chemistry

Many authors reported on the chemical composition and the variability of the main components in different species of *Hypericum* L. (Michaluk et al. 1956; Mathis and Ourisson 1963; Giese 1980; Meruelo et al. 1988; Kartnig et al. 1989; Southwell and Campbell 1991; Maisenbacher and Kovar 1992; Thomas et al. 1992; Yazaki and Okuda 1994; Aramaki et al. 1995; Girzu 1995; Cakir et al. 1997; Seidler et al. 1999; Schmidt et al. 2000). However, the chemical study of this genus began with *Hypericum perforatum* L. in 1830 with the pioneering isolation of hypericin by Bruchner, who named the compound "hypericum red". About one century later, in 1911, the compound was identified and renamed hypericin by Cerny, who also isolated other similar constituents without a proven structure determination, because the correct structure of hypericin was established in 1953 by Brockman and Sanne (Michaluk et al. 1956; Mathis and

[1]Department of Plant Physiology, Facultad de Farmacia, Universidad de La Laguna, 38200 La Laguna, Tenerife, Canary Islands, Spain

Biotechnology in Agriculture and Forestry, Vol. 51
Medicinal and Aromatic Plants XII (ed. by T. Nagata and Y. Ebizuka)
© Springer-Verlag Berlin Heidelberg 2002

Ourisson 1963; Bombardelli and Morazzoni 1995). Hypericin [4, 5, 7, 4′, 5′, 7′-hexahydro-2,2′-dimethyl-mes-o-naphthodianthrone] is a natural photosensitizing polycyclic aromatic dione which can be extracted from certain plants of the *Hypericum* L. genus. It is a photodynamic dark-red pigment and was the target of all the analytical methods reported in the literature (Araya and Ford 1981; Knox and Dodge 1985; Hölz and Ostrowski 1987; Kartnig et al. 1989; Lenard et al. 1993; Carpenter et al. 1994; Papageorgiou et al. 1996; Sureau et al. 1996). Moreover, the structural variation among such miscellaneous plant quinones was reviewed in 1979 by Thomson (Harborne 1980).

1.2 Medicinal Importance of *Hypericum* L.

Species of the genus *Hypericum* L. are used as medicinal remedies throughout the world (Mathis and Ourisson 1963; Gunatilanka et al. 1982; Kikuchi et al. 1985; Darias et al. 1986; Pérez de Paz and Medina 1988; Sakar et al. 1988; Jayasurikab et al. 1991; Southwell and Campbell 1991; Carin et al. 1992; Degar et al. 1992; Carpenter et al. 1994; Bombardelli and Morazzoni 1995; Girzu 1995; Jensen et al. 1995; Herrera et al. 1996; Cakir et al. 1997; Willis et al. 1997; Brumeton 1999). *Hypericum* L. species are an abundant source of pharmaceutical products and a significant interest has been concentrated on the activity, among others, of the naphthodianthrones hypericin and its derivatives (Vanhaelen and Vanhaelen-Fastre 1983; Duran and Song 1986; Meruelo et al. 1988; Kartnig et al. 1989; Lavie et al. 1989, 1990; Schinazi et al. 1990; Liebes et al. 1991; Thomas and Pardini 1992; Sureau et al. 1996). From current information it is clear that hypericin is only found in certain species belonging to the genus *Hypericum* L.. This photodynamic pigment is considered to be one of the most important compounds from a pharmacological point of view (Towers 1980; Freytag 1984). It displays considerable antiviral activity against several viruses (Vanhaelen and Vanhaelen-Fastre 1983; Meruelo et al. 1988; Takahashi et al. 1989; Kraus et al. 1990; Lavie et al. 1989, 1990; Andersen et al. 1991; López-Bazzocchi et al. 1991), including the human immunodeficiency virus type I (Schinazi et al. 1990; Hudson et al. 1991, 1993; Degar et al. 1992; Lenard et al. 1993; Carpenter et al. 1994). Thus, it is now considered to be a potentially valuable agent for the photodynamic therapy of cancer and other diseases (Towers 1980; Knox and Dodge 1985; Duran and Song 1986; Carin and Pardini 1992; Henderson and Dougherty 1992; Thomas and Pardini 1992; Diwu and Lown 1994). On the other hand, other authors reported that hypericin has antidepressant effects which were determined by inhibition of monoamino oxidase (MAO) in vitro. Since hypericin is considered to be an inhibitor of both types A and B of MAO, this result would confirm the supposition that it is the active constituent in *Hypericum* L. extract (Suzuki et al. 1984; Bombardelli and Morazzoni 1995).

2 *Hypericum canariense* L.

2.1 Botany

Hypericum canariense L. is a shrub species endemic to the Canary Islands, Spain, and belongs to the Hypericaceae family. This very interesting medicinal plant is widely distributed in these islands [Gran Canaria, Tenerife, Gomera, Hierro and La Palma] (Bramwell and Bramwell 1974, 1990; Kunkel 1992). In our islands, it is common and even abundant in moist, sunny areas of the Laurisilva (Laurel forest), and/or Fayal-Brezal and/or Thermophile Forest (Ceballos and Ortuño 1976; Santos 1980, 1983; Pérez de Paz 1982; Hansen and Sunding 1985; Kunkel 1971, 1972; Rivas-Martínez et al. 1993). *Hypericum canariense* L. is a widely used medicinal plant, known as Hiperico or Granadillo in Canary Islands folk medicine (Bramwell and Bramwell 1974; Kunkel 1971). This shrub species is glabrous or sometimes glaucescent. It flowers in spring and summer and reproduces from seeds with a low rate of viability. This species is abundantly branched and grows to between 1 and 3 m in height (Fig. 1). The stems are erect, branched at the top and 20–90 cm long. The leaves are smooth, simple, opposite, lanceolate or elliptic-lanceolate, 1–3 cm long and decussate (Fig. 2, top). Its large, very beautiful golden yellow flowers grow in groups of not very thick terminal inflorescences. Each flower has five lanceolated petals some 2 cm in length, with numerous stamens which form a crown around an overdeveloped pistil (Fig. 2, bottom). The ovary is ovoid. The capsules are oval-shaped, hard and reddish-brown and contain a large number of very tiny seeds (Cronquist 1981; Bramwell and Bramwell 1974, 1990; Kunkel 1992).

2.2 Importance of *Hypericum canariense* L.

Hypericum canariense L. is considered an important source of pharmaceutical compounds which are contained in the aerial parts of this plant. *Hypericum canariense* L. extracts used in folk therapy today are prepared by extracting the aerial parts of this species with mixtures of ethanol and water. This folk medicine has been recognised by the scientific, medical and commercial communities (González 1976, 1982; Darias et al. 1986; Pérez de Paz and Medina 1988; Rabanal Gallego 1996; Pérez de Paz and Hernández Padrón 1999). This has resulted in attempts to isolate the medicinally active agents and determine their range of application. Obviously, further investigation into alternative methods to obtain hypericin is needed. In vitro culture of *Hypericum canariense* L. is considered a very promising approach towards providing a long-term source of this valuable compound (Mederos et al. 1996).

Fig. 1. Natural populations of *Hypericum canariense* L. in flowering stage. (Photos were taken by Mederos)

Fig. 2. *Top* Branches of *Hypericum canariense* L. collected for the establishment of culture; *bottom* golden yellow flowers of this species. (Photos were taken by Mederos)

3 In Vitro Culture Studies of *Hypericum* Species

Tissue culture studies of the genus *Hypericum* L. have been reviewed by Yazaki and Okuda (1994) and Cellárová et al. (1995) and several papers about in vitro tissue culture of different *Hypericum* species have been published (Roest and Bokelmann 1973; Yazaki and Okuda 1990, 1994; Mederos 1991a,b,c, 1994; Mederos and Rodríguez 1991; Mederos et al. 1996; López et al. 1991; Berger-Büter et al. 1994; Brutovská et al. 1994; Cellárová et al. 1992, 1994, 1995, 1997; Dias and Fernández-Ferreira 1995; Cellárová and Bruñakova 1996; Beerhues et al. 1997; Halusková and Cellarová 1997; Schmidt et al. 2000). The regeneration of shoots is possible from the different types of explants (shoot, leaf, stem, root, callus) of *Hypericum* L. species which have proliferated (Roest and Bokelmann 1973; Yazaki and Okuda 1990, 1994; Cellárová et al. 1994, 1995, 1997). In contrast, the advantage of the insertion of a callus phase in the in vitro propagation lies in the possibility of greatly extending the base for the formation of adventitious shoots. However, it is possible that the morphogenetic capacity of the callus is lost during extended proliferation by repeated subcultures (Bajaj 1990; Herman 1995; George 1996). On the other hand, since the presence of callus during the multiplication stage increases the risk of off-types due to somaclonal variation, it is desirable that callus development be minimised or eliminated during micropropagation (Hartmann et al. 1990; George 1996; Herman 2000). In retrospect, the in vitro propagation of *Hypericum* L. species is attracting increasing economical interest. It potentially enables *Hypericum* L. species to be cloned, in a large number, under highly controlled conditions (Yazaki and Okuda 1990, 1994; Mederos 1991a,c, 1994; Mederos and Rodríguez 1991; Mederos et al. 1996; López et al. 1991; Cellárová et al. 1992, 1995, 1997; Berger-Büter et al. 1994; Brutovská et al. 1994; Cellárová and Bruñakova 1996; Pinzaru and Lupu 1999).

3.1 Micropropagation of *Hypericum canariense* L.

Micropropagation is a specific aspect of plant tissue culture dealing with the aseptic vegetative multiplication of in vitro plants. In this species the protocols were divided into several stages: stock plant selection and preparation should be called stage 0; initiation and establishment of an aseptic culture (stage 1); multiplication (stage 2); in vitro rooting and conditioning (stage 3) and acclimatisation to ex vitro environment (stage 4). Stages three and four may be combined in this species (Mederos, unpubl. data). To the best of our knowledge, no reference to in vitro mass propagation of adult plants of *Hypericum* L. genus is to be found in the literature. Micropropagation of *Hypericum* L. species is generally possible if explants from juvenile organs (leaf, stem, root) or very young seedlings are used (Yazaki and Okuda 1994; Cellárová et al. 1992, 1995, 1997). On the other hand, only seed germinations are used on a large scale for vegetative propagation of species of the genus *Hypericum canariense* L. in the Canary Islands (González-Martín 1995). Due to the limited seed viability and slow growth of *Hypericum canariense* L. in its natural habitat, the in vitro culture of this species resulted in micropropagation (Mederos 1991a,b,c; 1994; Mederos and Rodríguez 1991; Mederos et al. 1996) with a view to producing hypericin, although this technique should also be used to increase the natural populations. The potential of this species to be economically exploited as ornamental shrubs (González-Martín 1995) should also be considered.

3.1.1 *Adequate Starting Materials and the Preparative Stage*

It is important in a breeding programme when selected plant material is used to have as little genotypic variation as possible and therefore vegetative propagation methods should be used. For *Hypericum canariense* L. the starting material was selected for some particular phenotypic characteristics: these specific plants are vigorous and produce many flowers annually and a very high percentage of these flowers become fruit. On the other hand, these plants were disease-resistant. The plant material was collected in Tenerife and Gran Canaria, Canary Islands, between March and September from the annual sprouts of mature plants (23–33 years old) and for 7 years. They were stored in plastic bags in a refrigerator (temperature approximately 4°C), within 12h after collection.

The use of apical and/or axillary shoot explants minimises the possibility of recovering somaclonal variants, since it does not require dedifferentiation and redifferentiation, but only regeneration of a plant from an established, organised meristem (Bajaj 1990; Herman 1993, 1995, 1996, 2000; Gamborg and Phillips 1995; George 1996). Somaclonal variants were undesirable in this context because the objective was to have large-scale cloning of plantlets from selected mother plants. The explants used by us consisted of shoots excised from branches collected from the lower half of the shrubs because shoots collected from apical and media branches of the shrubs have a lesser influence on culture results. Overall, different seasonal effects were observed on the behaviour of *Hypericum canariense* L. plants, although the highest response was obtained with material collected in May and June. The behaviour of the explants collected between July and September during the stage of initiation of culture was detrimental and, moreover, the rate of survival of in vitro shoots was lowest (between 17 and 33%). The higher rate of explant viability and growth rate in May and June represents the normal active growth period in the local natural stand. The low explant viability and growth rate in late summer could be associated with the commencement of the dormant period and similar results are reported in other species (Fox et al. 1999; Mederos-Molina 2001a,b,c). On the other hand, after collecting the plant material of *Hypericum canariense* L., the explants were immediately sterilised. Explants were usually infected by fungi and contamination was normally visible after 8 days of culture. To avoid contamination, the stem has to be defoliated and cleaned prior to surface sterilisation. We found that the most effective sterilising agent was mercuric chloride (Mederos 1994; Mederos et al. 1996, 1997a). However, sodium hypochlorite (NaOCl) or calcium hypochlorite [$Ca(OCl)_2$] was detrimental for sterilising these explants, more than 73% of which became necrotic (Mederos et al. 1996). These results are in agreement with other investigations (Mederos et al. 1998a; Mederos and Trujillo 1999a,b). In addition, a seasonal effect was observed on explant viability and the degree of contamination. Contamination was significantly higher from late July to September. In agreement with this study, Kupila-Ahvenniemi (1977), Hohtola (1988), Mederos (1991d), George (1996), Herman (1993, 1995, 1966, 2000), Mederos and Amaro Luis (1997) and Mederos et al. (1998a,b) also noted seasonal changes in explant viability and/or contamination of tissue cultures of different woody plant species.

3.1.2 In Vitro Establishment of Shoots Stage

The purpose of the initial culture stage is to initiate axenic cultures. The development of a protocol for apical and axillary bud culture is a prerequisite for the micropropagation of *Hypericum canariense* L. One of the problems encountered at the moment of establishment of culture of this species was the presence of browning exudate arising from the cut tissue; this browning exudate seemed to lead to the death of the explants. On day two, explants necrosed in culture media used without antioxidants. Explant browning was relatively more advanced on liquid media with respect to solid media and the explants necrosed between 2 and 4 days of culture. For this purpose, an efficient method for eradication of browning exudate was developed (Mederos 1991a,c; Mederos and Amaro Luis 1997; Mederos et al. 1997a). Browning exudate was eliminated by rosmanol treatments; this compound is a powerful diterpenic antioxidant isolated from *Salvia canariensis* L. (Mederos 1991a; Luis et al. 1992; Mederos and Amaro Luis 1997; Mederos et al. 1997a). Another alternative to prevent browning exudates was the incubation of the culture at three different photosynthetic photon flux densities (PPFD) of 5, 10 and 15 $\mu m\ m^{-2}\ s^{-1}$ for 8, 16 and 24 days, respectively. The response to these light conditions at the early stage of development removed the browning from the explants and culture media (Mederos, unpubl. data). On the other hand, shoots can be induced from apical and axillary explants with a wide variety of media: [MS] Murashige and Skoog (1962), [LS] Linsmaier and Skoog (1965); [B-5] Gamborg et al. (1968); Gamborg and Phillips (1995); Woody Plant Medium [WPM] (Lloyd and McCown 1981) and modified Quoirin and Lepoivre (1977) [QL.4] media (Mederos 1991b; Mederos et al. 1994, 1997b; Mederos and Schöbert 1995; Mederos and Trujillo 1999a,b). However, after browning exudates were eliminated by several treatments, good establishment of shoot organogenesis potential was achieved in the Murashige and Skoog [MS] (1962) and modified Quoirin and Lepoivre [QL.4] media (Mederos 1991d) supplemented with

6-benzylaminopurine (BAP) and naphthaleneacetic acid (NAA) (Fig. 3, top; Mederos et al. 1997a). In other studies that involved only a few species of the Hypericaceae family with the purpose of growing axenic shoots, the results demonstrated that MS macroelement and microelement solutions (1962) supplemented with BAP and NAA (Mederos 1991a,c; Arello and Pinto 1993; Mederos et al. 1996) or basal medium consisting of Linsmaier and Skoog salts (1965), combined with BAP (Cellárová et al. 1992, 1994), or BAP plus 3-indoleacetic acid (IAA) (Yazaki and Okuda 1990) were used, resulting in good formation of shoots. Macroelement solutions have proved to be a vital factor in in vitro organogenesis of *Hypericum canariense* L. as shown by the fact that during this stage shoots developed well in an MS or QL.4 salts solution supplemented with BAP combined with NAA. However, significant differences have been observed when shoots developed in LS, B5 and WPM media supplemented with BAP combined with NAA or IAA because the shoots had stopped growing on the 17-day-old cultures. After 25 days of culture, irrespective of the macroelement solution in use [LS, B5 or WPM], all the shoots showed signs of localised leaf chlorosis (Mederos, unpubl. data).

3.1.3 Multiplication of Shoots Stage

The main objective of this stage is to produce the maximum number of shoots. Among the different methods for achieving propagule multiplication of plants from the Hypericaceae family, only axillary shoot proliferation is covered in this chapter. The proliferation rate of this method provides genetic stability and is achievable by most medicinal plant species (Hartmann et al. 1990; Mederos and Schöbert 1995; George 1996; Hermann 1996, 2000; Mederos and Amaro Luis 1997; Mederos et al. 1998a;). The number of developed *Hypericum canariense* L. shoots per explant is partially controlled by the amount of plant growth regulator added to the MS or QL.4 culture medium. Addition of BAP and NAA to the basal MS culture medium promoted the proliferation of many shoots (Fig. 3, bottom). The maximum amount of shoot multiplication, 13 shoots per shoot transferred, occurred in full-strength Murashige and Skoog (1962) inorganic salts and exhibited no chlorosis. However, preliminary experiments corroborated that shoot multiplication and elongation decreased in the QL.4 medium when compared to full-strength MS salts (Mederos et al. 1996). Linsmaier and Skoog (1965), Gamborg et al. (1968), Gamborg (1970),Gamborg and Phillips (1995)or Woody Plant Medium (Lloyd and McCown 1981) macroelements with MS full-strength microelements all gave poor shoot multiplication and elongation, and heavy chlorosis. On the other hand, the effective concentration of exogenous cytokinin required to enhance the branching of lateral buds from leaf axils varies with the species of the Hypericaceae family used (Cellárová et al. 1992; Arello and Pinto 1993), although the number of axillary shoots initiated in a medium supplemented with kinetin or zeatin was limited when compared to the number initiated in a medium supplemented with an equivalent amount of BAP (Mederos, unpubl. data). Thus, in this stage,

Fig. 3. *Top* In vitro establishment of shoots; *bottom* shoot multiplication and elongation in MS medium supplemented with BAP plus NAA. Photos were taken after 23 days of culture respectively. (Photos were taken by Mederos)

some plantlets could be obtained from only one in vitro shoot cultured in the optimum culture medium mentioned before, but supplemented with BAP and phloroglucinol (Mederos, unpubl. data). Exogenous hormones influenced growth, elongation and proliferation of both roots and shoots. The effect of the apex could be suppressed by exogenous hormones which promoted growth of axillary buds, but with time apical dominance was restored. This may be due to degradation of BAP within the medium or the production of endogenous hormones (Miller and Drew 1990; Sudriá et al. 1999; Mok et al. 2000). Thus, different clones of *Hypericum canariense* L. have been multiplied successfully for more than 7 years (Fig. 3, bottom).

3.1.4 *In Vitro Rooting Stage*

Length of *Hypericum canariense* L. shoots is a critical factor in root initiation. Poor root initiation was achieved with shoots shorter than 15 mm. Root induction was achieved readily when exogenous auxins were added to the culture medium. These in vitro plantlets were developed in modified Murashige and Skoog [MS] (1962) media (Mederos et al. 1994, 1997b; Mederos and Schöbert 1995) supplemented with BAP alone or combined with NAA at different concentrations. However, the best percentage and quality of roots were obtained in half-strength MS medium supplemented with IBA at low concentrations (Fig. 4). The cultures were maintained in a growth room under 20 $\mu m\, m^{-2} s^{-1}$ (Mederos et al. 1996). Thus, in every stage, direct organogenesis was achieved from apical and axillary cuttings without callus production. However, the development of plantlets is possible in half-strength MS medium supplemented with phloroglucinol (Mederos, unpubl. data). Phloroglucinol was suitable for rooting as well as for the growth of buds from medicinal woody trees and shrubs (George 1996; Herman 1995, 2000). On the other hand, in some *Hypericum* L. species phloroglucinol and derivatives are present (Decosterd et al. 1990). Moreover, the influence of medium volume and culture vessels on the growth yields of *Hypericum canariense* L. was determined after 3 weeks of incubation and high positive correlations occurred between the culture vessels capacity and fresh weight (fwt), leaf number, shoot number (stages 1 and 2) and root number (stage 3) (Mederos, unpubl. data). These results are in accordance with the observations in other medicinal plants (Herman 2000; Tisserat and Silman 2000).

4 Hypericin from Clonal Propagation of *Hypericum canariense* L.

Concerning the isolation of naphthodianthrone hypericin, there also exist several *Hypericum* L. species propagated in vitro which produce this econom-

Fig. 4. Proliferation and rooting in half-strength MS culture medium supplemented with different concentrations of IBA. Photo was taken after 23 days of culture. (Photo was taken by Mederos)

ically important metabolite (Cellárová et al. 1994, 1997; Mederos et al. 1996; Halusková and Cellárová 1997). From a cross-section of leaves from in vitro and in situ *Hypericum canariense* L. plantlets, a spongy parenchyma containing dark red secretory glands was observed (data not shown). These observations are in accordance with the results of other authors (Larsen 1963; Curtis and Lerste, 1991; Cellárová et al. 1992). However, Cellárová et al. (1992) and Fornasiero et al. (1998) found that in vitro regenerated shoots showed some interesting differences in the anatomical structure of leaves, probably because they achieved multiple shoots from different initial explants. To the best of our knowledge, extraction of hypericin from in vitro shoots and in vitro plantlets of *Hypericum canariense* L. was achieved for the first time and described in Mederos et al. (1996). The in vitro shoots and plantlets (3.5 months old), previously dried and powdered, were macerated at room temperature and the technique for hypericin extraction was described earlier (Mederos et al. 1996). For this purpose, the effect of plant growth regulators on different culture media was studied for proliferation of shoots (BAP plus NAA) and rooting (IBA) stages (Mederos et al. 1996). The amount of hypericin isolated from in vitro shoots showed no significant differences at the 1% level from that extracted from the aerial parts of in vitro plantlets (Mederos et al. 1996). Several authors have also reported that hypericin content of in vitro regenerated plants of *Hypericum perforatum* L. were not affected by the concentration of 6-benzylaminopurine which promoted shoot differentiation (Cellárová et al.

1994, 1997; Umek et al. 1999). However, comparison of the mean values of hypericin content in diploid, triploid and tetraploid somaclones and their R1 and R2 progenies indicated its dependence on the ploidy (Cellárová et al. 1997). Furthermore, in our case, the mean hypericin content was 90.7 $\mu g\, g^{-1}$ dry weight (in vitro shoots) and 293.0 $\mu g\, g^{-1}$ dry weight (aerial parts of in vitro plantlets) and these hypericin levels are in agreement with other research results (Cellárová et al. 1994, 1997). However, the mean hypericin content in vegetative growth of *Hypericum* L. plants cultivated in the greenhouse depends on the biotypes. Jensen et al. (1995) found that the mean hypericin level in vegetative growth of 24 biotypes was different and there was also significant variation in hypericin content of *Hypericum perforatum* L. biotypes due to their seed origin. Differences in hypericin levels between different locations show that location and biotype differences may occur within a small geographical area. At that time, hypericin was produced by seedlings of *Hypericum perforatum* L. genotypes (Jensen et al. 1995) and by different organs from adult plant material of some *Hypericum* L. species (Hölz and Ostrowski 1987; Kartnig et al. 1989; Bombardelli and Morazzoni 1995). On the other hand, further experiments are in progress in plant tissue culture because recent studies have characterised the relationship between environmental factors and levels of hypericin in callus and plant regeneration from *Hypericum canariense* L. (Mederos et al., unpubl. data).

5 Future Prospects

Primary metabolism in plants is essential for growth and development of cells, tissue and organs. Until recently, secondary metabolism was generally regarded as a non-essential process that produced by-products or plant wastes. However, there is evidence that secondary metabolites such as phenols, terpenoids and alkaloids may contribute to plant defence against attacks by herbivores and pathogenic micro-organisms (Kingsbury 1964; Bomme 1997; Galitzer and Oehme 1978; Giese 1980; Hildebrand and Jensen 1991; Ryu and Furosaki 1994; Jensen et al. 1995; Van der Meijden 1996; Liu et al. 1998). In addition, the cytotoxic nature of some secondary metabolites has made them medicinally important (Schripsema and Verpoorte 1995; Liu et al. 1998). Examples included the naphthodianthrone hypericin and its derivatives which have especially been used, due to their antiviral and anticancer activities (Carpenter et al. 1994; Diwu and Lown 1994). A general increase in interest in use of Hypericaceae medicinal plants as natural products for pharmaceutical applications has swept the world over the past years (Seidler et al. 1999; Umek et al. 1999). As the demand for hypericin products expands, the efforts of plant collectors have sought other alternatives. For this purpose, plant cell and tissue culture techniques have developed into a branch of biotechnology with attractive prospects for agriculture as well as for the chemical and pharmaceutical industries. Due to the presence of the therapeutically valuable cytotoxic and antidepressant naphthodianthrones hypericin and derivatives in some *Hypericum* L. species, these medicinal plants became one of the major fields of interest in modern plant cell biotechnology. The high demand for these chemical compounds was one of the major motives to study the possibilities for their production by tissue culture techniques (Cellárová et al. 1994, 1995, 1997; Cellárová and Bruñakova 1996; Mederos et al. 1996). In the last 10 years, perspectives have changed and considerable progress has been achieved in the therapeutic approach with hypericin extracts. On the other hand, the production of hypericin is still far from being economically feasible. However, considerable improvement in the efficiency of plant cell and tissue culture may be expected. To the best of our knowledge, no patents have been registered for the commercial production of hypericin. Furthermore, more experience with large-scale culturing of in situ and in vitro plant material is needed for this purpose.

6 Conclusion

An overview is given of some aspects of the history, principles and applications of this medicinal plant endemic to the Canary Islands. The main applications discussed in the review are related to micropropagation, in vitro plant breeding and production of hypericin. In spite of the fact that in vitro propagation of mature *Hypericum canariense* L. often proved difficult, satisfactory micropropagation was achieved (Mederos 1991a,b,c, 1994; Mederos et al. 1996). Thus, the cultivation of this endemic medicinal species is discussed with reference to other *Hypericum* L. species. In conclusion, as described above, shoot multiplication and rooting of *Hypericum canariense* L., a very interesting medicinal plant, was achieved in vitro using shoot tips and nodal segments as explants for the purpose of obtaining the chemical product hypericin. The possibility to increase the concentration of this compound by tissue culture is presently being considered (Mederos et al., unpubl. data).

7 Protocol

To illustrate the complete procedure of micropropagation of *Hypericum canariense* L., a protocol of apical and axillary shoots culture of this species is provided for the production of hypericin.

7.1 Obtaining Plant Materials

1. Annual sprouts of mature plants.
2. Seasonal effects: the highest response was obtained with material collected in May and June.
3. Branches collected from the lower half of the shrubs have a good influence on culture results.

7.2 Surface Disinfection

1. Branches have to be defoliated and cleaned with water prior to surface sterilisation.
2. Immerse the branches in a systemic fungicide: 0.1% benlate for 10 min.
3. Dip branches in 50 or 70% ethanol for 2 or 1 min respectively.
4. Dissect explants: apical and axillary shoots. Soak these explants in 0.1 Cl_2Hg with 1 ml Tween 80 between 5 and 10 min.
5. Rinse four times in sterile distilled water.

7.3 In Vitro Establishment of Shoots Stage

1. Culture apical and axillary shoots explants onto an MS (Murashige and Skoog 1962) or Ql.4 (a modified Quoirin and Lepoivre culture medium 1977; Mederos 1991d; Mederos et al. 1997b) solid culture media with the following additives: 100 $mg\,l^{-1}$ myo-inositol, 20 $g\,l^{-1}$ sucrose, 6 $g\,l^{-1}$ Difco-bacto agar (Sigma), 10^{-7} M BAP and 10^{-8} M NAA (Mederos et al. 1996).
2. To prevent browning exudates, use the culture media mentioned before supplemented with 10^{-5} M rosmanol antioxidant (Mederos 1991a; Mederos and Amaro Luis 1997; Mederos et al. 1997a). Another alternative to prevent browning exudates was the incubation of the culture at a photosynthetic photon flux densities (PPFD) of 5, 10 and 15 $\mu mol\,m^{-2}\,s^{-1}$ for 8, 16 and 24 days respectively (Mederos, unpubl. data).
3. Incubate cultures under 16 h photoperiod and at a temperature of 24 °C. The PPFD may be equal or lower at 15 $\mu mol\,m^{-2}\,s^{-1}$ (see item 2, above) and 20 $\mu mol\,m^{-2}\,s^{-1}$ after 24 days of culture.

7.4 In Vitro Multiplication of Shoots Stage

1. Transfer each shoot onto solid MS culture medium (Murashige and Skoog 1962) with 10^{-7} M BAP and 10^{-8} M NAA (Mederos 1991a; Mederos et al. 1996).
2. Incubate shoots under 16-h photoperiod with an intensity (PPFD) of 20 $\mu mol\,m^{-2}\,s^{-1}$ and at a temperature of 24 °C.

3. At this stage, it is possible to achieve plantlets if shoots are cultured in the optimum culture medium mentioned before with phloroglucinol and without NAA (Mederos, unpubl. data).

7.5 In Vitro Rooting Stage

1. Transfer shoots into a half-strength MS medium (Murashige and Skoog 1962) supplemented with 10^{-8} M IBA (Mederos, 1991a; Mederos et al. 1996).
2. Incubate shoots under 16-h photoperiod with a PPFD of 20 μmol $m^{-2} s^{-1}$ and at a temperature of 24 °C.
3. Finally, at these stages, lights were provided by Philips cool white fluorescent lamps.

7.6 Hypericin for Clonal Propagation of *Hypericum canariense* L.

1. Take 25 g fresh wt. of shoots.
2. Take 25 g fresh wt. of aerial plantlets.
3. Take 12 g fresh wt. of roots.
4. Dry weight of shoots, aerial plantlets and roots were 4.2, 4.3 and 2 g respectively.
5. Macerate with distilled Cl_2CH_2 at room temperature for 2-day periods.
6. Treat repeatedly to obtain an extract of 76.8 mg from shoots and aerial part of plantlets and 28.2 mg from roots.
7. Extracts were chromatographed on a silica gel column with a solution of methylene chloride/acetone/methanol (75/15/10 vol%) as eluant.
8. Finally, hypericin was isolated from shoots and from the aerial part of plantlets of this species endemic to the Canary Islands, Spain (Mederos et al. 1996).

Acknowledgements. This review is dedicated to the memory of Professor Bajaj, an eminent and highly respected biologist in the field of Plant Cell, Tissue and Organ Culture. I would also like to take the opportunity to contribute with this article to Biotechnology in Agriculture and Forestry. Thanks are also due to Mrs. P. Agnew for correcting the English text. A part of the in vitro plant material of *Hypericum canariense* L. was destined to CICYT Projects I + D FAR 88-501.

References

Andersen D, Weber N, Wood S, Hughes B, Murray B, North J (1991) In vitro viral activity of selected anthraquinones and anthraquinone derivatives. Antivir Res 16:185–196

Aramaki Y, Chiba K, Tada M (1995) Spiro-lactones, hyperolactone A–D from *Hypericum chinense*. Phytochemistry 38(6):1419–1421

Araya O, Ford E (1981) An investigation of the type of photosensitization caused by ingestion of St. John's Wort (*Hypericum perforatum*) by calves. J Comp Pathol 91:135–141

Arello EF, Pinto JE (1993) In vitro propagation of *Kielmeyera coriaceae* Martius. I. Effects of 6-benzylaminopurine and naphtaleneacetic acid on shoot multiplication. Pesq Agropec Bras Brasília 28(1):25–31

Bajaj YPS (ed) (1990) Somaclonal variation in crop improvement. In: Bajaj YPS (ed) Biotechnology in agriculture and forestry, vol 11. Springer, Berlin Heidelberg New York, pp 27–63

Beerhues L, Barillas W, Peters S, Schmidt W (1997) Xanthone biosynthesis in plant cell cultures. Recent Res Dev Phytochem 1:69–76

Berger-Büter K, Büter B, Schaffner W (1994) In vitro propagation of *Hypericum perforatum* L.: impact of thidiazuron and 6-benzylaminopurine. Book of abstract of VIIIth Int Congr Plant Tissue Cell Culture, Firenze, June, pp 62

Bombardelli E, Morazzoni P (1995) *Hypericum perforatum*. Fitoterapia LXVI 1:43–68

Bomme U (1997) Technology of field production of perforated St. John's Wort (*Hypericum perforatum* L.). Z Arznei Gewürzpflanzen 2(3):127–134

Bramwell D, Bramwell Z (1974) Flores Silvestres de las Islas Canarias. Ed. Ecxmo. Cabildo Insular de Gran Canaria and Stanley Thornes, Cheltenham, UK [Span]

Bramwell D, Bramwell Z (1990) Flores Silvestres de las Islas Canarias. 2a ed, ed Rueda, Madrid, Spain [Span]

Bruneton J (1999) Pharmacognosy, phytochemistry, medicinal plants.2nd ed. Ed Agricell Report, P.O. Box 255, Schrub Oak, NY 10588, USA, pp 1131

Brutovská R, Cellárová E, Davey M, Power JB, Lowe K (1994) Stimulation of multiple shoot regeneration from seedling leaves of *Hypericum perforatum* L. by Pluronic F-68. Acta Biotechnol 14:347–353

Cakir A, Duru M, Harmandar M, Ciriminna R, Passannanti S, Piozzi F (1997) Comparison of the volatile oils of *Hypericum scabrum* L. and *Hypericum perforatum* L. from Turkey. Flavour Fragance J (UK) 12(4) 285–287

Calvert J (1930) St. John's wort – botanical classification. J Council Sci Ind Res 3:189–190

Campbell MH (1985) Germination, emergence and seedling growth of *Hypericum perforatum* L. Weed Res 25:259–266

Campbell MH, Delfose E (1984) The biology of Australian weeds, *Hypericum perforatum* L. J Aust Inst Agric Sci 50:63–73

Carin T, Pardini R (1992) Oxygen dependence of hypericin-induced phototoxicity to EMT6 mouse mammary carcinoma cells. Photochem Photobiol 55:831–837

Carin T, MacGill R, Miller G, Pardini R (1992) Photoactivation of hypericin generates singlet oxygen in mitochondria and inhibits succinoxidase. Photochem Photobiol 55:47–53

Carpenter S, Fehr M, Kraus G, Petrich J (1994) Chemiluminescent activation of the antiviral activity of hypericin: a molecular flashlight. Proc Natl Acad Sci USA 91:12273–12277

Ceballos L, Ortuño F (1976) Estudio sobre la vegetación y flora forestal de las Islas Canarias occidentales. Aula de Cultura de Santa Cruz de Tenerife, Islas Canarias [Span]

Cellárová E, Bruñakova K (1996) Somaclonal variation in *Hypericum perforatum* (St. John's Wort). In: Bajaj YPS (ed) Biotechnology in agriculture and forestry, vol 36. Somaclonal variation in crop improvement. Springer, Berlin Heidelberg New York, pp 267–279

Cellárová E, Kimáková K, Brutovská R (1992) Multiple shoot formation and phenotypic changes of Ro regenerants in *Hypericum perforatum* L. Acta Biotechnol 12:445–452

Cellárová E, Daxnerová Z, Kimáková K, Halusková J (1994) The variability of hypericin content in the regenerants of *Hypericum perforatum*. Acta Biotechnol 14:267–274

Cellárová E, Kimáková K, Daxnerová Z, Mártonfi P (1995) *Hypericum perforatum* (St. John's Wort) In vitro culture and the production of hypericin and other secondary metabolites. In: Bajaj YPS (ed) Biotechnology in agriculture and forestry, vol 33. Medicinal and aromatic plants. Springer, Berlin Heidelberg New York, pp 261–275

Cellárová E, Brutovská R, Daxnerová Z, Brunáková K, Weigel R (1997) Correlation between hypericin content and the ploidy of somaclones of *Hypericum perforatum* L. Acta Biotechnol 17: 83–90

Crompton C, Hall I, Jensen K, Hildebrandt P (1988) The biology of Canadian weeds. *Hypericum perforatum* L. Can J Plant Sci 68:149–162

Cronquist A (1981) An integrated system for classification of flowering plants. Colombia University Press, New York

Curtis JD, Lersten NR (1991) Internal secretory structures in *Hypericum*: *H. perforatum* L and *H. balearicum* L. New Phytol 114:571–580

Darias V, Bravo L, Barquín E, Martín Herrera D, Fraile C (1986) Contribution to the ethnopharmacological study of the Canary Islands. J Ethnopharmacol 15:169–193

Degar S, Prince A, Pascual D, Lavie G, Levin B, Mazur Y, Lavie D, Ehrlich L, Carter C, Meruelo D (1992) Inactivation of the human immunodefficiency virus by hypericin: evidence for photochemical alterations of p24 and a block in uncoating. AIDS Res Human Retroviruses 9:1929–1936

Decosterd L, Hoffmann E, Kyburz R, Bray D, Hostettmann K (1990) A new phloroglucinol derivative from *Hypericum calycinum* with antifungal and in vitro antimalarial activity. Planta Med 57:548–551

Dias ACP, Fernández-Ferreira M (1995) In vitro cultures of tissues and cell suspensions of *Hypericum androsaerum*. First Meeting of the Portuguese chapter of International Association for Plant Tissue Culture, November, Lisbon, Portugal

Diwu Z, Lown J (1994) Photoperiodic potential of alternative photosensitizers to porophyrins. Pharm Ther 63:1–35

Duran N, Song PS (1986) Hypericin and its photodynamic action. Yearly review. Photochem Photobiol 43:677–680

Fornasiero R, Bianchi A, Pinetti A (1998) Anatomical and ultrastructural observations in *Hypericum perforatum* L. J Herbs Spices Plants 5(4):21–33

Fox LR, Ribeiro SP, Brown VK, Masters GJ, Clarke IP (1999) Direct and indirect effects of climate change on St John's wort, *Hypericum perforatum* L. (*Hypericaceae*). Oecologia 120(1):113–122

Freytag W (1984) Bestimmung von Hypericin und Nachweis von Pseudohypericin in *Hypericum perforatum* durch HPLC. Dtsch Apoth Ztg 124:2383–2386

Galitzer S, Oehme F (1978) Photosensitization: a literature review. Vet Sci Commun 2:217–230

Gamborg OL (1970) The effects of amino acids and ammonium on the growth of plant cells in suspension culture. J Plant Physiol 45:372–375

Gamborg OL, Phillips G (1995) Plant cell, tissue and organ culture. Fundamental Methods. Springer Lab Manual, Berlin Heidelberg New York

Gamborg OL, Miller R, Ojima K (1968) Nutrient requirements of suspension cultures of soybean root cells. Exp Cell Res 50:151–158

George EF (1996) Plant propagation by tissue culture, parts I and 2. Exegetic Ltd, UK

Giese AC (1980) Hypericism. Photochem Photobiol Rev 5:229–255

Girzu M (1995) Composition en acides gras de la sommité fleurie du millepertuis (*Hypericum perforatum* L., Hypericaceae). Oleagineux Corps Gras Lipides (Fr) 2(4):317–318

González AG (1976) Natural products isolated from plants of the Canary Islands. Dr W Junk, The Hague, pp 297–326

González AG (1982) Productos aislados de plantas de las Islas Canarias entre 1975 y 1981. Instituto de Estudios Canarios en su cincuenta aniversario, Tomo I, pp 151–178 [Span]

González-Martín M (1995) Germination and light requirements of species of genus *Hypericum* in the Canary Islands. Bot Macaronésica Spain 21:43–49 [Span]

Gunatilanka A, De Silva A, Sotheesearan S (1982) Studies on medicinal and related plants of Sri Lanka. Part 6 Minor xanthones of *Hypericum mysorense* H. Phytochemistry 21:1751–1753

Halusková J, Cellárová E (1997) RFLP analysis of *Hypericum perforatum* L. somaclones and their progenies. Euphytica 95:229–235

Hansen A, Sunding P (1985) Flora of Macaronesia. Checklist of vascular plants. 3a rev Sommerfeltia (ed) Univ Oslo

Harbone JB (1980) Plant phenolics: secondary plant products. In: Bell E, Charlwood B (eds) Bioactive compounds from plants. Springer, Berlin Heidelberg New York, pp 329–402

Hartmann H, Kester D, Davies F (1990) Principles of tissue culture for micropropagation. In: Hartmann H, Kester D, Davies F (eds) Plant propagation: principles and practices, 5th edn. Prentice Hall, Englewood Cliffs, pp 459–495

Henderson BW, Dougherty TJ (1992) How does photodynamic therapy work? Photochem Photobiol 26:67–74

Herman EB (1993) Recent advances in plant tissue culture, vol 2. Secondary metabolite production 1988–1993. Agritech Consultants , Inc P.O. 255, Shrub Oak, NY 10588, USA

Herman EB (1995) Recent advances in plant tissue culture, vol 3. Regeneration and micropropagation: techniques, systems and media (1991–1995). Agritech Consultants , Inc P.O. Box 255, Shrub Oak, NY 10588, UK

Herman EB (1996) Microbial contamination of plant tissue cultures, vol 4. Recent advances in plant tissue culture. Agricell Report , Inc P.O. 255, Shrub Oak, NY 10588, USA

Herman EB (2000). Regeneration and micropropagation: techniques, systems and media between 1997 and 1999. Agricell Report

Herrera RM, Pérez M, Martín-Herrera DA, López-García R, Rabanal R (1996) Antimicrobial activity of extracts from plants endemic to the Canary Islands. Phytother Res 10:364–366

Hildebrand P, Jensen K (1991) Potential for the biological control of St. John's-wort (*Hypericum perforatum*) with an endemic strain of *Colletotrichum gloeosporioides*. Can J Plant Physiol 13:60–70

Hölz J, Ostrowski H (1987) Johanniskraut (*Hypericum perforatum* L.). HPLC-Analyse der wichtigen Inhaltsstoffe und deren Variabilität in einer Population. Dtsch Apoth Ztg 127:1227–1230

Hohtola A (1988) Seasonal changes in explant viability and contamination of tissue cultures from mature Scots pine. Plant Cell Tissue Organ Cult 15:211–222

Hudson J, López-Bazzocchi I, Towers G (1991) Antiviral activities of hypericin. Antiviral Res 15:101–112

Hudson J, Harris L, Towers G (1993) The importance of light in the anti-HIV effect of Hypericin. Antivir Res 20:173–178

Jayasurikab H, Clark A, Machesney J (1991) New antimicrobial filicinic acid derivatives from *Hypericum drummondii*. J Nat Prod 54:1314–1320

Jensen K, Gaul S, Specht E, Doohan D (1995) Hypericin content of Nova Scotia biotypes of *Hypericum perforatum* L. Can J Plant Sci 75:923–926

Kartnig TH, Gruber A, Sauer H (1989) Comparative phytochemical investigations of *Hypericum* species. Planta Med 55:215–218

Kikuchi T, Kadto S, Matsuda S, Tanaka K, Namba T (1985) Studies on the constituents of medicinal and related plants in Sri Lanka. II. Isolation and structures of y-pirone and related compounds from *Hypericum mysorense* Heyne. Chem Pharm Bull 33:557–564

Kingsbury J (1964) Poisonous plants in the United States and Canada. Prentice-Hall, Englewood Cliffs

Knox JP, Dodge AD (1985) Isolation and activity of the photodynamic pigment hypericin. Plant Cell Environ 8:19–25

Kraus G, Pratt D, Tossberg J, Carpenter S (1990) Antiretroviral activity of synthetic hypericin. Plant Cell Environ 8:19–25

Kunkel G (1971) Bibliografía Botánica: índice de obras sobre la flora y vegetación del Archipiélago Canario, publicadas entre 1960 y 1969. Cuadernos de Botánica Canaria supl 1 [Span]

Kunkel G (1972) Enumeración de las plantas vasculares de Gran Canaria. Monogr Biol Canaria Supl 3, pp 86 [Span]

Kunkel G (1992) Flora y Vegetación del Archipiélago Canario: Tratado Florístico, 2a parte. Edirca (ed), Gran Canaria. [Span]

Kupila-Ahvenniemi S (1977) Winter and spring changes in the microsporogeneous cells of the Scotch pine as revealed by Feulgen photometry and Auramin fluorometry. Can J Bot 55: 1434–1442

Larsen K (1963) Contribution to the cytology of the endemic Canarian element. Bot Notiser 116: 409–424

Lavie G, Valentine F, Leviu B, Mazur Y, Gallo G, Lavie D, Weiner D, Meruelo D (1989) Studies of the mechanisms of action of the antiretroviral agents hypericin and pseudohypericin. Proc Natl Acad Sci USA 86:5963–5967
Lavie G, Mazur Y, Lavie D, Levin B, Ittah Y, Meruelo D (1990) Hypericin as an antiretroviral agent. Ann New York Acad Sci 616:556–562
Lenard J, Rabson A, Vanderoef R (1993) Photodynamic inactivation of infectivity of human immunodeficiency virus and other enveloped viruses using hypericin and rose bengal: inhibition of fusion and syncytia. Proc Natl Acad Sci USA 90:158–162
Liebes L, Mazur Y, Freeman D, Lavie G, Kudler N, Mendoza S, Levin B, Hochster H, Meruelo D (1991) A method for the quantitation of hypericin an antiviral agent, in biological fluids using high performance liquid chromatography. Anal Biochem 195:77–85
Linnaeus C (1753) Species *Hypericum* (1967–1959). Facsim Ray Soc , London
Linsmaier E, Skoog F (1965) Organic factor requirement of tobacco tissue culture. Physiol Plant 18:100–127
Liu Z, Carpenter S, Bourgeois W, Yu Y, Constantin R, Falcon M, Adams J (1998) Variations in the secondary metabolite camptothecin in relation to tissue age and season in *Camptotheca acuminata*. Tree Physiol 18:265–270
Lloyd G, McCown B (1981) Commercially feasible micropropagation of mountain laurel, *Kalmia latifolia* by use of shoot-tip culture. Combined Proc Int Plant Propagat Soc 30:421–426
López L, Rodríguez B, Mederos S (1991) Influencia del aporte amonio/nitrato y de ciertas hormonas sobre la micropropagación de *Hypericum canariense* L. En: Actas de la IX Reunión de la Soc. Española de Fisiología Vegetal y II Congreso Hispano-Luso de Fisiología Vegetal. Septiembre. Madrid, España, pp 230 [Span]
López-Bazzocchi I, Hudson J, Towers G (1991) Antiviral activity of the photoactive plant pigment hypericin. Photochem Photobiol 54:95–98
Luis J, González A, San Andrés L, Mederos-Molina S (1992) Diterpenes from in vitro-grown *Salvia canariensis* L. Phytochemistry 31:3272–3275
Maisenbacher P, Kovar K (1992) Analysis and stability of *Hyperici oleum*. Planta Med 58:351–354
Mathis C, Ourisson G (1963) Étude chimio-taxonomique du genre *Hypericum*. I. Repartition de l´ hypericine. Phytochemistry 2:157–171
Mederos S (1991a) Micropropagación del *Hypericum canariense* L. In: Frazao S, Jacobson E, Carmo M (eds) Book of abstract II, p-2, das II Jornadas Ibéricas de Plantas Medicinais e Oleos Essenciais. Marzo. Lisboa, Portugal [Span]
Mederos S (1991b) In vitro growth and multiplication of *Hypericum canariense* L. In: Mascherpa JM, Moncousin Ch (eds) International Symposium Plant biotechnology and its contribution to plant development multiplication and improvement. April. Palexpo, Geneva, Switzerland
Mederos S (1991c) In vitro growth and multiplication of *Hypericum canariense* L. Acta Hortic 289:133–134
Mederos S (1991d) Multiplicación vegetativa in vitro de especies leñosas de interés agronómico: *Pistacia atlantica* Desf. y *Pistacia vera* L. cv. Mateur. Ph. D. Thesis, Universidad de La Laguna, Tenerife, Canary Islands, Spain, 395 pp [Span]
Mederos S (1994) Micropropagación del *Hypericum canariense* L. In: Franzao S, Jacobson E Carmo M (eds) Proceedings das II Jornadas Ibéricas de Plantas Medicinais e Oleos Essenciais, vol I. Lisboa, Portugal, pp 84–88 [Span]
Mederos-Molina S. (2001a). Influence of supporting substrates on survival and growth in vitro plantlets of Tagasaste. In: Economou A (ed) Book of Abstracts of 1st International Symposium on Acclimatization and Establishment of Micropropagated Plants, September, Sani-Haldikiki, Greece, 111 pp
Mederos-Molina S. (2001b). In vitro tissue culture of *Maytenus canariensis* (Loes) Kund & Sund Mature Plants. I. Macroelement solution requirement for apical meristem cultures. Plant Tissue Culture (in press)
Mederos-Molina S. (2001c). Micropropagation of *Maytenus canariensis* (Loes) Kund & Sund Mature plants through apical meristem cultures. II. Plant growth regulators requirement. Plant Tissue Culture (in press)

Mederos-Molina S, Amaro Luis (1997) In vitro mass propagation of *Salvia canariensis* by axillary shoots. Acta Soc Bot Polon 66 (3–4):351–354

Mederos S, Rodríguez R (1991) Effect of plant hormones on the development of apical and axillary buds cultured in vitro of *Hypericum canariense* L. In: Altman A, Thörpe T (eds) In vitro plant. In vitro cellular and developmental biology, 27 (3, part 2), Int Assoc for Plant Tissue Cult & Biotech (IAPTC&B) Anaheim, California, 384 pp

Mederos-Molina S, Schöbert CH (1995) Micropropagation of *Ricinus communis* L. J Plant Physiol 147:270–272

Mederos-Molina S, Trujillo MI (1999a) Techniques for in vitro germination in *Pistacia* species. S Afr J Bot 65(2):149–152

Mederos-Molina S, Trujillo MI (1999b) Elimination of browning exudate and in vitro development of shoots in *Pistacia vera* L. cv. Mateur and *Pistacia atlantica* Desf. culture. Acta Soc Bot Polon 68(1):1–24

Mederos S, Trujillo I, López I (1994) Micropropagation of pistachio: *Pistacia vera* cv. Mateur. Plant Tissue Culture 4:111–116

Mederos S, López-Bazzocchi I, Ravelo A, González AG (1996) Hypericin for clonal propagation of *Hypericum canariese* L. Plant Tissue Culture 6:7–13

Mederos S, San Andrés L, Luis J (1997a) Rosmanol controls explants browning of *Hypericum canariense* L. during the in vitro establishment of shoots. Acta Soc Bot Polon 66(3–4):347–349

Mederos S, Trujillo I, López I. (1997b) Effects of nutrient media on the in vitro establishment of shoots of *Pistacia atlantica* Desf. seedlings. Tr J Agricult For 21:345–348

Mederos S, Martín C, Navarro E, Ayuso M J (1998a) Micropropagation of a medicinal plant, *Plantago major* L. Biol Plant 40(3):465–468

Mederos-Molina S, Trujillo MI, Gutiérrez Jerez F (1998b) In vitro Propagation of *Vitis vinifera* L. cv listán negro. Plant Tissue Cult 8(2):165–172

Meruelo D, Lavie G, Lavie D (1988) Therapeutic agents with dramatic antiretroviral activity and little toxicity at effective doses: aromatic polycyclic diones: hypericin and pseudohypericin. Proc Natl Acad Sci USA 85:5230–5234

Michaluk A, Brunarska Z, Beonarska D (1956) Tannins and flavones in different species of *Hypericum*. Diss Pharm 8:47–62

Miller RM, Drew RA (1990) Effect of explant type on proliferation of *Carica papaya* L. in vitro. Plant Cell Tissue Organ Culture 21:39–44

Milne-Redhead E (1953) *Hypericaceae*. In: Turrill WB, Milne-Redhead E (eds) Flora of Tropical East Africa. Cambridge Univ Press, Cambridge, 97 pp

Mok MC, Martin RC, Mok DWS (2000) Cytokinins: biosynthesis, metabolism and perception. In Vitro Plant 36(2):102–107

Murashige T, Skoog F (1962) A revised medium for rapid growth and bioassays with tobacco tissue cultures. Physiol Plant 15:473–497

Papageorgiou G, Alygizaki-Zorba A, Loukas S, Brody S (1996) Photodynamic effects of hypericin on photosynthetic electron transport and fluorescence of *Anacystis nidulans* (*Synechococcus* 6301). Photosyn Res 48:221–226

Pérez de Paz PL (1982) Perspectiva histórica de los últimos 50 años (1932–1982) de la Botánica en Canarias. Instituto de Estudios Canarios (ed). La Laguna, Tenerife [Span]

Pérez de Paz PL, Hernández Padrón CE (1999) Plantas medicinales o útiles en la flora canaria. In: Lemus F (ed) Aplicaciones populares. La Laguna, Tenerife, Canary Islands [Span]

Pérez de Paz PL, Medina I (1988) Catálogo de las plantas medicinales de la Flora Canaria. Aplicaciones populares. Gobierno de Canarias e Instituto de Estudios Canarios. La Laguna, Tenerife, Canary Islands [Span]

Pinzaru G, Lupu C (1999) Research concerning the determination of the best germination temperature for hardy (*Hypericum perforatum* L.) seeds. Cercetari Agron Moldova 31(3,4): 137–140

Quoirin M, Lepoivre P (1977) Improved medium for in vitro culture of *Prunus* sp. Acta Hortic 78:437–442

Rabanal-Gallego R (1996) Plantas medicinales canarias (ciclo de conferencias). Vicerrectorado de Extensión Universitaria de la Universidad de La Laguna, Tenerife, Islas Canarias [Span]

Rivas-Martínez S, Wildpret de la Torre W, Del Arco Aguilar M, Rodríguez O, Pérez de Paz PL, García Gallo A, Acebes Ginovés JR, Díaz González TC y Fernández González F (1993) Las comunidades vegetales de la Isla de Tenerife (Islas Canarias). Itinera Geobot 7:169–374 [Span]

Robson N (1958) The genus *Hypericum* in Africa south of the Sahara, Madagascar and the Mascarenes. Kew Bull 12:443–446

Robson N (1968) *Guttiferales*. In Flora Europea. Tutin T, Heywood V, Burges N (eds) vol 2. Cambridge University Press, Cambridge, pp 261–269

Robson N (1981) Studies in the genus *Hypericum* L. (*Guttiferae*). Sections: *Brathys* (part 2). Sections: *Trigynobrathys*. Bull Br Mus Nat Hist Bot 20:1–151

Robson N (1990) Studies in the genus *Hypericum* L. (*Guttiferae*). Characters of the genus. Bull. Br. Mus. Nat. Hist. Bot. 8:55–226

Robson N (1993) Studies in *Hypericum*: validation of new names. Bull Nat Hist Mus Lond (Bot) 23 (2):67–70

Roest S, Bokelmann G (1973) In vitro culture of *Hypericum perforatum*. Sci Horticult 1:120–122

Ryu D, Furosaki S (1994) Advances in plant biotechnology. Elsevier Academic Press, Amsterdam

Sakar M, Tamer A, Tokur S (1988) Antimicrobial activities of some *Hypericum* species growing in Turkey. Fitoterapia 59:49–52

Sampson A, Parker K (1930) St. John's wort on rangelands of California. Calif Agric Exp Stn Bull Berkeley 503:88

Santos A (1980) Contribución al conocimiento de la flora y vegetación de la Isla del Hierro (Islas Canarias). Fund J March Ser Univ Madrid, España [Span]

Santos A (1983) Vegetación y Flora de La Palma. Interinsular Canaria, Santa Cruz [Span]

Schinazi R, Chu CH, Babu J, Oswald B, Saalmann V, Cannon D, Eriksson B, Nasr M (1990) Anthraquinones as a new class of antiviral agents against human immunodeficiency virus. Antivir Res 13:265–272

Schmidt W, Peters S, Beerlmes L (2000) Xanthone 6-hydroxylase from cell cultures of *Centaurium erythraea* Rafn. and *Hypericum androsaeroum* L. Phytochemistry 53 (4):427–431

Schripsema J, Verpoorte R (1995) Primary and secondary metabolism of plant cell cultures III. Kluwer, Dordrecht

Seidler Lozykowska K, Dabrowska J, Zygmunt B (1999) Content of active substances in herb St. John´s wort (*Hypericum perforatum*) cv. Topaz in different growth phases. Herba Polon 45(3):169–172

Southwell I, Campbell M (1991) Hypericin content variation in *Hypericum perforatum* in Australia. Phytochemistry 30:475–478

Sudriá C, Piñol M, Palazón J, Cusidó R, Vila R, Morales C, Bonfill M, Cañiguera S (1999) Influence of plant growth regulators on the growth and essential oil content of cultured *Lavandula dentata* plantlets. Plant Cell Tissue Organ Culture 58(3):177–184

Sureau F, Miskovsky P, Chinsky L, Turpin PY (1996) Hypericin-induced cell photosensitization involves an intracellular pH decrease. J Am Chem Soc 118:9484–9487

Suzuki O, Katsumata Y, Oya M, Bladt S, Wagner H (1984) Inhibition of monoamine oxidase by hypericin. Planta Med 50:272–274

Takahashi I, Nakanishi S, Kobayashi E, Nakano H, Suzuki K, Tamaoki T (1989) Hypericin and pseudohypericin specifically inhibit protein kinase C.: possible relation to their antiretroviral activity. Biochem Biophys Res Commun 165:1207–1212

Thomas C, Pardini R (1992) Oxygen dependence of hypericin-induced phototoxicity to EMT6 mouse mammary carcinoma cells. Photochem Photobiol 55:831–837

Thomas C, Macgill R, Miller G, Pardini R (1992) Photoactivation of hypericin generates singlet oxygen in mitochondria and inhibits succinoxidase. Photochem Photobiol 55:47–53

Tisserat B, Silman R (2000) Interactions of culture vessels, media volume, culture density and carbon dioxide levels on lettuce and spearmint shoot growth in vitro. Plant Cell Rep 19(5): 464–471

Towers GH (1980) Photosenzitizers in plants and their photodynamic action. Prog Phytochem Ecol 40:54–62

Umek A, Kreft S, Kartnig T, Heydel B (1999) Quantitative phytochemical analyses of six *Hypericum* species growing in Slovenia. Planta Med 65(4):388–390

Van der Meijden E (1996) Plant defence, an evolutionary dilemma: contrasting effects of (specialist and generalist) herbivores and natural enemies. Entomol Exp Appl 80:307–310

Vanhaelen M, Vanhaelen-Fastre R (1983) Hypericin and pseudohypericin specifically inhibit protein kinase C: possible relation to their antiretroviral activity. J Chromatogr 281:263–271

Willis A, Groves R, Ash J (1997) Seed ecology of *Hypericum gramineum*, an Australian Forb. Aust J Bot 45:1009–1022

Yazaki K, Okuda T (1990) Procyanidins in callus and multiple shoot cultures of *Hypericum erectum*. Planta Med 56:490–491

Yazaki K, Okuda T (1994) *Hypericum erectum* Thunb. (St. John's Wort). In vitro culture and the production of procyanidins. In: Bajaj YPS (ed) Biotechnology in agriculture and forestry, vol 26. Medicinal and aromatic plants. Springer, Berlin Heidelberg New York, pp 167–178

7 Production of Quinolizidine Alkaloids in in Vitro Cultures of Legumes

MICHAEL WINK[1]

1 Introduction

1.1 Occurrence of Quinolizidine Alkaloids

Quinolizidine alkaloids (QA), of which more than 170 structures have been reported (overviews in: Mears and Mabry 1971; Kinghorn and Balandrin 1984; Salatino and Gottlieb 1980; Wink 1993), are distinct from other alkaloids in that they contain at least one quinolizidine ring system. According to the degree of substitution at least seven different groups can be distinguished: (1) the lupinine-type, (2) the *Leontidine*-type, (3) the sparteine/lupanine/multiflorine-type, (4) the alpha-pyridone-type, (5) the matrine-type, (6) the *Ormosia*-type, and (7) piperidine and dipiperidines, which often occur together with QA (Fig. 1).

Quinolizidine alkaloids are mainly distributed within the Leguminosae, the third largest family of flowering plants after the Asteraceae and Orchidaceae. Leguminosae consist of about 750 genera and 18,000 species. QA-accumulating tribes include Sophoreae, Dalbergieae, Euchresteae, Thermopsidae, Genisteae, Bossiaeae, Brongniartieae, Podalyrieae, Liparieae and Crotalarieae (Fig. 2; Kinghorn and Balandrin 1984; Kaess and Wink 1995, 1997). However, restricted occurrences have been reported from a number of other families which are unrelated to the Leguminosae, such as Berberidaceae (*Leontice, Caulophyllum*), Solanaceae, Ranunculaceae (*Cimicifuga*), Chenopodiaceae (*Anabasis*) and Rubiaceae (*Readea*). Parasitic or hemiparasitic plants tap phloem and xylem sap of their host plants. If the host plant contains QA, an alkaloid transfer is plausible, as was observed for a number of species of the Scrophulariaceae: *Orobanche* (Wink et al. 1982), *Castilleja* (Stermitz and Harris 1983; Adler and Wink 2001),and *Pedicularis* (Stermitz et al. 1985), but also in other families such as in *Viscum* (Loranthaceae), *Cuscuta* (Cuscutaceae) (Czygan et al. 1988) and *Osyris* (Santalaceae) (Le Scao-Bogaert et al.1978; Moges and Wink 2002).

Using plant cell suspension cultures of *Daucus carota* (Umbelliferae), *Conium maculatum* (Umbelliferae), *Atropa belladonna* (Solanaceae),

[1]University of Heidelberg, Institute for Pharmaceutical Biology, Im Neuenheimer Feld 364, 69120 Heidelberg, Germany

Biotechnology in Agriculture and Forestry, Vol. 51
Medicinal and Aromatic Plants XII (ed. by T. Nagata and Y. Ebizuka)
© Springer-Verlag Berlin Heidelberg 2002

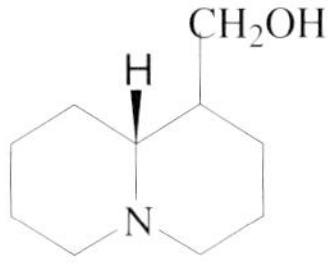

Lupinine (1)

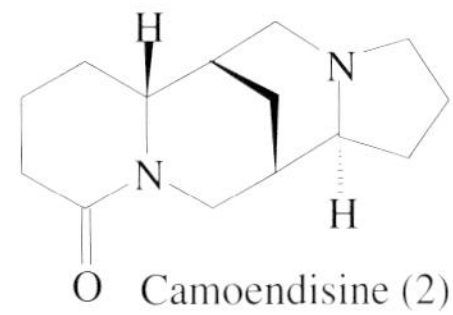

Camoendisine (2)

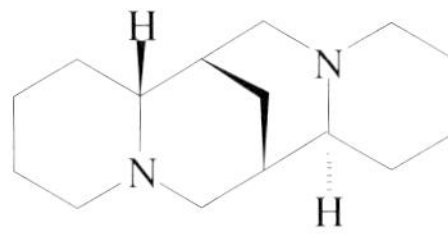

Sparteine (3a)

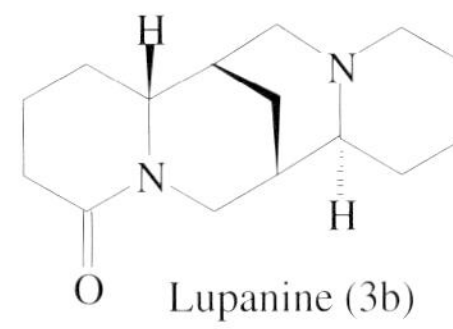

Lupanine (3b)

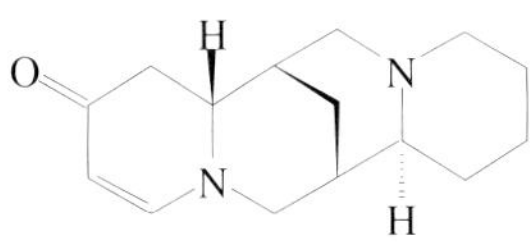

Multiflorine (3c)

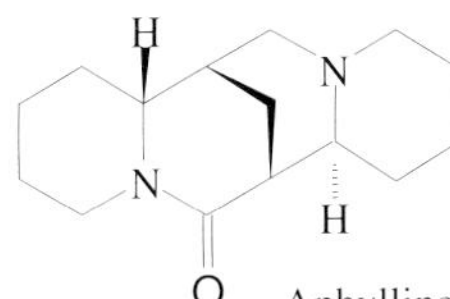

Aphylline (3d)

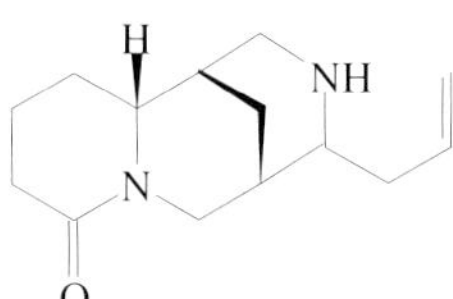

Angustifoline (3e)

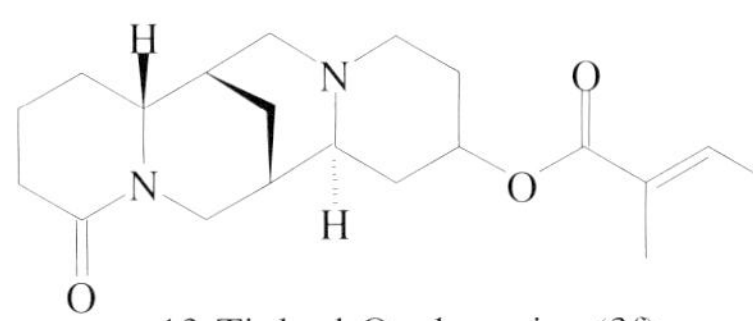

13-Tigloyl-Oxylupanine (3f)

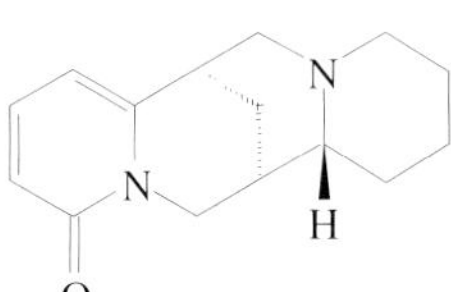

Anagyrine (4)

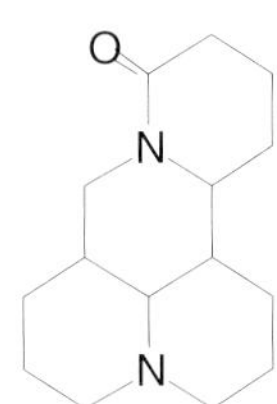

Matrine (5)

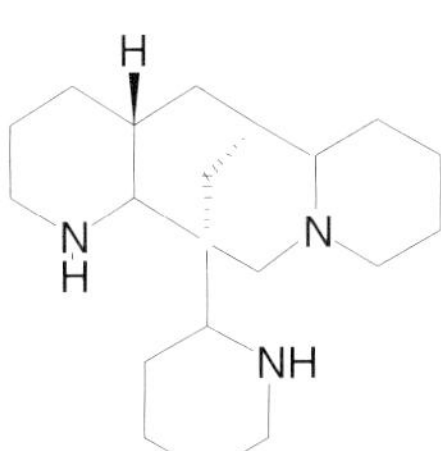

Piptanthine (6)

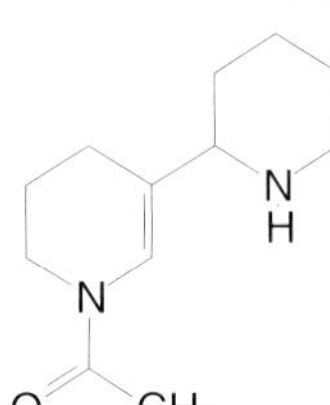

Ammodendrine (7)

Fig. 1. Structures of quinolizidine alkaloids

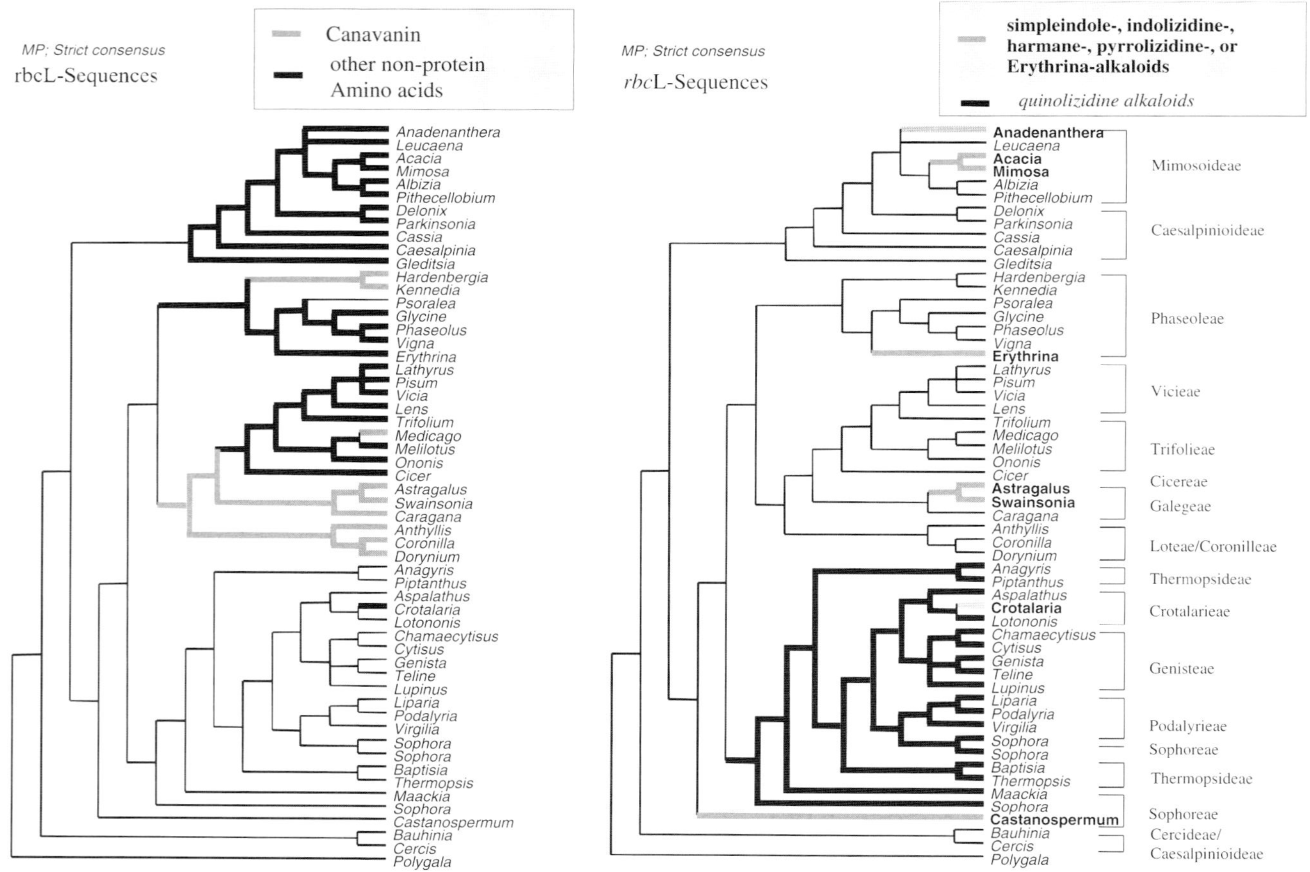

Fig. 2. Molecular phylogeny of the Leguminosae and distribution of QA. The phylogeny was inferred from nucleotide sequences of the *rbc*L gene (after Wink and Waterman 1999). *Left* Distribution of non-protein amino acids. *Right* Distribution of alkaloids. Note that the occurrences of non-protein amino acids and alkaloids are almost mutually exclusive. In the tribe Crotalarieae pyrrolizidine alkaloids and not quinolizidine alkaloids are present which would have been expected if alkaloids were a strict taxonomic marker

Chenopodium rubrum (Chenopodiaceae), *Spinacia oleracea* (Chenopodiaceae) and *Symphytum officinale* (Boraginaceae), we observed that these cultures produce minute amounts of lupanine, especially when challenged with elicitors (Wink and Witte 1983). Using a very sensitive radio immunoassay (RIA) for lupanine we found traces of lupanine in ca 50 out of 100 plants studied (Wink 1987b). The next step will be to determine the nucleotide sequences of the corresponding genes, to see whether and how the genes and plants are related. These surprising findings suggest that the genes of QA biosynthesis might be widely distributed within the plant kingdom, but are generally "silent". Mainly in the Leguminosae the genes for biosynthesis, transport and storage are switched on, since here the alkaloids serve for chemical defence. The distribution of QAs is only partly consistent with the phylogeny of legumes (Fig. 2), indicating that QAs represent an adaptive trait which tells more about the defence strategies of respective plants than their taxonomic relationships (Wink and Waterman 1999).

1.2 Biosynthesis of Quinolizidine Alkaloids in Plants

QA biosynthesis from lysine via cadaverine to sparteine/lupanine takes place in the aerial green parts of legumes. Its intracellular site is the chloroplast (Wink and Hartmann 1982a,b). Biosynthesis is regulated by light, thus QA concentrations display a diurnal rhythm, with a stimulated production period during the day (Wink and Witte 1984). QA are translocated by the phloem all over the plant (Wink et al. 1982; Wink and Witte 1984, 1991), and they accumulate predominantly in epidermal and subepidermal cell layers (Wink et al. 1984; Wink 1987d). The subcellular site of QA storage is the vacuole, into which the alkaloids are pumped by a selective carrier system (Mende and Wink 1987; Wink and Mende 1987). Especially rich in alkaloids are the seeds, which can store up to 8% of their dry weight as alkaloids. During germination the alkaloids are translocated from the cotyledons to the newly formed tissues, where the alkaloids are partly degraded, obviously serving as nitrogen storage compounds (Wink and Witte 1985). QA also disappear from senescing leaves during the vegetation period, indicating that they are not end products but metabolically mobile compounds (Wink and Witte 1985).

1.3 Function of Quinolizidine Alkaloids

It could be shown experimentally that pure QA or QA mixtures are biologically active (for review of the older literature, see Kinghorn and Balandrin 1984), in that they inhibit the multiplication of potato-X virus (Wink 1987a), the growth of gram-negative and gram-positive bacteria (Wink 1984a; Tyski et al. 1988), and of certain fungi (Wink 1984a; Wippich and Wink 1985). QA deter or repel feeding of a number of herbivores (nematodes, bees, caterpillars, beetles, aphids, locusts, snails, rabbits, cows) or are directly toxic or mutagenic for them (Keeler 1976; Bentley et al. 1984; Dreyer et al. 1985; Johnson and Bentley 1988; Wink 1984b, 1985a, 1987a, 1988; Matsuda et al. 1989; for review see Wink 1992; Wink and Twardowski 1992). The toxic effects are probably due to inhibition of Na^+ and K^+ ion channels (Paolisso et al. 1985; Koerper et al. 1998), to interactions with the nicotinic and muscarinic acetylcholine receptor (Schmeller et al. 1994, Wink et al. 1998), and to inhibition of protein synthesis (Korcz et al. 1987; Wink and Twardowski 1992). Thus QA are broad-range allelochemical defence compounds.

The concentration of QAs in the plant is sufficiently high to perform their inhibitory effects observed in vitro. In addition, QA contents can be increased by wounding: this effect was highest under greenhouse conditions, but also measurable in the field (Wink 1983a, 1985a; Johnson et al. 1989). In this context, QA localisation in epidermal tissues can be interpreted as a strategically important adaptation, since this tissue has to ward off small herbivores and pathogens in the first instance.

Plant breeders have selected lupin mutants which produce only minute amounts of alkaloids. These so-called sweet lupins have a dramatically reduced fitness under natural conditions as compared to their "bitter" wild forms. They are preferentially eaten by rabbits, leaf miners, aphids or beetles and are vulnerable to other pathogens (Cantot and Papineau 1983; Wink 1985a, 1987a; review Wink 1988, 1992). Thus, it is seems well established that the ecological function of QA is that of defence against herbivores, but also against microorganisms. Some evidence suggests activity against other competing plants (Wink 1983b; Wink and Twardowski 1992).

1.4 Pharmacological Properties of Quinolizidine Alkaloids

In addition to toxic and repellent properties, lupin alkaloids have a number of pharmacological activities (Kinghorn and Balandrin 1984; Wink 1992, 2000): Sparteine, lupanine and other QA exhibit antiarrhythmic properties. Since only sparteine can be easily isolated from broom (*Cytisus scoparius*), it is the only lupin alkaloid that is commercially available and exploited in medicine as an antiarrhythmic drug. However, about 10% of all patients are unable to metabolise sparteine and suffer from sparteine intoxication. Because of these side effects and the availability of more reliable synthetic heart drugs, the use of sparteine in modern medicine is declining and restricted. The utilisation of sparteine as a uterus-contracting drug has been abandoned for the same reasons. Sparteine, lupanine and 13-hydroxylupanine have hypotensive and CNS-depressant properties and furthermore, are hypoglycemic, i.e., they reduce the level of blood glucose and might thus be interesting as antidiabetic drugs. In addition to QA, the alkaloid-free fraction of *L. albus* seeds also seems to have antidiabetic activities. Matrine and cytisine are amoebicidal, matrine and 17-oxolupanine effective anti-inflammatory compounds. Some of these pharmacological properties can be explained through activation of the acetylcholine receptors and inhibition of K^+ and Na^+ channels (reviewed in Wink 1995, 2000).

2 Quinolizidine Alkaloid Formation in in Vitro Cultures

2.1 Biosynthesis and Accumulation

QA metabolism has been studied in the cell and organ cultures of more than 14 legume species (Wink et al. 1983; Wink 1987b,c, 1989,1990; Saito et al. 1989): Whereas the differentiated plants produce a variety of QA, the alkaloid patterns of the cell cultures of *Lupinus, Cytisus, Laburnum, Genista, Baptisia,* and *Sophora* were very simple (Table 1) in that lupanine was always the main alkaloid irrespective of the alkaloid patterns of the intact plant (Wink et al. 1983; Wink and Hartmann 1980, 1982a,b). Callus cultures of *Sophora flavescens* accumulate matrine, and those of *Thermopsis lupinoides,* lupanine as the main product (Saito et al. 1989). Alkaloid levels in cell cultures were about two to three orders of magnitude lower than those of the differentiated plant (Wink et al. 1983; Table 2).

Alkaloid formation is correlated with light and greening of the cell culture (Wink et al. 1982b,c; Saito et al. 1989). In organised tissues, QA production can be observed in shoots and leaves but not in roots (Table 2; Wink 1987d; Saito et al. 1989). In aseptically grown roots and shoots of *L. polyphyllus, L. succulentus, L. subcarnosus, Cytisus scoparius* and *Laburnum anagyroides,* QA were only produced by shoots and not by roots. The alkaloid profiles of shoots resembled those of the intact plant (leaves). Feeding experiments with the biogenetic precursor cadaverine confirmed the inactivity of the roots towards QA biosynthesis (Wink 1987b). We conclude that the genes of QA metabolism are correctly expressed in differentiated shoot cultures but silent in roots, i.e. QA genes are regulated in a tissue-specific way, as is the case with many other cellular phenomena (Wink 1988, 1989,1990, 1999a,b).

Most of our physiological and biochemical studies have been performed with cell cultures of *Lupinus polyphyllus* (review Wink 1985c, 1987b,c) and of another North American lupin, *L. hartwegii* (Wink et al. 1992). Besides the biosynthesis of QA, their cellular storage and degradation (see below) appear as critical factors (Mende and Wink 1987; Wink and Mende 1987; Wink 1987d).

The kinetics of alkaloid accumulation in relation to physiological parameters in the cell culture medium and the cells during the same growth period, such as sucrose and phosphate contents, osmolarity and conductivity, hydrogen ion concentration, protein content and RNA formation and chlorophyll content have been documented for *L. polyphyllus, L. hartwegii* and *Cytisus scoparius* (Wink et al. 1980, 1982, 1992; Wink and Hartmann 1980).

Table 1. Profiles of QA in plants and cell suspension cultures (after Wink et al. 1983; and unpubl. data)

Alkaloid	*Lupinus polyphyllus*		*L. luteus*		*L. hartwegii*		*Cytisus scoparius*		*Cytisus purpureus*		*Genista canariensis*		*Genista pilosa*		*Laburnum alpinum*		*Baptisia australis*		*Sophora japonica*	
	cc	l	cc	l	cc	l	cc	l	cc	l	cc	l	cc	l	cc	l	cc	l	cc	l
Lupinine	–	–	–	19	–	–	–	–	–	–	–	–	–	–	–	–	–	–	–	–
Sparteine	1–5	<1	<1	66	<5	–	<40	62	–	–	–	–	–	70	–	–	–	<2	–	–
Ammodendrine	–	<1	–	<1	–	–	–	<1	–	–	–	–	–	–	–	39	–	–	–	–
N-Methylcytisine	–	–	–	–	–	–	–	–	–	64	–	19	–	5	–	45	–	74	–	tr
Cytisine	–	–	–	–	–	–	–	–	–	1	–	39	–	<1	–	2	–	13	–	tr
Epiaphylline/ aphylline	–	–	–	–	–	72	–	–	–	–	–	–	–	–	–	–	–	–	–	–
17-Oxosparteine	<1	<1	–	<1	3	–	–	19	–	–	–	–	–	10	–	<1	–	tr	–	–
Lupanine	70–90	60	95	<1	80	23	50–70	7	95	6	95	<1	95	5	95	3	98	<1	95	30
Anagyrine	–	–	–	–	–	–	–	<1	–	26	–	40	–	–	–	<1	–	7	–	–
13-Hydroxylupanine	<1	7	–	–	–	–	–	3	–	–	–	–	–	–	–	–	–	–	–	–
Hydroxylupanine esters	tr	26	–	–	–	–	–	–	–	–	–	–	–	–	–	–	–	–	–	–
Lupinine esters	–	–	–	13	–	–	–	–	–	–	–	–	–	–	–	–	–	–	–	–
Alkaloid content (mg/g fresh wt.)	<0.015	5	<0.005	1.2	<0.01	0.9	<0.012	5.5	<0.001	0.1	<0.001	0.1	<0.001	0.1	<0.001	1	<0.001	2.5	0.001	<0.1

cc, Cell suspension culture; l, leaf; data are % of total alkaloids.

Table 2. Expression of quinolizidine alkaloid biosynthesis in cell cultures, organ cultures and *in planta*

Type	Alkaloid content (μg/g fresh wt.)		
	L. polyphyllus	*Cytisus scoparius*	*Laburnum anagyroides*
Cell cultures			
Callus	0.01–1	0.01–5	0.1–1[a]
Suspension	0.01–5	0.8–12	0.1–1[a]
Organ cultures			
Roots	0–4	0	0
Shoots	76	64	21
Intact plants			
Leaves	500–5000	100–5500	100–900
Seeds	35,000	210	10,000

[a] *Laburnum alpinum.*

After administration of the precursor cadaverine to *L. hartwegii* cell suspension cultures, which is formed after decarboxylation of lysine and figures as the sole precursor of the tetracyclic lupin alkaloids (review: Wink 1987b), QA formation was significantly enhanced during the initial 3 days and up to 33 times higher than untreated controls. Most of the alkaloids produced were released into the cell culture medium as had been observed before with cell suspension cultures of *L. polyphyllus* (Wink and Witte 1983; Wink and Hartmann 1982a,b; Wink 1985b). After ca. 4 days, almost all alkaloids produced previously had disappeared which is probably due to QA degradation (Wink 1985b; Wink and Witte 1985). When cadaverine was added at different growth stages, it was observed that QA formation is most active during the initial lag period and to a smaller degree when the cells have reached the final stationary growth phase. In cell suspension cultures of *L. polyphyllus,* both alkaloid formation and the activity of the biosynthetic enzymes, lysine decarboxylase and oxosparteine synthase, showed similar trends (Wink and Hartmann 1982a).

QA formation can be stimulated by a number of chemical or biological elicitors (Wink 1985c), as is the case with many other secondary metabolites. During the last decade, it became evident that many of these compounds or processes induce the formation of jasmonic acid (after wounding or herbivore attack) or salicylic acid (after microbial infection) (reviewed in Wink 1999a,b) which in turn stimulate gene expression. Agents which can induce QA formation in cell cultures included several alkaloids (unrelated to QA), polyamines, cAMP and bacterial cell walls (Wink 1985b). Preliminary evidence indicates that the activation involves an induction of the respective alkaloid genes. The increase of alkaloid formation is transient and after 3–5 days the alkaloids formed disappear due to degradation (Wink 1985c, 1994).

2.2 Alkaloid Uptake and Degradation

Quinolizidine alkaloids are stored in vacuoles (Mende and Wink 1987). Since QA are charged molecules under physiological conditions they cannot diffuse freely across biomembranes. We could show in earlier experiments that the transport of QA into vacuoles is achieved by a specific carrier process. Furthermore, uptake was found to be fuelled by ATP and a K^+ gradient (Mende and Wink 1987). Whereas vacuoles were highly selective in QA uptake (only vacuoles of QA-producing plants accumulated QA), protoplasts showed a lesser degree of specificity, i.e. also protoplast from different species were able to sequester QA. The exact mechanism of QA transport across the plasma membrane remains to be elucidated. Experimental evidence suggests that carrier proteins should be involved (Mende and Wink 1987; Wink and Mende 1987).

Sparteine and lupanine were administered to suspension-cultured cells of *L. polyphyllus*. Alkaloid uptake showed an optimum during the log phase of growth during days 7–9. Cells in the stationary phase were almost inactive. Sparteine and lupanine uptake is influenced by the hydrogen ion concentration in the external medium. At higher pH values, sparteine uptake is enhanced. This is partly due to diffusion since the amount of the lipophilic base increased with alkaline pH values. Using an Arrhenius plot, we could determine an activation energy of 66 kJ/mol for sparteine and 63 kJ/mol for lupanine uptake. Since diffusion is characterised by values <20 kJ/mol, these data support the idea of a carrier-mediated process. The phenomenon seen in protoplasts that alkaloid uptake was rather unspecific was also found at the level of suspension-cultured cells. Cells of *L. polyphyllus* also accumulate alkaloids of different classes and cells from different species take up sparteine and the tropane alkaloid hyoscyamine.

Our experimental data thus indicate that QA uptake is achieved to some part by diffusion (at higher pH values) but also by a carrier-mediated process, as was found for protoplasts and vacuoles (Mende and Wink 1987; Wink and Mende 1987). However, since two membrane systems, i.e. the plasmalemma and the tonoplast and the cell wall, which also can bind alkaloids, are involved in QA sequestration, the kinetic data are less clear than those obtained with isolated vacuoles (Mende and Wink 1987). In the intact plant, a preferred site of alkaloid storage is the epidermis. We have shown previously that QA uptake into these cells is also achieved by a carrier-mediated process (Wink and Mende 1987). As compared to this system, alkaloid uptake by suspension-cultured cells is remarkably reduced by almost 70 to 90% (Wink 1987b; Wink et al. 1992). We assume that in epidermal cells, more carrier protein molecules are expressed.

In vitro cultures of lupins and brooms are able to degrade exogenously applied alkaloids such as sparteine and lupanine. When sparteine is administered at 200 μM, degradation and transformation into other alkaloids, such as 11,12-dehydrosparteine, 17-oxosparteine and lupanine (Wink et al. 1980), were most pronounced. When 200 μM (= 2400 μg) sparteine or lupanine was added to the culture medium just after cell transfer, we observed a rapid uptake from the medium into the cells and concomitantly, an active degradation; at day 22 only 10 or 40%, respectively, of the initially applied alkaloid was still present. The pronounced degradation of sparteine has been observed in other cell culture systems; some cell cultures even remained viable on media with sparteine as the sole nitrogen source for several months, whereas controls on nitrogen-free media died after ca. 4 weeks. These results indicate that cells can virtually degrade QA and can utilise their nitrogen for amino acid metabolism (Wink and Witte 1985). A similar situation can be seen in lupin seedlings which degrade the alkaloids present in the seeds after germination (Wink and Witte 1985; Wink 1984c).

QA are often found in the medium, especially after induction. Also, many hydrolysing or oxidising enzymes are released by suspension-cultured cells into the medium (Wink 1985b,c, 1994). Usually, secondary metabolites and these enzymes are sequestered in the vacuole which serves as a storage and lytic compartment. We have suggested that in cell cultures the extracellular space is almost unlimited and that the cell culture medium therefore forms a kind of secondary storage ("SINK") and lytic compartment (Wink 1985b,c, 1994). A hypothetical scheme of these processes is shown in Fig. 3.

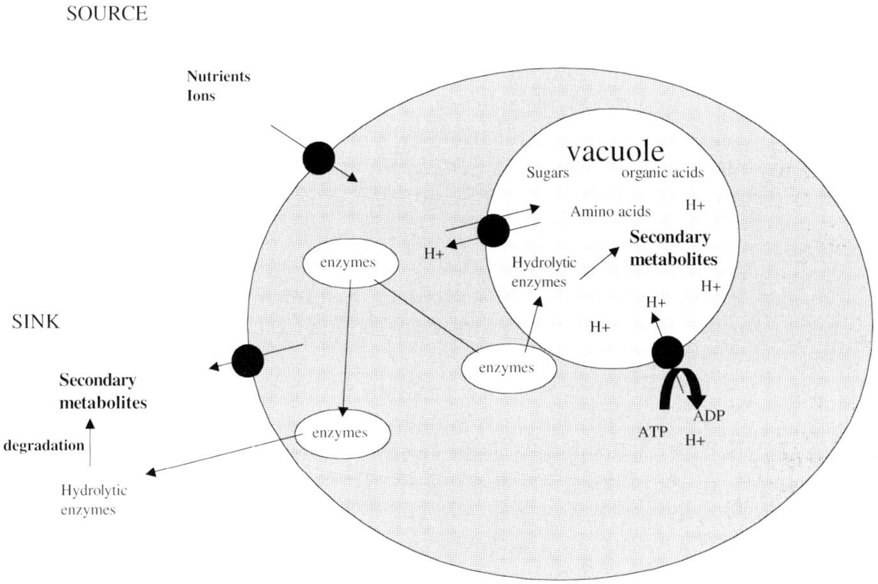

Fig. 3. Schematic illustration of alkaloid accumulation, uptake, release and degradation in suspension-cultured cells. *Filled circles* represent carrier proteins

3 Why Is Quinolizidine Alkaloid Production Reduced in Cell Suspension Cultures?

We have analysed cell cultures of more than 14 different legume species (Table 1) and after more than 2 years in vitro, all cultures produced detectable amounts of QA. However, QA levels were two to three orders of magnitude lower than those of the differentiated plants (Table 2). If we consider alkaloid storage in legume seeds, which contain up to 8% dry weight alkaloid, this difference is even larger. The question is then: why is QA production reduced in cell cultures?

Production is a complex process, consisting of a number of interrelated events, such as biosynthesis, intra- and intercellular transport, storage and degradation, which additionally may differ in space and time.

Biosynthesis: The activities of the main biosynthetic enzymes, lysine decarboxylase and oxosparteine synthase (review: Wink and Hartmann 1985; Wink 1987b) are reduced in cell cultures by one to two orders of magnitude (Wink and Hartmann 1980, 1982c). Since QA biosynthesis takes place in the chloroplast, it is not surprising that in green cell cultures, i.e. those with functional chloroplasts, QA formation is positively correlated with greening (Wink and Hartmann 1980, 1982b,c; Saito et al. 1989). Light is also important and triggers QA biosynthesis, which ceases in the dark. Both in the intact plant and cell cultures, we observed a diurnal fluctuation of QA formation (Wink and Hartmann 1982a,b; Wink and Witte 1984).

Transport: In plants, QA are exported from the chloroplast and translocated via the phloem all over the plant, especially to the stems and fruits (seeds). Thus QA do not accumulate at the site of synthesis. This separation of synthesis and storage cannot occur in cell cultures. However, we find that part of the alkaloids produced, especially after precursor feeding or elicitation (Wink 1985c) are secreted into the culture medium, which also sequesters a variety of hydrolysing and oxidating enzymes (Wink 1985c, 1994).

Storage: QA are not stored in a similar fashion in all cell types, but preferentially in epidermal cells. As mentioned above, suspension-cultured cells have only a reduced capacity for QA storage, which may be thus another limiting step in QA production (Wink 1987b; Wink and Mende 1987).

Degradation: QA are obviously not end products of metabolism, but undergo degradation, both diurnally and during seed germination (Wink and Witte 1984, 1985). Whereas capacities for QA biosynthesis and QA storage are reduced in suspension-cultured cells, degradation seems to be enhanced. One reason is the overproduction of peroxidase in lupin cell cultures (Wink 1985c; Perrey et al. 1989, 1991), an enzyme which can degrade QA. This and other enzymes are released into the culture medium into which QA are also sequestered (Fig. 3). As a consequence, a rapid QA degradation can be observed.

4 Gene Expression in Suspension-Cultured Cells of *L. polyphyllus*

The data discussed previously imply that QA production is strongly reduced in suspension-cultured cells, since the necessary degree of tissue differentiation is not realised in undifferentiated cells. Further evidence comes from the observation that in vitro-cultured shoots of lupins, but not roots, show enhanced alkaloid formation (Table 2; Wink 1987d). Thus QA production is regulated in a tissue-specific way, similar to the situation of many other cellular events of the differentiated plants, such as flowering, germination etc.

To further analyse this hypothesis we have started to analyse gene expression in suspension-cultured cells (a transcriptomic approach) as compared to differentiated tissues, such as stems, leaves and roots. Since this approach is not documented in detail in most systems, we decided to consider it more holistically and not only with respect to QA formation, i.e. we would generally like to know which genes are turned on in suspension-cultured cells which look morphologically undifferentiated.

In a proteomic approach, we compared the protein patterns of suspension-cultured cells as analysed by SDS and 2D-polyacrylamid electrophoresis (Fig. 4; Wink 1985a) with those of stems, leaves and roots (Schneider 1990). Obviously, suspension-cultured cells have a different pattern of expressed proteins and many of them are unique for the cells. These proteins have molecular weights between 20 and 50 kDa and isoelectric points vary over the whole pH range. In the next step, we need to identify those proteins whose activity and expression status changed under these conditions, in order to understand the process of dedifferentiation and stress of suspension-cultured cells.

In a second approach, we established a cDNA-library in pUC19 constructed from mRNA that was isolated from rapidly growing cells of *L. polyphyllus* (Perrey and Wink 1991; Perrey et al. 1991). More than 30 cDNA clones were isolated and their DNA-sequence analysed (Perrey and Wink 1991; Perrey et al. 1990a,b,c, 1991; Warskulat et al. 1990). Some of the isolated clones represent interesting genes, such as those encoding peroxidase (Perrey and Wink 1991), heatshock protein HSP70, HSP60, ubiquitin (Perrey et al. 1990b,c), and several ribosomal proteins (Warskulat et al. 1991; Maercker and Wink 1992).

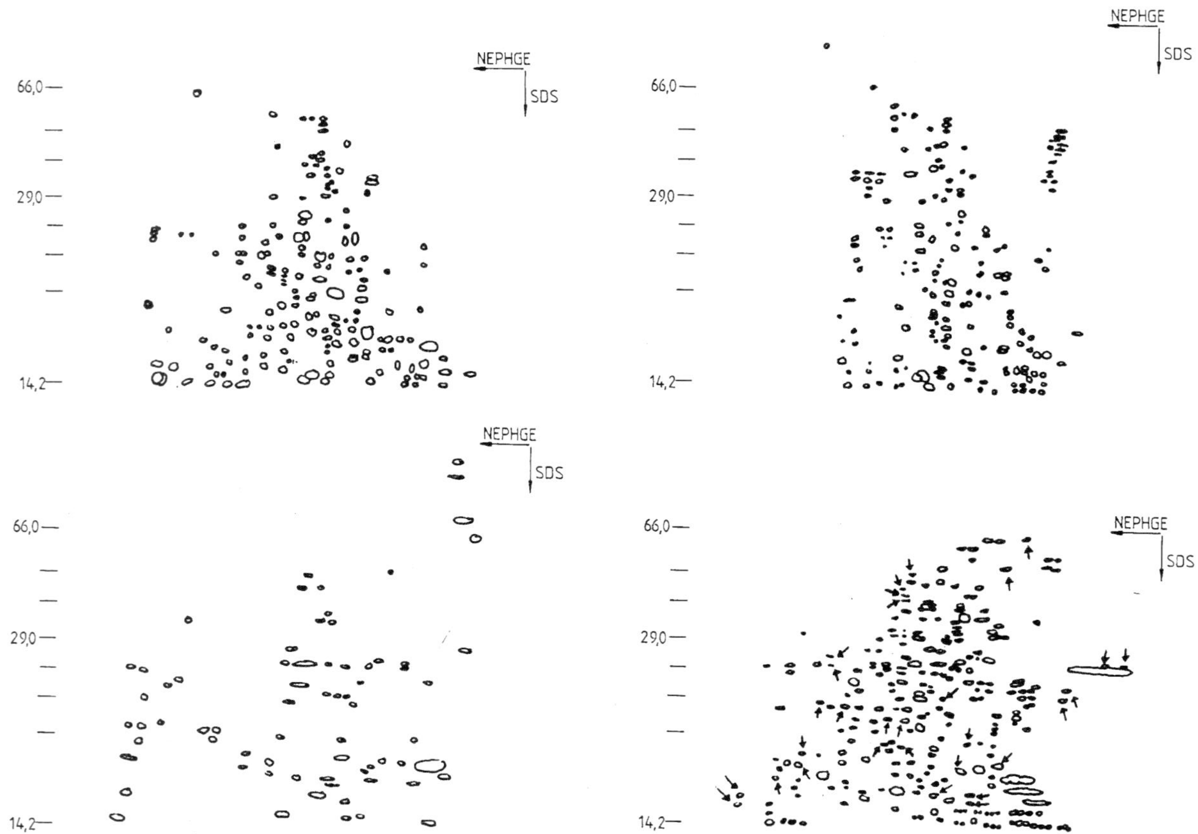

Fig. 4. *Top left* Analysis of proteins from suspension-cultured cells, *top right* stems , *bottom left* leaves and *bottom right* roots of *L polyphyllus* by 2D-PAGE and visualised by silver staining. (After Wink et al. 1992)

Table 3. Expression of cDNA clones from a cDNA library of cell suspension cultures of *L polyphyllus* as compared to callus cultures and leaves, stems and roots of intact plants. Results of a Northern or dot-blot analysis. (After Wink et al. 1992)

cDNA-Clone	Size bp	Expression in cell cultures				
		Suspension	Callus	Leaves	Stems	Roots
pPLZ02	875	++	nd	++	++	++
pPLZ09[a]	588	++	nd	++	+	++
pPLZ10	600	+	+	+	+	+
pPLZ12	790	++	nd	++	+	++
pPLZ15[a]	650	+	++	+	+	+
pPLZ22	800	++	++	++	+	+
pPLZ24	600	+	++	+	+	+
pPLZ26	380	+	++	++	++	+
pPLZ17[c]	1005	++	++	–	–	–
pPLZ06	623	++	nd	–	++	++
pPLZ23	1000	++	+	–	++	+
PPLZ11[b]	701	++	nd	++	–	++
pPLZ19	700	+	++	+	+	–
pPLZ27	900	++	++	++	++	–
pPLZ18[d]	1500	++	++	+	+	–

[a] Gene code for ribosomal proteins.
[b] Gene code for peroxidase.
[c] Gene code for HSP 70.
[d] Gene code for HSP60.

We have compared the expression of these cDNA clones in suspension-cultured cells, callus cultures, and roots, stems and leaves of intact plants by Northern or dot-blot analysis. As can be seen from Table 3, gene expression in these systems does not follow a simple rule:

1. Some genes are expressed in all systems, such as clone pPLZ02, pPLZ09, pPLZ10,pPLZ12, pPLZ15, pPLZ22, pPLZ24, and pPLZ26. These can be considered as "house-keeping genes", necessary for all cells. This group includes genes encoding ribosomal proteins.
2. A few genes are only expressed in suspension-cultured cells, such as pPLZ15. This clone represents the gene for HSP70. Whereas it is constitutively expressed in cell culture, it is heat-inducible in the intact plant (Perrey et al. 1991).
3. A third group of cDNA-clones is expressed in most systems, but not in leaves, roots or stems. Since the corresponding genes are new and the function of the respective proteins unknown, we cannot explain the pattern of differential gene expression at present.

So far, we have analysed only a small sample of cDNA-clones, but it appears that suspension-cultured cells display a unique pattern of gene activation. Especially astonishing is the fact that heat shock proteins HSP70 and 60 are actively expressed in cell culture without prior heat shock (Perrey et al. 1991).

We assumed that the culture conditions produce a stressful environment (i.e., osmotic and shear stress), resulting in the permanent expression of these genes which are known to respond to a variety of stress factors besides heat shock (Nover et al. 1989). It has been shown repeatedly that heat shock turns off the expression of most other genes including those encoding enzymes of secondary metabolism (review: Nover et al. 1989). We speculate that maybe a similar consequence results in our lupin system, where HSP genes are activated permanently. Further analysis, especially of the genes involved in QA biosynthesis, transport and degradation (which are under investigation in our laboratory), is needed to explain the reduced expression of QA production in cell cultures, i.e. whether HSP gene activation or differential and tissue-specific gene expression play a more important role.

In conclusion, genes of eukaryotes are regulated in a tissue-, organ- or development-specific process. It is likely, and already it has been shown in several instances, that the genes of plant secondary metabolism are regulated in a similar way. As outlined in this review, the accumulation of alkaloids is the result of a complex network involving biosynthesis, transport, degradation and storage which often show tissue and organ specificity. Considering the molecular background of gene expression, it is not surprising that it is difficult to obtain correct production of secondary metabolites in undifferentiated cell suspension cultures, in which most genes should be expected to be turned off, or at least no longer under specific control. The failure of production or the instability of compound formation in many cell cultures could be explained in this way. We need a much better understanding of the molecular, biochemical and regulatory basis of secondary metabolite formation in plants and plant cell cultures to overcome the current obstacles in plant biotechnology. There is hope, however, that the new disciplines of genomics, functional genomics, transcriptomics and proteomics might help to elucidate the underlying principles.

Acknowledgements. Our research was supported by grants of the Deutsche Forschungsgemeinschaft. A number of coworkers and technical assistants have contributed to the results discussed in this review. My thanks are due to C. Theuring, U. Schade, M. Weyerer, H. Wurm, U. Dostal, H. Staudter, K. von Borstel, P. Mende, R. Perrey, M. Schneider, U. Warskulat and D. Wandel.

References

Adler L, Wink M (2001) Transfer of quinolizidine alkaloids from hosts to hemiparasites in two *Castilleja – Lupinus* associations: analysis of floral and vegetative tissues. Biochem Syst Ecol 29(6):551–561

Bentley MD, Leonard DE, Reynolds EK, Leach S, Beck AB, Murakoshi I (1984) Lupine alkaloids as larval feeding deterrents for spruce budworm, *Choristoneura fumiferana* (Lepidoptera: Tortricidae). Ann Entomol Soc Am 77:398–400

Cantot P, Papineau J (1983) Discrimination des lupins basse teneur en alcaloides par les adultes de *Sitonia lineatus* L (Col.; Curculionidae). Agronomie 3:937–940

Czygan F-C, Wessinger B, Warmuth K (1988) *Cuscuta* und ihre Fähigkeit zur Aufnahme und Speicherung von Alkaloiden der Wirtspflanzen. Biochem Physiol Pflanz 183:495–501

Dreyer D, Jones KC, Molyneux RJ (1985) Feeding deterrency of some pyrrolizidine, indolizidine, and quinolizidine alkaloids towards pea aphid (*Acyrthosiphon pisum*) and evidence for phloem transport of the indolizidine alkaloid swainsonine. J Chem Ecol 11:1045–1051

Johnson ND, Bentley BB (1988) Effects of dietary protein and lupine alkaloids on growth and survivorship of *Spodoptera eridania*. J Chem Ecol 14:1391–1403

Johnson ND, Rigney L, Bentley BL (1989) Short-term induction of alkaloid production in lupines. Differences between N_2-fixing and nitrogen-limited plants. J Chem Ecol 15:2425–2434

Käss E, Wink M (1995) Molecular phylogeny of the Papilionoideae (family Leguminosae): RbcL gene sequences versus chemical taxonomy. Bot Acta 108:149–162

Käss E, Wink M (1997) Phylogenetic relationships in the Papilionoideae (family Leguminosae) based on nucleotide sequences of cpDNA (rbcL) and ncDNA (ITS1 and 2). Mol Phylogenet Evol 8:65–88

Keeler RF (1976) Lupin alkaloids from teratogenic and nonteratogenic lupins :III Identification of anagyrine as the probable teratogen by feeding trials. J Toxicol Environ Health 1:887–898

Kinghorn AD, Balandrin MF (1984) Quinolizidine alkaloids of the Leguminosae: structural types, analysis, chemotaxonomy, and biological activities. In: PelletierWS (ed) Alkaloids: chemical and biological perspectives, vol 2. Wiley, New York, pp 105–148

Körper S, Wink M, Fink RA (1998) Differential effects of alkaloids on sodium currents of isolated single skeletal muscle fibres. FEBS Lett 436:251–255

Korcz A, Markiewicz M, Pulikowska J, Twardowski T (1987) Species-specific inhibitory effects of lupine alkaloids on translation in plants. J Plant Physiol 128:433–442

Le Scao-Bogaert F, Faugeras G, Paris RR (1987) Sur la presence d'alcaloides quinolizidinique chez l' *Osyris alba* (Santalacees). Plant Med Phytother 12:315–318

Maercker D, Perrey R, Schneider M, Wink M (1992) Genexpression in Lupinenzellkulturen. In: Wink M (ed) Lupinen 1991 – Forschung, Anbau und Verwertung, pp 166–174

Matsuda K, Kimura M, Komai K, Hamada M (1989) Nematicidal activities of (–)-*N*-methylcytisine and (–)-anagyrine from *Sophora flavescens* against Pine wood nematodes. Agric Biol Chem 53:2287–2288

Mears JA, Mabry TJ (1971) Alkaloids in the Leguminosae. In: Harborne JB, Boulter D, Turner BL (eds) Chemotaxonomy of the Leguminosae. Academic Press, London, pp 73–178

Mende P, Wink M (1987) Uptake of the quinolizidine alkaloid lupanine by protoplasts and vacuoles of *Lupinus polyphyllus* cell suspension cultures. J Plant Physiol 129:229–242

Moges G, Wink M (2002) Concomitant occurrence of pyrrolizidine and quinolizidine alkaloids in the hemiparasite *Osyris alba* L (Santalaceae). Biochem Syst Ecol (in press)

Nover L, Neumann D, Scharf KD (1989) Heat shock and other stress response systems in plants. Springer, Berlin Heidelberg New York

Paolisso G, Nenquin M, Schmeer W, Mathot F, Meissner HP, Henquin JC (1985) Sparteine increases insulin release by decreasing K^+ permeability of the B-cell membrane. Biochem Pharmacol 34:2355–2361

Perrey R, Wink M (1991) Constitutive expression and molecular characterization of a cDNA clone encoding a partial HSP70 gene in cell suspension cultures of *Lupinus polyphyllus*. J Plant Physiol 137:744–748

Perrey R, Hauser M T, Wink M (1989) Cellular and subcellular localization of peroxidase isoenzymes in plants and cell suspension cultures from *Lupinus polyphyllus*. Z Naturforsch 44c:931–936

Perrey R, Warskulat U, Wink M (1990a) Molecular cloning of a lupin-specific gene from a cDNA library of suspension-cultured cells of *Lupinus polyphyllus*. Plant Mol Biol 15:175–176

Perrey R, Warskulat U, Wink M (1990b) Molecular cloning of a cDNA for the ubiquitin gene of *Lupinus polyphyllus*. Nucleic Acids Res 18:6428

Perrey R, Schneider M, Wink M (1990c) Tissue-specific expression of a cDNA clone from cell suspension cultures of *Lupinus polyphyllus*. Plant Mol Biol 14:1055–1056

Perrey R, Warskulat U, Wink M (1991) Molecular cloning of a *Lupinus polyphyllus* cDNA encoding a basic peroxidase isoenzyme of cell suspension cultures. J Plant Physiol 137:537–540

Saito K, Yamazaki M, Yamakawa K, Fujisawa S, Takamatsu S, Kawaguchi A, Murakoshi I (1989) Lupine alkaloids in tissue culture of *Sophora flavescens* var. *angustifolia*. Greening induced production of matrine. Chem Pharm Bull (Tokyo) 37(11):3001–3004

Salatino A, Gottlieb OR (1980) Quinolizidine alkaloids as systematic markers of the Papilionoideae. Biochem Syst Ecol 8:133–147

Schmeller T, Sauerwein M, Sporer F, Müller WE, Wink M (1994) Binding of quinolizidine alkaloids to nicotinic and muscarinic receptors. J Nat Prod 57:1316–1319

Schneider M (1990) Vergleich der Genexpression in Pflanzen und Zellsuspensionskulturen von *Lupinus polyphyllus*. Master Thesis, University Mainz

Stermitz FR, Harris GH (1983) Chemistry of the Scrophulariaceae. 10. Transfer of pyrrolizidine and quinolizidine alkaloids to *Castilleja* hemiparasites from composite and legume host plants. J Chem Ecol 13:1917–1925

Stermitz FR, Belofsky GN, Ng D, Singer MC (1985) Quinolizidine alkaloids obtained by *Pedicularis semibarbata* from *Lupinus fulcratus* fail to influence the specialist herbivore *Euphydryas editah* (Lepidoptera). J Chem Ecol 15:2521–2530

Tyski S, Markiewicz M, Gulewicz K, Twardowski T (1988) The effect of lupin alkaloids and ethanol extracts from seeds of *Lupinus angustifolius* on selected bacterial strains. J Plant Physiol 133:240–242

Warskulat U, Perrey R, Wink M (1991) Molecular cloning of a cDNA from *Lupinus polyphyllus* cell cultures encoding a ribosomal protein (rps 16). Plant Mol Biol 16:739–740

Wink M (1983a) Wounding-induced increase of quinolizidine alkaloid accumulation in lupin leaves. Z Naturforsch 38c:905–909

Wink M (1983b) Inhibition of seed germination by quinolizidine alkaloids. Aspects of allelopathy in *Lupinus albus* L. Planta 158:365–368

Wink M (1984a) Chemical defence of Leguminosae. Are quinolizidine alkaloids part of the antimicrobial defence system of lupins? Z Naturforsch 39c:548–552

Wink M (1984b) Chemical defense of lupins. Mollusc-repellent properties of quinolizidine alkaloids. Z Naturforsch 39c:553–558

Wink M (1984c) Evidence for an extracellular lytic compartment of plant cell suspension cultures. Naturwissenschaften 71:635

Wink M (1985a) Chemische Verteidigung der Lupinen: zur biologischen Bedeutung der Chinolizidinalkaloide. Plant Syst Evol 150:65–81

Wink M (1985b) Composition of the spent culture medium: time course of ethanol formation and the excretion of hydrolytic enzymes into the medium of suspension-cultured cells of *Lupinus polyphyllus*. J Plant Physiol 121:287–293

Wink M (1985c) Metabolism of quinolizidine alkaloids in plants and cell suspension cultures. In: Neumann KH, Barz W, Reinhard E (eds) Primary and secondary metabolism of plant cell cultures. Springer, Berlin Heidelberg New York, pp 107–116

Wink M (1987a) Chemical ecology of quinolizidine alkaloids. ACS Symp Ser 330:524–533

Wink M (1987b) Quinolizidine alkaloids: biochemistry, metabolism, and function in plants and cell suspension cultures. Planta Med 53:509–514

Wink M (1987c) Why do lupin cell cultures fail to produce alkaloids in large quantities? Plant Cell Tissue Organ Cult 8:103–111

Wink M (1987d) Site of lupanine and sparteine biosynthesis in intact plants and in vitro organ cultures. Z Naturforsch 42c:868–872

Wink M (1988) Plant breeding: importance of plant secondary metabolites for protection against pathogens and herbivores. Theor Appl Genet 75:225–233

Wink M (1989) Genes of secondary metabolism: differential expression in plants and in vitro cultures and functional expression in genetically transformed microorganisms. In: Kurz WGW (ed) Primary and secondary metabolism of plant cell cultures. Springer, Berlin Heidelberg New York, pp 239–251

Wink M (1990) Physiology of secondary product formation in plants. In: Charlwood BV, Rhodes MJC (eds) Secondary products from plant tissue culture. Clarendon Press, Oxford, pp 23–41

Wink M (1992) The Role of quinolizidine alkaloids in plant insect interactions. In: Bernays EA (ed) Insect-plant interactions, vol IV. CRC-Press, Boca Raton, pp 133–169
Wink M (1993) Quinolizidine alkaloids. In: Waterman P (ed) Methods in plant biochemistry, vol 8. Academic Press, London, pp 197–239
Wink M (1994)The cell culture medium-a functional extracellular compartment of suspension cultured cells. Plant Cell Organ Culture 38:307–319
Wink M (1995) Biological activities and potential application of lupin alkaloids. In: Neves-Martin JM, Beirao da Costa ML (eds) Advances in lupin research. International Lupin Association, Lisbon, pp 161–178
Wink M (1999a) Biochemistry of plant secondary metabolism. Annual plant reviews, vol 2. Sheffield Academic Press, Sheffield, 358 pp
Wink M (1999b) Function of plant secondary metabolites and their exploitation in biotechnology. Annual plant reviews, vol 3. Sheffield Academic Press, Sheffield, 362 pp
Wink M (2000) Interference of alkaloids with neuroreceptors and ion channels. In: Atta-Ur-Rahman (ed) Bioactive natural products, vol 11. Elsevier, Amsterdam, pp 3–129
Wink M, Hartmann T (1980) Production of quinolizidine alkaloids by photomixotrophic cell suspension cultures: biochemical and biogenetical aspects. Planta Med 40:149–155
Wink M, Hartmann T (1982a) Localization of the enzymes of quinolizidine alkaloid biosynthesis in leaf chloroplasts of *Lupinus polyphyllus*. Plant Physiol 70:74–77
Wink M, Hartmann T (1982b) Diurnal fluctuation of quinolizidine alkaloid accumulation in legume plants and photomixotrophic cell suspension cultures. Z Naturforsch 37c:369–375
Wink M, Hartmann T (1982c) Physiological and biochemical aspects of quinolizidine alkaloid formation in cell suspension cultures. In: Fujiwara A (ed) Plant tissue culture. Japanese Association for Plant Tissue Culture, Tokyo, pp 333–334
Wink M, Hartmann T (1985) Enzymology of quinolizidine alkaloid biosynthesis. In: Zalewski RI, Skolik JJ (eds) Natural products chemistry 1984. Elsevier, Amsterdam, pp 511–520
Wink M, Mende P (1987) Uptake of lupanine by alkaloid-storing epidermal cells of *Lupinus polyphyllus*. Planta Med 53:465–469
Wink M, Twardowski T (1992) Allelochemical properties of alkaloids. Effects on plants, bacteria and protein biosynthesis. In: Rizvi SJH, Rizvi V (eds) Allelopathy: basic and applied aspects. Chapman and Hall, London, pp 129–150
Wink M, Waterman P (1999) Chemotaxonomy in relation to molecular phylogeny of plants. In: Wink M (ed) Biochemistry of plant secondary metabolism. Annual plant reviews, vol 2, Sheffield Academic Press, Sheffield, pp 300–341
Wink M, Witte L (1983) Evidence for a wide-spread occurrence of the genes of quinolizidine alkaloid biosynthesis. Induction of alkaloid accumulation in cell suspension cultures of alkaloid-"free" species. FEBS Lett 159:196–200
Wink M, Witte L (1984) Turnover and transport of quinolizidine alkaloids: diurnal variation of lupanine in the phloem sap, leaves and fruits of *Lupinus albus*. L-Planta 161:519–524
Wink M, Witte L (1985) Quinolizidine alkaloids as nitrogen source for lupin seedlings and cell suspension cultures. Z Naturforsch 40c:767–775
Wink M, Witte L (1991) Storage of quinolizidine alkaloids in *Macrosiphum albifrons* and *Aphis genistae* (Homoptera: Aphididae). Entomol Gener 15:237–254
Wink M, Hartmann T, Witte L (1980) Biotransformation of cadaverine and potential intermediates of lupanine biosynthesis by plant cell suspension cultures. Planta Med 40:31–39
Wink M, Hartmann T, Witte L, Rheinheimer J (1982) Interrelationship between quinolizidine alkaloid producing legumes and infesting insects: exploitation of the alkaloid-containing phloem sap of *Cytisus scoparius* by the broom aphid, *Aphis cytisorum*. Z Naturforsch 37c:1081–1086
Wink M, Witte L, Hartmann T, Theuring C, Volz V (1983) Accumulation of quinolizidine alkaloids in plants and cell suspension cultures: Genera *Lupinus, Cytisus, Baptisia, Genista, Laburnum*, and *Sophora*. Planta Med 48:253–257
Wink M, Heinen HJ, Vogt H, Schiebel HM (1984) Cellular localization of quinolizidine alkaloids by laser desorption mass spectrometry (LAMMA 1000). Plant Cell Rep 3:230–233

Wink M, Perrey R, Schneider M, Warskulat U, von Borstel K, Mende P (1992) Alkaloid metabolism and gene expression in cell suspension cultures of *Lupinus polyphyllus* and *L hartwegii*. In: Oono K, Hirabayashi T, Kikuchi S et al. (eds) Plant tissue culture and gene manipulation for breeding and formation of phytochemicals.National Institute of Agricultural Resources, Tsukuba Science City, pp 101–119

Wink M, Schmeller T, Latz-Brüning B (1998) Modes of action of allelochemical alkaloids: interaction with neuroreceptors, DNA and other molecular targets. J Chem Ecol 24:1881–1937

Wippich C, Wink M (1985) Biological properties of alkaloids. Influence of quinolizidine alkaloids and gramine on the germination and development of powdery mildew, *Erysiphe graminis f.sp.hordei*. Experientia 41:1477–1478

8 *Malva* sp. (Mallow): In Vitro Culture and the Production of Secondary Metabolites

S.E. Kintzios[1]

1 Introduction

1.1 Distribution and Importance of the Plant

Although about 1000 species are designated with the common name of mallow, approximately 30 species belonging to the genus *Malva* (of the Malvaceae family) are known for their medicinal value, mostly in a traditional sense.

The common (blue or high) mallow (*Malva sylvestris* L.) is a biennial to short-lived perennial with prostrate to semi-erect stems (10–80 cm long) and long-stalked rounded leaves with a heart-shaped base and five to seven broad shallow-toothed lobes (Fig. 1). The leaves of *M. sylvestris* var. *incanescens* Gris are hairy. The flowers (appearing from May to September) are pale lilac to bright mauve-purple and the seeds are flat button-like nutlets. The plant is found naturally in marginal or waste lands, hedgerows and roadsides and is approximately 1 m high, with stalked, roundish, five- to seven-lobed leaves (Grieve 1994; Podlech 1996). Plant parts abound with a mild mucilage.

Malva aegyptia (Egyptian mallow) is an annual species, endemic in the Mediterranean countries, 20–50 cm high with purple-blue flowers.

Malva cretica (Crecian mallow) is another Mediterranean species, which is an annual, 10–30 cm high with rose-coloured leaves.

Malva ambigua Guss (*M. sylvestris* var. *ambigua*) is a perennial with hairy roots and purple leaves.

Malva neglecta is also known as common mallow or 'cheeses'. It has white to pale lilac flowers (appearing from May to September) with five slightly notched petals. Leaves are wavy up and down, their edges having rounded teeth. Plants (10–100 cm high) are covered with hairs and are branched at the base (Gleason and Cronquist 1991). *M. neglecta* has similar pharmacological properties to common mallow. The dwarf mallow (*Malva rotundifolia* L.) is a similar but smaller species with prostrate stems and pale lilac flowers (Grieve 1994).

Malva pussila is an annual weed with white-purple flowers.

[1]Agricultural University of Athens, Faculty of Agricultural Biotechnology, Laboratory of Plant Physiology, Iera Odos 75, 11855 Athens, Greece

Biotechnology in Agriculture and Forestry, Vol. 51
Medicinal and Aromatic Plants XII (ed. by T. Nagata and Y. Ebizuka)
© Springer-Verlag Berlin Heidelberg 2002

Fig. 1. Common mallow (*Malva sylvestris* L.)

Malva parviflora is also an annual, with white-blue flowers. It is a poisonous plant for various animal species (Chopra and Badhwar 1940)

The musk mallow (*Malva moschata* L.) is a related perennial species with light-green leaves and rose-coloured flowers, which are considerably larger than those of the common mallow. It is one of the few mallow species that are cultivated, even on a marginal level.

Mallows are abundantly distributed in temperate and tropical regions and virtually worldwide.

Common mallow has been traditionally used both as a food source and a remedy for various diseases. The young leaves and shoots of this plant (gathered in July and August) have been eaten since at least the 8th century B.C. Pliny called common mallow 'the food of the poor'. The seeds are also edible and the boiled foliage can be consumed as a vegetable, making a good soup (Grieve 1994; Podlech 1996; Scroumbis 1998). *M. neglecta* is used for the preparation of fresh or cooked 'panerak' in Pakistan (Iqbal et al. 1994). Country people have traditionally used flowers for decorative purposes.

The medicinal value of the plant is reflected in the Spanish adage, 'A kitchen garden and mallow, sufficient medicines for a home'. The leaves and flowers are the main plant parts used, although from musk mallow the white root is mostly used. Traditional uses of the species include the treatment of coughs and throat infections and other bronchial problems, as well as stomach and intestinal irritations (Podlech 1996). The flowers and leaves are emollient and used for the softening of sensitive areas of the skin. It is applied as a poultice to reduce swelling and draw out toxins. Taken internally, the leaves reduce gut irritation and have a laxative effect. Combined with eucalyptus, it makes a good remedy for coughs and other chest ailments. The mucilage has an anti-complementary activity (Tomoda et al. 1989). *M. rotundifolia* is a potential vasodilator (i.e., a blood pressure-lowering agent) (Kratzer and Greif 1969).

Based on ethnomedical information gathered from surveys at clinics and hospitals as well as interviews with traditional healers and rural dwellers, Grierson and Afolayan (1999) revealed that *M. parvifolia* is among the most commonly used plants for the treatment of wounds in the eastern Cape Province, South Africa, although extracts from this species did not show any antibacterial activity.

Huang et al. (1998) used short-term in vivo and in vitro tests to further confirm the anti-tumour activities of *Malva crispa* powder (MCP), a vegetable powder, prepared from *Malva crispa* L. In cancer cell culture systems, the MCP fat-soluble extract revealed inhibitory effects on the growth and proliferation of the human hepatoma and the gastric cancer cells in a dose-response manner. In the colony formation test, MCP also altered the morphology of human gastric cancer cells. It was suggested that MCP could be consumed not only by healthy subjects for cancer prevention but also by patients with cancer as a supplementary treatment in combination with anti-carcinogenic drugs such as 5-FU or cyclophosphamide (CP).

In Table 1, reported pharmacological properties of different common mallow species are presented cumulatively, along with indicative dosages of herbal preparations.

1.2 Chemical Constituents and Structure-Related Activities

Several secondary metabolites, belonging to various chemical groups have been isolated from various parts of the plant.

1.2.1 Mineral Elements, Vitamins and Minor Organic Compounds

The leaves of *M. sylvestris* are rich in ascorbic acid (240–2.855 ppm), β-carotene (33–390 ppm), calcium (900–10,715 ppm), iron (37–440 ppm), phosphorus (360–5000 ppm), vitamins (niacin, riboflavin, thiamin) and fibre (9000–107,000 ppm). Tossi et al. (1995) have isolated the coumarin scopoletin from aerial

Table 1. Reported pharmacological properties of common mallow and indicative dosages of herbal preparations

Pharmacological property	Indicative dose
Analgesic	5–10 g/day
Antiarthritic	1 g/day
Antiasthmatic	1000 mg/day
Anticataract	350 mg/day
Anticold	1–2 g/day
AntiCrohn's	50–100 mg/day
Antidepressant	2000 mg/day
Antieczemic	3.5–5 g/day
Antihaemorrhagic	1 g/day
Antihepatitic	2–6 g/day
Antiherpetic	1–5 g/day
Antioxidant	100 ppm
Antiosteoporotic	500 ppm
Antiviral	1–5 g/day
Diuretic	700 mg/day
Hypocholesterolemic	300–1000 mg/day
Hypotensive	1000 mg/day
Immunostimulant	180 mg/day
Mucolytic	1 g/day

Sources: Pizzorno and Murray (1985); Davies and Stewart (1990); Simon (1990); Duke (1992b), Werbach (1993).

parts of Italian mallow plants. *M. neglecta* is particularly rich in choline and tannin (flowers). *M. parviflora* contains considerable amounts of iron (454–4745 ppm), calcium (3120–23,650 ppm), phosphorus (670–4891 ppm), vitamins and fibre (15,000–109,000 ppm) (Duke 1992a,b).

1.2.2 Anthocyanins

Mallow flowers contain considerable amounts of anthocyanins and leucoanthocyanins, as well as flavonoid glycosides (malvin, malvidin) (Pourrat et al. 1990). Farina et al. (1995) analysed aqueous alcoholic mallow (*M. sylvestris*) flower extracts by both HPTLC densitometry in the reflectance mode at 530 nm and reversed-phase HPLC with gradient elution. The best chromatographic resolution was obtained by HPLC, which revealed only two main compounds, confirmed by fast atom bombardment–mass spectrometry (FAB–MS): malvidin 3,5-*O*-diglucoside (malvin) and malvidin 3-*O* (6″ *O*-malonylglucoside)-5-*O*-glucoside.

Mas et al. (2000) measured the DNA triplex stabilisation property of seven natural anthocyanins (five monoglucosides and two diglucosides) by triplex thermal denaturation experiments. They noted a difference between the diglucosides that do not modify this melting temperature and the monoglucosides, such as the 3-*O*-β-D-glucopyranoside of malvidin, which was extracted from dried petals of *M. sylvestris* and presents a weak but significant stabilising effect. It appeared clearly that the difference between the two series could be due to the supplementary sugar moiety at the 5 position for the diglucosylated compounds, that would make them too crowded to allow interaction with the triplex. Among the monoglucoside series, the most active compounds are the only ones to embody a catechol B-ring in their structure that could be important for such an interaction.

1.2.3 Carbohydrates and Polysaccharides

Mucilage is found in the whole plant (Gibbs 1974; Tomoda et al. 1990; Hasler et al. 1990; Duke 1992a,b), while various carbohydrates (D-galactose, L-arabinose, L-rhamnose) are found in the leaves. Aqueous extracts of mallow, as in many other polysaccharide-containing plants, are widely used in therapy for irritated mucus membranes in the pharynx region. In order to prove the existence of mucilaginous effects of polysaccharide hydrocolloids on epithelia, Schmidgall et al. (2000) developed an ex vivo system based on porcine buccal membranes. Polysaccharides from *M. moschata* showed only moderate bioadhesion to epithelial tissue. The adhesive effects were concentration-dependent. Histological studies of membranes, incubated with a fluorescence-labelled rhamnogalacturonan, indicated the presence of distinct polysaccharide layers on the apical membrane surface. With these results, adsorption effects of certain polysaccharides on mucus membranes were shown for the first time. Such effects suggest that this may account, at least in part, for the therapeutic effects of mucilage-containing plants in the treatment of irritated buccal membranes.

Gonda et al. (1990) isolated a main acidic polysaccharide preparation from the leaves of *Malva sylvestris* L. var *mauritiana* Mill. It was composed of L-rhamnose, D-galactose, D-galacturonic acid, and D-glucuronic acid in the molar ratio of 22:6:22:11, and it contains 7.7% peptide. It was homogeneous by electrophoresis and gel chromatography, which gave a value of 11,000 as molecular weight. The structure of the polysaccharide component was elucidated by methylation analysis and partial hydrolysis. The substance showed considerable anti-complementary activity.

Classen and Blaschek (1998) isolated high molecular weight acidic polysaccharides (HMWAPs) from the mucilage of leaves and flowers of *M. sylvestris*, with a molecular weight in the range of $1.3–1.6 \times 10^6$ daltons. HMWAPs were found to be composed mainly of glucuronic acid, galacturonic acid, rhamnose and galactose.

Capek et al. (1997) isolated an acidic heteropolysaccharide composed of L-rhamnose, D-galactose, L-arabinose, D-xylose, and uronic acids in 1.0:0.5:0.1:0.1:1.8 mole proportion from the flowers of *M. mauritiana* L. The results of chemical and spectroscopic analyses indicated a branched structure with the backbone made up of alternating sequences of two- and four-linked rhamnogalacturonan-type units with linear segments (DP = 2, 4, 12), segments bearing on *O*-3 of D-GalA monomeric D-GlcA randomly distributed along the chain (DP = 7, 10, 19, 31), and segments having each D-GalA substituted (DP = 15). Moreover, the results of methylation analysis indicated that about 28% of L-rhamnose residues were branched through *O*-4.

1.3 Conventional Practices for Mallow Propagation – Significance and Utilisation in Agriculture – Phytopathology

Plants are propagated by seed or scions transplanted in autumn or spring. Plant spacing varies from 40 to 50 cm × 60 to 70 cm (Scroumbis 1998). *Malva sylvestris* and *M. meschata* are cross-pollinated (insect-visited), while *M. rotundifolia* is self-fertilising (Grieve 1994). Kozlowski et al. (1989) have shown that the seed germination of mallow in laboratory is irregular and not so good (10–50%), only occasionally it reaches 70%.

Mallow species are still underutilised and only plants from the wild are used and incorporated into various cosmetic and homeopathic products. The species is not systematically cultivated and is rather considered a broad-leaf weed. *M. sylvestris* is a common weed in olive and wine cultures, but is efficiently controlled by chemical herbicides. The round-leaved mallow (*M. pusilla*) interferes with spring wheat and flax (Friesen et al. 1992). *M. pusilla* is controlled by the pathogenic fungi *Colletotrichum gloesporiodes* f. sp. *malvae* which is also known as a bioherbicide under the trade name BioMal (Makowski and Mortensen 1999). Consequently, no commercial mallow varieties exist.

Ostrowski and Daczewska (1981, 1984) studied the potential use of *Malva verticillata* L. as a fodder plant. Crop yield and structure, nutrient content in different vegetation periods and suitability for preservation were evaluated. There were linear relationships between the content of dry matter and crude fibre, and in vitro digestibility of dry matter on one side and the number of days of growth on the other. The regression between the in vitro digestibility of dry matter and the content of fibre in green forage was also linear (r = −0.948x). Cattle, sheep, swine, horses and rabbits readily ate green forage of *Malva verticillata*. In vitro digestibility of the dry substance of young plants was satisfactory but was low for the older ones. The content of dry substance in the plants was relatively low. A large content of slime can have dietetic effects. Seong et al. (1995) studied the photosynthesis and effects of shading and harvesting methods on the growth and yield of this species.

Although not a crop plant, common mallow has been the focus of some interesting phytopathological investigations. Felklova (1965) studied the *Puccinia malvacearum* Mont. disease on *Malva sylvestris* L. ssp. Mauietanica Thell. Marco (1975) studied the occurrence of alfalfa mosaic virus and *Malva* vein clearing virus on weed members of Malvaceae in Israel.

Extracellular pectic lyase and polygalacturonase activities of *Colletotrichum gloeosporioides* f. sp. *malvae* were detected in broths containing mallow cell wall extract, pectin or glucose as the carbon source (Shih et al. 2000). The initial pH of the broth as well as the carbon source had major influences on pectinase enzyme activities. In the host, only pectic lyase activity was detected, which began at the end of the biotrophic phase and increased in the necrotrophic phase of infection. Two full-length pectate lyase cDNAs, pel-1 and pel-2, were cloned from the fungus. Both genes showed similar patterns of expression when the fungus was grown in mallow cell-wall extract and pectin medium, and the only major difference in expression in culture was that only pel-2 was expressed in glucose broth. Expression of pel-1 and pel-2 was also affected by the initial pH of the medium. Expression of pel-2, but not pel-1, was detected during infection of the host, round-leaved mallow, *Malva pusilla*. Transcripts of pel-2 were first detectable during the necrotrophic phase of infection approx. 24 h after the first detection of pectic acid (pectale) lyase enzyme activity.

In addition, Jin et al. (1999) cloned two actin genes, actA from the hemibiotrophic anthracnose fungus, *Colletotrichum gloeosporioides* (Penz.) Penz. & Sacc. f. sp. *malvae*, and act1 from its host, *Malva pusilla* (Sm.) from a cDNA library developed from infected host tissue. The actin gene, actA, of *C. gloeosporioides* f. sp. *malvae*, which is similar to that of other euascomycetes, appears to be expressed constitutively. The actin gene of *M. pusilla* is most similar to one of the actin genes of *Arabidopsis thaliana* that is unique in being responsive to environmental stimuli such as wounding. Expression of actA was used to follow the growth of the fungus in the plant tissue. Low actA expression occurred until 72–96 h after inoculation and then increased rapidly, corresponding with the timing of the shift from slower biotrophic fungal growth to much more rapid necrotrophic growth. In contrast, expression of act1 approximately doubled during the biotrophic phase and then rapidly declined during the necrotrophic phase. Increased host actin expression could be due to host cytoskeleton rearrangement in response to biotrophic infection, and the subsequent decrease in host actin expression could be due to host cell disruption resulting from tissue maceration during necrosis. This was the first report of a host actin gene that can increase in expression during a compatible plant-pathogen interaction.

2 The In Vitro Culture of *Malva* sp.

Due to the yet limited commercial significance of mallow species, there have been few attempts to utilise *Malva* sp. in modern agrobiotechnology. In particular, the establishment of efficient protocols for callus induction from common mallow, plant regeneration via organogenesis or somatic embryogenesis and clonal selection could potentially help towards considerably reducing the time needed for the introduction of newly selected, outstanding mallow lines. In addition, mallow tissue cultures could be used as cell factories for the in vitro production of various bioactive substances, such as polysaccharides with anti-tumour properties.

In a general sense, only a few species of the Malvaceae have been studied in vitro, with the most representative example being cotton (*Gossypium hirsutum* L.) and kenaf (*Hibiscus cannabinus* L.), both being used as sources of fibres. In a recent example, Zapata et al. (1999) described an improved protocol for the regeneration of plants from the shoot apices of three different cultivars of cotton and kenaf. Hasson and Poljakoff-Mayber (1995) also reported the establishment of callus cultures from another Malvaceae species, *Kosteletzkya virginica* L. Presl. There are even fewer reports on the tissue culture of *Malva* sp. These are largely focused on experiments concerning the effect of medium composition and environmental factors on the establishment of common mallow callus cultures and the induction of somatic embryogenesis thereof. Some preliminary information also exists on the in vitro production of secondary metabolites. The results of these studies are summarised in the following.

2.1 Experimental Approaches for Establishing Tissue Cultures from Common Mallow

2.1.1 Plant Material – Explant Preparation

Until now, the common mallow *M. sylvestris* has been the only species from which cell and callus cultures have been successfully established (Dexiang et al. 1985; Katsouri 1997). Donor plants (bearing six to eight leaves) collected from the wild at both the flowering and non-flowering stage have been used as the source of explants in all experiments conducted so far.

2.1.2 Culture Media

All types of culture media consisted of Murashige and Skoog (MS) (Murashige and Skoog 1962) basal medium solidified with 0.8% agar and supplemented with 3% sucrose. The media were supplemented with either α-naphthaleneacetic acid (NAA) and benzyladenine (BA) as the plant growth regulators (PGRs) at different concentrations and in factorial combinations. In addition, the effect of casein hydrolysate was studied. Culture incubation took place under a photosynthetic photon flux density (PPFD) of either $250\,\mu mol\,m^{-2}\,s^{-1}$ (16/8, from cool white fluorescent lamps) or $50\,\mu mol\,m^{-2}\,s^{-1}$ at 25 °C.

2.1.3 Assessment of Culture Development

In order to determine the kinetics of mallow tissue growth and development in vitro, various parameters have been studied in different reports. These include fresh and dry weight, chlorophyll content, number of globular embryos per explant, macronutrient accumulation, cell metabolism, cell division rate and histological differentiation.

2.2 Initiation and Development of Mallow Tissue Cultures

So far, the establishment of in vitro cultures from *Malva sylvestris* has been achieved only from stem or petiole explants. Leaf explants are entirely non-responsive under any of the aforementioned treatments and eventually decline by developing brown necrotic areas. The initiation and development of mallow tissue culture depends significantly on the complex interaction of the culture medium composition and the intensity of light.

Explant response to the culture treatment is very fast: within only 3 days from inoculation and irrespective of the particular PGR combinations, stem and petiole explants form either dedifferentiated callus masses and/or numerous globular somatic embryos (proembryos; Fig. 2). Embryo formation is observed exclusively on both cut surfaces of the explants, however a polar predisposition is evident: remarkably more embryos are formed on the surface of the explant that is distal to the root (Christodoulopoulou 2000).

2.2.1 Factors Affecting Culture Growth and Embryo Proliferation

2.2.1.1 Plant Growth Regulators

Culture growth and the proliferation of the globular embryos are significantly ($p < 0.0001$) affected by the combination of growth regulators at a particular concentration and are more intensive at 9 μM NAA + BA (Kintzios et al. 1998b). Tissue growth is negatively affected by higher PGR concentrations. This observation is in accordance with the results of previous studies on culture initiation from other Malvaceae species, such as cotton (Gossett et al. 1994; Rajasekaran 1996), kenaf (Mc Lean et al. 1992) and *Hibiscus syriacus* (Zhao et al. 1991).

2.2.1.2 Casein

Casein addition to the culture medium (at a maximum concentration of 0.5 g/l) has remarkable effects on the induction of somatic embryogenesis from mallow cultures. While only a few embryogenic calli can be observed on a

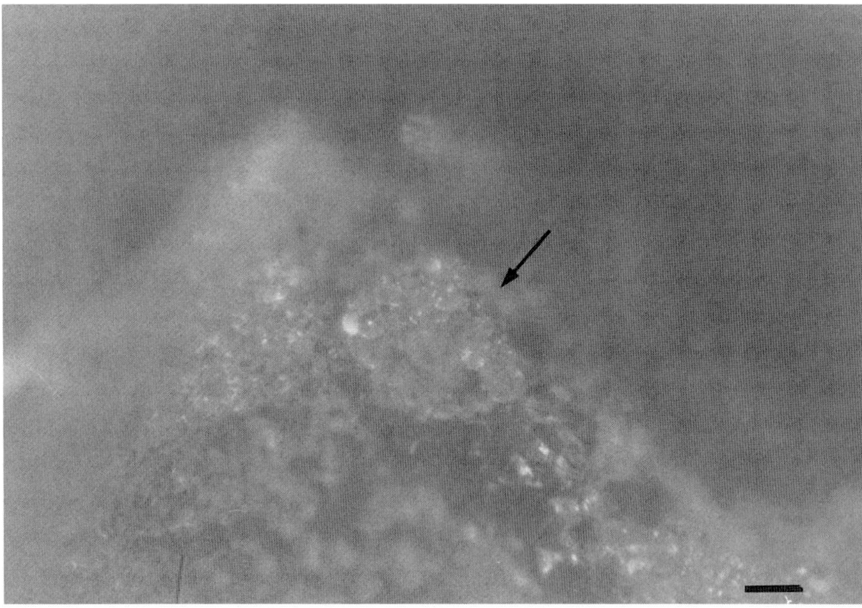

Fig. 2. Somatic proembryos (globular-stage embryos; *arrow*) on *M. silvestris* stem explants (*scale bar* 0.5 mm)

casein-free medium, the number of globular embryos increases in parallel with casein concentration on casein-containing culture media (Fig. 3). In addition, casein is associated with the expression of direct somatic embryogenesis, i.e., without the simultaneous formation of callus. On the other hand, the effect of casein on callus induction and growth is not yet defined: increasing casein concentration in the culture medium is related to a decline of callus dry weight (Fig. 4) and chlorophyll content (Fig. 5). This effect depends on light intensity and culture duration. Indeed, under conditions of high light intensities (150–250 μmol m^{-2} s^{-1}), explant fresh and dry weight were negatively correlated with total chlorophyll accumulation ($r^2 = -0.95109$ and -0.96594 respectively) (Kintzios et al. 2000; Papagiannakis 2000). Cell proliferation is favoured by either low (0.1 mg/l) or high (0.5 mg/l) casein concentrations (Fig. 6). Finally, maximum explant metabolism (as indicated by the dehydrogenase activity of cultured cells) was observed at 100 mg l^{-1} casein.

The beneficial effect of casein on the induction of somatic embryogenesis from mallow petiole explants was verified by microscope observations on explant sections derived 10 days after culture initiation (Kintzios et al. 2000; Papagiannakis 2000): abundant globular embryos were formed in cortex parenchyma. As far as the effects of individual amino acids are concerned, addition of tyrosine gave the best embryogenic response, followed by histidine and isoleucine. This response was lower than with casein addition, however it indicated a possible implication of various biochemical pathways (such as the

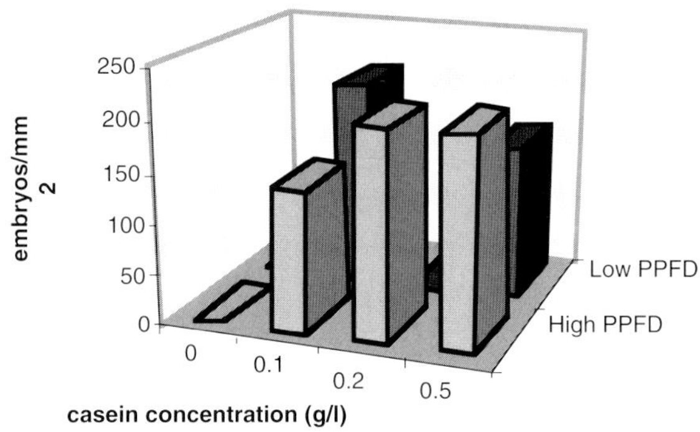

Fig. 3. Effect of casein and light intensity on the proliferation of globular somatic embryos from *M. silvestris* stem explants on MS medium supplemented with 9 μM NAA + 9 μM BA

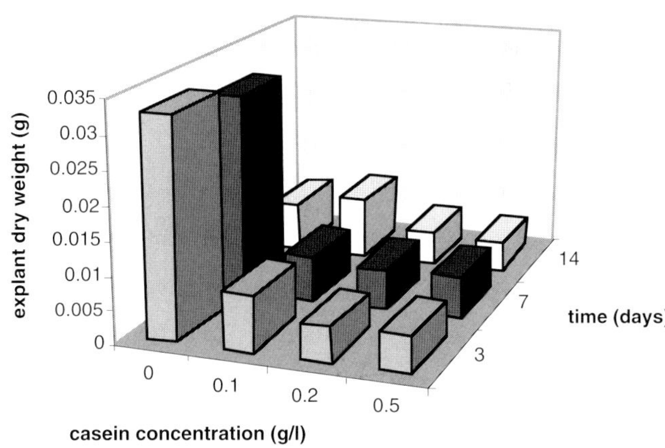

Fig. 4. Effect of casein on the dry weight of cultures derived from *M. silvestris* stem explants on MS medium supplemented with 9 μM NAA + 9 μM BA

biosynthesis of auxins and phenylpropanoids, and the stimulation of photosynthesis) to the process of embryogenic induction.

2.2.1.3 Light

Increasing light intensities are associated with increased induction and proliferation of mallow somatic embryos (at the globular stage; Fig. 3), however callus induction is adversely affected (Fig. 7; Christodoulopoulou 2000;

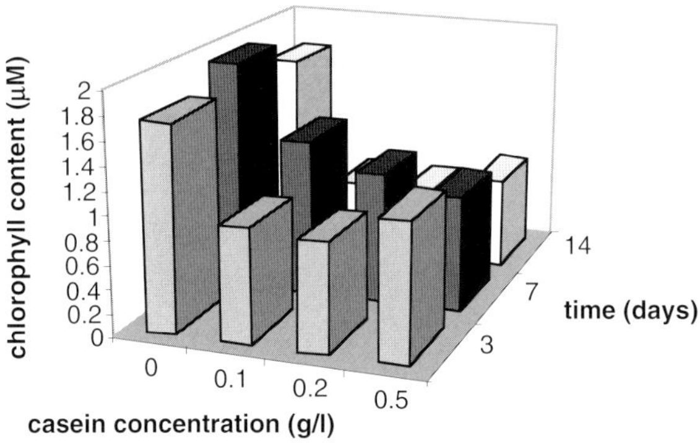

Fig. 5. Effect of casein on the chlorophyll content of cultures derived from *M. silvestris* stem explants on MS medium supplemented with 9μM NAA + 9μM BA

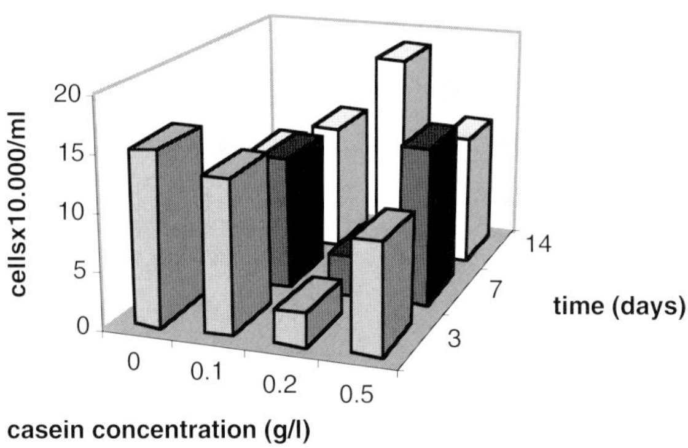

Fig. 6. Effect of casein on the cell division rate of cultures derived from *M. silvestris* stem explants on MS medium supplemented with 9μM NAA + 9μM BA

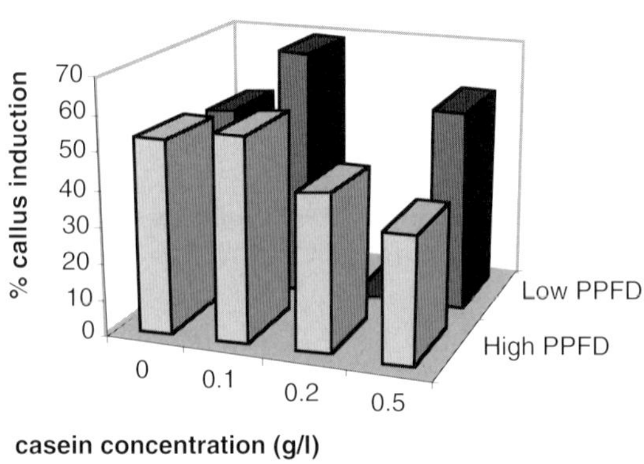

Fig. 7. Effect of casein and light intensity on callus induction from *M. silvestris* stem explants on MS medium supplemented with 9μM NAA + 9μM BA

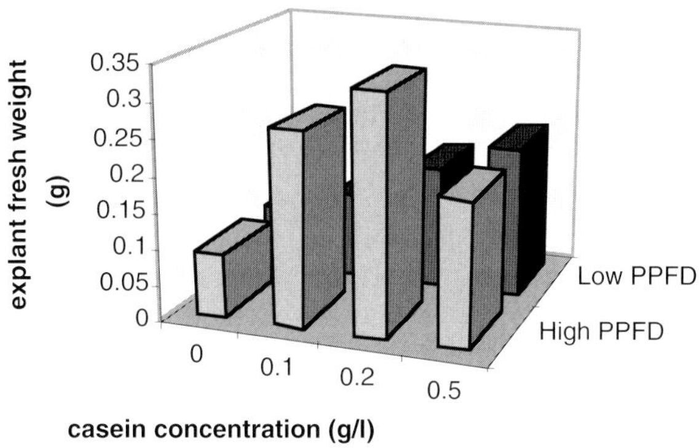

Fig. 8. Effect of casein and light intensity on the fresh weight of cultures derived from *M. silvestris* stem explants on MS medium supplemented with 9 μM NAA + 9 μM BA

Kintzios et al. 2000). On the other hand, callus growth, expressed as callus fresh weight, is favoured by higher PPFD values, but only if the culture medium contains 0.1–0.2 mg/l casein (Fig. 8). In all other cases, addition of casein seems to contradict the effect of increasing light intensities on mallow tissue cultures. A beneficial effect of light on embryogenic tissue growth and somatic embryogenesis has been previously reported for other species, such as melon, rose, squash, pepper and gardenia (Kintzios and Taravira 1997; Kintzios et al. 1998a).

2.2.2 Somatic Embryo Development

Globular somatic embryos on tissues which were allowed to continue growing on the induction medium supplemented with 1.8–18 μM NAA and BA were able to further develop on the same medium until heart- and torpedo-shaped forms (1–1.5 mm long, one to two embryos per stem explant tissue) appeared (Fig. 9). In contrast, globular embryos subcultured on growth regulator-free medium failed to develop after that stage. The PPFD value had no effect on the embryo development process. Histological studies revealed that those embryos appeared normal in structure. However, none of the fully developed embryos, which were transferred to either growth regulator-free solid MS medium or solid MS medium with 1 mg/l GA_3, germinated into plants. Embryo germination ('conversion') also failed on MS media supplemented with various growth regulators, such as gibberellic acid and abscisic acid. Thus no regenerants were recovered from the investigated cultures.

2.2.3 Macronutrient Accumulation in Tissue Cultures of Mallow

It has been previously reported (Kintzios et al. 2000) that potassium accumulation in stem-derived mallow cultures increased with casein concentration.

Fig. 9. Embryogenic *M. silvestris* tissue with heart-shaped somatic embryo on MS + BA (*scale bar* 1 mm)

Moreover, somatic embryo induction was highly positively correlated with K^+ accumulation ($r^2 = 0.930687$),. Potassium is the most abundant cation in the cytoplasm. It promotes the photosynthetic activity of plant cells by establishing the transmembrane pH gradient necessary for the synthesis of ATP in chloroplasts. It also facilitates RuBP carboxylase activity, respiration and starch accumulation. Thus, both K^+ and chlorophyll accumulation at higher casein concentrations could have promoted somatic embryogenesis in mallow cultures through an increase of their metabolic and photosynthetic activity.

2.3 Use of Mallow Tissue Cultures for the Production of Secondary Metabolites – Preliminary Experiments

In order to estimate the potential use of mallow tissue culture for the in vitro production of various bioactive products, Kintzios et al. (1998a,b) assessed the secondary metabolite accumulation in common mallow tissue cultures over a period of 4 weeks. Total secondary metabolite extraction and measurement were conducted according to the protocol of Afek et al. (1995), which is specific for phenolic derivatives. The constituents of the concentrated extracts were separated by thin-layer chromatography (9.5 toluene:0.5 ethyl acetate v/v) and their total concentration was determined spectrophotometrically at 257 nm.

Metabolite accumulation in cultured *M. sylvestris* explants started with culture establishment and increased progressively with the proliferation of somatic embryos and culture growth. Under a high photosynthetic proton flux density (250 μmol $m^{-2} s^{-1}$), maximum accumulation was observed during the third week, when the growth rate of the cultures was decreasing, and rapidly declined thereafter. Thus, when cultures were incubated under a high light intensity, secondary metabolite production was negatively correlated with culture growth (r^2 = –0.4278). This effect was not reproduced at low PPFD values, under which both tissue growth and secondary metabolite accumulation progressively increased with time (Fig. 10) and were, therefore, positively correlated (r^2 = 0.8550). However, cultures incubated under high light intensities demonstrated a remarkably increased secondary metabolite accumulation rate.

A total of five constituents were detected in tissues cultured in vitro, while only one constituent could be detected in stems received from plants growing in vivo. Differences were observed between tissues cultured under different

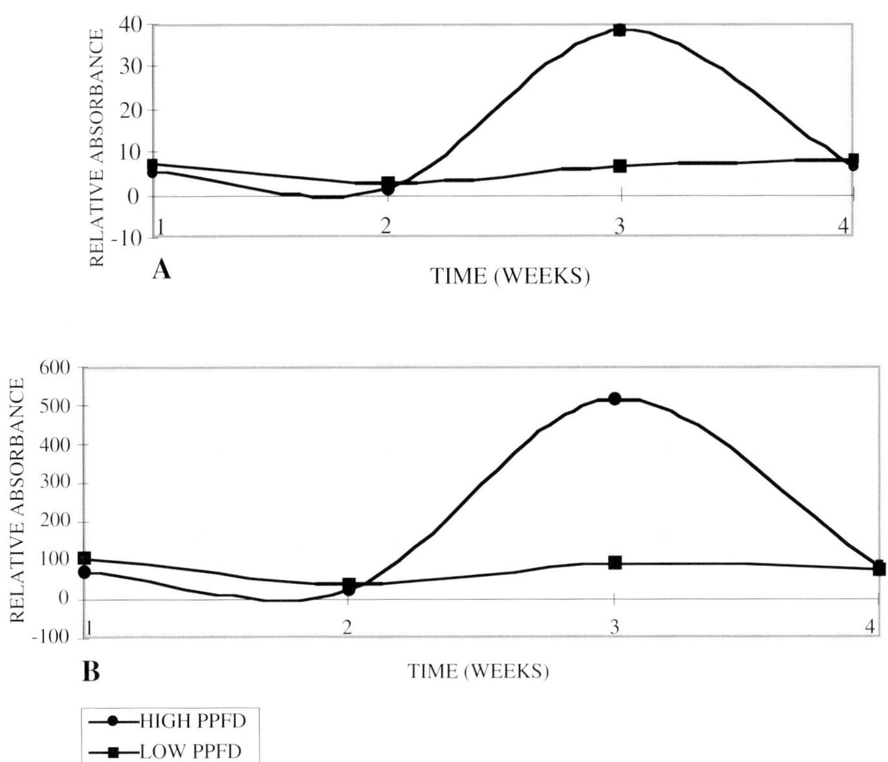

Fig. 10. Patterns of relative secondary metabolite accumulation (expressed as absorbance at 257 nm/g of **A** fresh or **B** dry culture weight) in embryogenic tissues of common mallow cultured at different PPFDs (*circles* high PPFD values, *squares* low PPFD values)

PPFDs: incubation under a high light intensity was associated with the appearance of spots on TLC only during the last 2 weeks of culture. Therefore, the observed pattern of in vitro secondary metabolite was very complex, since it was strongly dependent on the intensity of light under which cultures are incubated. Although the concentration of the different constituents which were detected in cultured tissues was too low to allow for a further chemical characterisation of them, it was demonstrated that tissue cultures of *Malva* sp. are biosynthetically active.

3 Conclusion and Prospects

Although documented research with tissue culture of *Malva* sp. is limited to a small number of reports, the successful induction of callus and somatic embryos from this species allows for a discernible potential for the further application of biotechnological techniques on this species for various purposes, such as the in vitro production of useful metabolites and the expression of genes related to the induction of somatic embryogenesis. Mallow petiole culture could be exploited as a reliable, fast-responding model system for investigating the effect of various chemical factors on the induction of callus or somatic embryogenesis from dicotyledonous plant species. Indeed, this system is being currently used for evaluating the growth regulatory potential of oligosaccharide fractions derived from various Malvaceae species (Research Project PENED 1999). Further progress in somatic embryogenesis and other advanced aspects of cell culture (e.g. protoplast culture and fusion) could lead to a significant involvement of biotechnology to the utilisation of mallow as a modern medicinal and/or fodder plant species.

4 Protocol

1. Surface sterilise 1-cm-long stem or petiole segments for 12 min in 1% (w/v) sodium hypochlorite solution, containing 1–2% Tween-80, then rinse three times in sterilised distilled water.
2. *Callus induction*: inoculate explants on MS + 9 μM NAA + 9 μM BA and incubate at 25 °C under 50 $\mu mol\, m^{-2} s^{-1}$.
3. *Somatic embryo induction and development*: inoculate explants on MS + 9 μM NAA + 9 μM BA + 0.2 g/l casein hydrolysate and incubate at 25 °C under 250 $\mu mol\, m^{-2} s^{-1}$.

Acknowledgements. I wish to thank my associates Eva Katsouri (now at the Laboratory of Pharmacognosy, Faculty of Pharmacy, University of Athens), Lito Christodoulopoulou and Emmanouil Papagiannakis for providing me with unpublished data from their research on mallow tissue culture.

References

Afek U, Aharoni N, Carmeli S (1995) Increasing celery resistance to pathogens during storage and reducing high-risk psoralen concentration by treatment with GA_3. J Am Soc Hortic Sci 129:562–565

Capek P, Matulova M, Kardosova A (1997) An acidic heteropolysaccharide from the flowers of *Malva mauritiana* L. J Carbohydrate Chem 16:1373–1391

Chopra RN, Badhwar RL (1940) Poisonous plants of India. School of Tropical Medicine, Calcutta

Christodoulopoulou L (2000) Study of the combined effect of casein hydrolysate, light intensity and duration of incubation on the induction of callus and somatic embryogenesis from common mallow (*Malva sylvestris* L.). Thesis, Faculty of Agricultural Biotechnology, Agricultural University of Athens, Athens, Greece

Classen B, Blaschek W (1998) High molecular weight acidic polysaccharides from *Malva sylvestris* and *Alcea rosea*. Planta Med 64(7):640–644

Davies S, Stewart A (1990) Nutritional medicine. Avon Books, New York

Dexiang N, Pifang Z, Gang C (1985) The effect of light quality on callus growth and root formation of *Malva sylvestris* L. in vitro. Acta Agric Shanghai 1(3):39–46

Duke J (1992a) Handbook of phytochemical constituents of GRAS herbs and other economic plants. CRC Press, Boca Raton

Duke J (1992b) Handbook of biologically active phytochemicals and their activities. CRC Press, Boca Raton

Farina A, Doldo A, Cotichini V, Rajevic M, Quaglia MG, Mulinacci N, Vincieri FF (1995) HPTLC and reflectance mode densitometry of anthocyanins in *Malva silvestris* L.: a comparison with gradient-elution reversed-phase HPLC. J Pharm Biomed Anal 14(1–2):203–211

Felklova M (1965) Contribution to the study of the *Puccinia malvacearum* Mont. disease on *Malva sylvestris* L. ssp. Mauietanica Thell. Preliminary report. Cesk Farm 14(8):409–413

Friesen LF, Nickel KP, Morrison IN (1992) Round-leaved mallow (*Malva pusilla*) growth and interference in spring wheat (*Triticum aestivum*) and flax (*Linum usitatissimum*). Weed Sci 40:448–454

Gibbs RD (1974) Chemotaxonomy of flowering plants, vol III. McGill-Queen's University Press, Montreal, pp 1443–1460

Gleason HA, Cronquist A (1991) Manual of the vascular plants of northeastern United States and adjacent Canada, 2nd edn. New York Botanical Garden, Bronx, New York

Gonda R, Tomoda M, Shimizu N, Yamada H (1990) Structure and anticomplementary activity of an acidic polysaccharide from the leaves of *Malva sylvestris* var. *mauritiana*. Carbohydr Res 198(2):323–329

Gossett DR, Millhollon EP, Lucas MC, Bank SW, Marney MM (1994) The effects of NaCl on antioxidant enzyme activities in embryogenic tissue tissues of salt-tolerant and salt-sensitive cotton cultivars (*Gossypium hirsutum* L.). Plant Cell Rep 13:498–503

Grierson DS, Afolayan AJ (1999) Antibacterial activity of some indigenous plants used for the treatment of wounds in the Eastern Cape, South Africa. J Ethnopharmacol 66(1):103–106

Grieve M (1994) Mallow. In: Leyel CF (ed) A modern herbal. Tiger Books International, London, pp 508–509

Hasler A, Meier B, Sticher O (1990) Quantitative HPLC analysis of flavonoid aglycones in different medicinal plants. Planta Med 56:575–579

Hasson E, Poljakoff-Mayber A (1995) Callus culture from hypocotyls of *Kosteletzkya virginica* (L.) seedlings Its growth, salt tolerance and response to abscisic acid. Plant Cell Tissue Organ Cult 43(3):279–285

Huang CY, Zeng LF, He T, Wang CJ, Hong JR, Zhang XQ, Hou YH, Peng SS (1998) In vivo and in vitro studies on the antitumor activities of MCP (*Malva crispa* L. Powder). Biomed Environ Sci 11(4):297–306

Iqbal B, Hussain T, Shah H, Manan SA (1994) Chemical composition of *Malva neglecta* grown in Kalam district Swat [Pakistan]. Sarhad J Agric 10:125–129

Jin S, Xu R, Wei Y, Goodwin PH (1999) Increased expression of a plant actin gene during a biotrophic interaction between round-leaved mallow, *Malva pusilla*, and *Colletotrichum gloeosporioides* f. sp. *malvae*. Planta 209(4):487–494

Katsouri E (1997) Tissue culture and in vitro secondary metabolite production from common mallow (*Malva sylvestris* L.). Thesis, Faculty of Agricultural Biotechnology, Agricultural University of Athens, Athens, Greece

Kintzios S, Taravira N (1997) Effect of genotype and light intensity on somatic embryogenesis and plant regeneration in melon (*Cucumis melo* L.). Plant Breed 116:359–362

Kintzios S, Hiureas G, Shortsianitis E, Sereti E, Blouhos P, Manos C, Makri O, Taravira N, Drossopoulos J, Holevas CD (1998a) The effect of light on the induction, development and maturation of somatic embryos from various horticultural and ornamental species. Acta Hortic 461:427–432

Kintzios S, Katsouri E, Peppes D, Koulocheri S (1998b) Somatic embryogenesis and in vitro secondary metabolite production from common mallow (*Malva sylvestris* L.) collected in Greece. Acta Hortic 457:173–180

Kintzios S, Papagiannakis EM, Aivalakis G, Konstas J, Bouranis D, Christodoulopoulou L (2000) The effects of casein and its constituents on the development of tissue culture and somatic embryogenesis from *Malva silvestris* L.: a preliminary study. J Herbs Spices Med Plants (in press)

Kozlowski J, Szczyglewska D, Formanowiczowa H (1989) Biology of germination of medicinal plants seeds, Part 14. Seeds of species from Malvaceae family: marsh mallow (*Althea officinalis* L.) and mallow (*Malva silvestris* L.). Herba-Polon 35(2–3):99–107

Kratzer KE, Greif JM (1969) *Malva rotundifolia*, a potential vasodilator. J Am Osteopath Assoc 68(10):1029–1035

Makowski RMD, Mortensen K (1999) Latent infections and residues of the bioherbicide agent *Collecottrichum gloeosporioides* f.sp. *malvae*. Weed Sci 47:589–595

Marco S (1975) Occurrence of alfalfa mosaic virus and Malva vein clearing virus on weed members of Malvaceae in Israel. Plant Dis Rep (USA) 59(1):34–36

Mas T, Susperregui J, Berke B, Cheze C, Moreau S, Nuhrich A, Vercauteren J (2000) DNA triplex stabilization property of natural anthocyanins. Phytochemistry 53:679–687

McLean KS, Lawrence GW, Reichert NA (1992) Embryogenic tissue induction and adventitious organogenesis of kenaf (*Hibiscus cannabinus* L.). Plant Cell Rep 11:532–534

Murashige T, Skoog F (1962) A revised method for rapid growth and bioassays with tobacco tissue cultures. Physiol Plant 15:472–497

Ostrowski R, Daczewska M (1984) Preliminary results of investigations on the economic and fodder value of feed mallow (*Malva verticillata*).In: Research and animal nutrition in Poland. Polska Akademia Nauk, Warszawa, Poland, pp 249–256

Ostrowski R, Daczewska M (1981) Investigations on mallow (*Malva verticillata* L.) as a fodder plant and on possibilities of its cultivation in the Wielkopolska region [Poland]. Pol J Anim Sci Technol 8(2):59–69

Papagiannakis EM (2000) Effect of casein hydrolysate on the induction of somatic embryogenesis from common mallow (*Malva sylvestris* L.). Thesis, Faculty of Agricultural Biotechnology, Agricultural University of Athens, Athens, Greece

Pizzorno JE, Murray MT (1985) A textbook of natural medicine. John Bastyr College Publications, Seattle, Washington

Podlech D (1996) Herbs and medicinal plants of Britain and Europe. Harper Collins Publishers, London, pp 92–93

Pourrat H, Texier O, Barthomeuf C (1990) Identification and assay of anthocyanin pigments in *Malva sylvestris* L. Pharm Acta Helv 3:65–68

Rajasekaran K (1996) Regeneration of plants from cryopreserved embryogenic cell suspension and embryogenic tissue cultures of cotton (*Gossypium hirsutum* L.). Plant Cell Rep 15:859–864

Research Project PENED (1999) Structural biotransformation of renewable hydrocarbons as a tool for producing bioactive materials with growth regulatory activity. General Secretary of Research and Technology

Schmidgall J, Schnetz E, Hensel A (2000) Evidence for bioadhesive effects of polysaccharides and polysaccharide-containing herbs in an ex vivo bioadhesion assay on buccal membranes. Planta Med 66:48–53

Scroumbis B (1998) Aromatic, medicinal and apiarian plants of Greece. Agrotypos Press, Athens, pp 172–173

Seong KC, Cho CR, Yu IC (1995) Studies on the photosynthesis and effects of shading and harvesting methods on the growth and yield of mallow (*Malva verticillata* L.). RDA J Agric Sci 37:349–355

Shih J, Wei Y, Goodwin PH (2000) A comparison of the pectate lyase genes, pel-1 and pel-2, of *Colletotrichum gloeosporioides* f.sp. *malvae* and the relationship between their expression in culture and during necrotrophic infection. Gene 243(1–2):139–150

Simon PW (1990) Carrots and other horticultural crops as a source of provitamin A carotenes. HortScience 25(12):1495

Tomoda M, Gonda R, Shimizu N, Yamada H (1989) Plant mucilages. XLII. An anti-complementary mucilage from the leaves of *Malva sylvestris* var. *mauritiana*. Chem Pharm Bull (Tokyo) 11:3029–3032

Tomoda M, Shimizu N, Gonda R, Kanari M, Yamada H, Hikino H (1990) Anti-complementary and hypoglycemic activities of the glycans from the seeds of *Malva verticilata*. Planta Med 56:168–173

Tossi B, Tirillini B, Donini A, Bruni A (1995) Presence of scopoletin in *Malva sylvestris*. Int J Pharmacogn 33:353–355

Werbach M (1993) Healing with food. Harper Collins, New York

Zapata C, Srivatanakul M, Park S-H, Lee B-M, Salas MG, Smith RH (1999) Improvements in shoot apex regeneration of two fiber crops: cotton and kenaf. Plant Cell Tissue Organ Cult 56(3): 185–191

Zhao Y-X, Yao D-Y, Harris PJC (1991) Isolation and culture of protoplasts from embryogenic tissue of *Hibiscus syriacus* L. Plant Cell Tissue Organ Cult 25:17–20

9 In Vitro Culture and the Production of Polyphenol and Volatile Compounds in *Ocimum*

K. Ishimaru[1], Y. Murakami[1], and K. Shimomura[2]

1 Introduction

The genus *Ocimum* of the mint family Labiatae, some of which are well known as 'basil', is found in temperate regions in the world. The plants are popularly used as culinary herbs and flower garden ornamentals. One of the major species of basil is *O. basilicum* L. (sweet basil, Fig. 1), which has some varieties such as *O. basilicum* 'Cinnamon' (cinnamon basil), 'Crispum' (lettuce-leaf basil), 'Minimum' (bush basil) and 'Purpurascens' (dark opal basil). It is an aromatic annual herb, native of tropical Asia, Africa and islands of the Pacific Ocean (Hikawa 1998). Holy basil (*O. sanctum*), frequently cultivated in warm parts of Asia (in conjunction with the Hindu religion), has a sweet fragrance and is popularly used as an ornamental.. The perennial *O. basilicum* 'Citriodorum' (lemon basil) has a strong lemon fragrance which is appropriate for kitchen vinegar. Tree basil (*O. suave*) of Africa, which is a many branched shrub and closely resembles *O. gratissimum* (tea bush) of India, is an important garden tree, growing in well-drained soil and a sunny location (Everett 1981).

In general, basil species contain a variety of essential oils (0.5–1.1%, in the aerial part) such as methylchavicol, linalool, cineol, geraniol, pinane etc., together with vitamin B1 (0.1–0.2 mg/100 g DW), vitamin C (150–250 mg/100 g DW), carotene (1.2–2.8 mg/100 g DW) and bitter ingredients. They are important materials which can be blended in tea to treat coughs, stomach disorders, or used as diuretics etc. (Hikawa 1998). The essential oils obtained from basil herbs are also used for soap, shampoo and perfume.

Phenolic constituents of *Ocimum* and various other genera in Labiatae have been well studied and some caffeic acid derivatives have been identified. In particular, rosmarinic acid (RA, Fig. 2) is one of the most common antioxidant phenolics occurring in Labiatae and some investigations on RA production in *Ocimum* species have been done using gas chromatography (Gerhardt and Schroter 1983) and thin-layer chromatography (Reschke 1983) methods.

---

[1]Department of Applied Biological Sciences, Faculty of Agriculture, Saga University, 1 Honjo, Saga 840-8502, Japan
[2]Tsukuba Medicinal Plant Research Station, National Institute of Health Sciences, 1 Hachimandai, Tsukuba, Ibaraki 305-0843, Japan

Biotechnology in Agriculture and Forestry, Vol. 51
Medicinal and Aromatic Plants XII (ed. by T. Nagata and Y. Ebizuka)
© Springer-Verlag Berlin Heidelberg 2002

Fig. 1. *O. basilicum* 'Sweet' field-grown plant. (Photo by Ishimaru 1996)

2 In Vitro Studies of *Ocimum*

Nine basil (*O. basilicum*) cultivars were studied for in vitro callus formation and plant regeneration (Matsubara et al. 1996). Callus formed best on Murashige-Skoog (MS) medium N (Murashige and Skoog 1962) supplemented with isopentenyladenine, among other cytokinins, but only 6-benzyladenine (BA) actually induced regeneration of plantlets from callus. Callus growth from 'lettuce basil' was three times higher than that from 'cinnamon basil' and adventitious buds regenerated only from 'sweet basil', 'dark opal basil' and 'bush basil'.

Dalton (1983, 1984) studied the effect of sugar supply rate on photosynthetic development of *O. basilicum* (sweet basil) cells and on chlorophyll production in fed-batch cultures of the species. When fructose in the feed was decreased, the specific rate of chlorophyll production increased and the biomass became greener with an increment of actual photosynthesis rate. The chlorophyll production was inhibited when glucose concentration in the cells was above a threshold of about 1.2% dry biomass.

In addition to the major phenolic constituent, such as RA in *Ocimum* plants, recently related large molecular weight phenolics (caffeic acid derivatives) of lithospermic acid (LA; Kelley et al. 1975) and lithospermic acid B (LAB; Tanaka

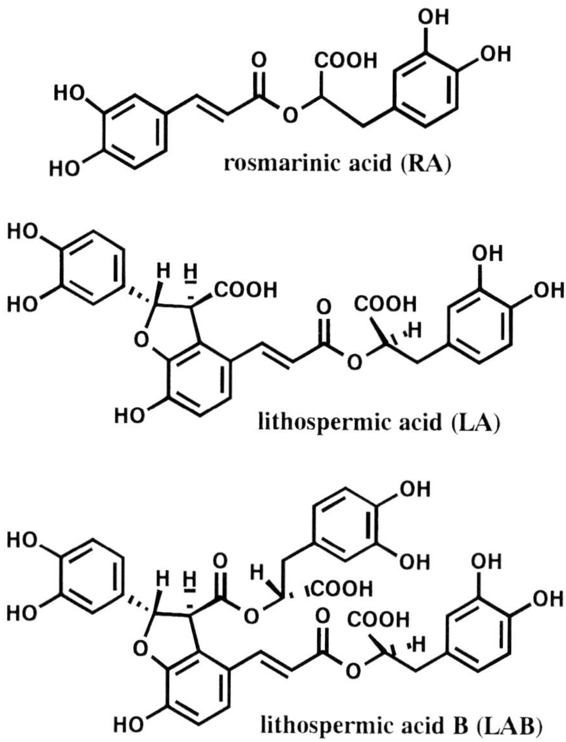

Fig. 2. Structures of RA, LA and LAB

et al. 1989; Fig. 2) have attracted attention with their interesting biological activities used for renal disease (Yokozawa et al. 1989, 1990a,b) and anti-HIV (Lee et al. 1992). This chapter describes the production of phenolic constituents (RA, LA and LAB) in some tissue cultures (shoot, callus, adventitious root and transformed root) of various *Ocimum* cultivars. Furthermore, volatile constituents in shoot cultures of *Ocimum* varieties were analyzed and the composition of volatile compounds in vitro was compared to that of the intact field-grown plants.

3 Tissue Culture and Phenolic Production of *Ocimum*

3.1 Shoot Culture

Shoot cultures of ten *O. basilicum* varieties were established on phytohormone-free MS solid medium (Fig. 3). All in vitro plants produced a high content of RA in both leaf and root portions, together with a lower amount of LA and

Fig. 3. Shoot cultures of **a** *O. basilicum* 'Aromata'; **b** *O. basilicum* 'Cinnamon'; **c** *O. basilicum* 'Citriodorum'; **d** *O. basilicum* 'Greek'; **e** *O. basilicum* 'Sweet'; **f** *O. basilicum* 'Purpureum'; **g** *O. basilicum* 'Purple Ruffles'; **h** *O. sanctum* on phytohormone-free MS medium. (Photo by Murakami 1997)

LAB in some varieties (Table 1). In particular, *O. sanctum* showed the maximum content of RA (5.15% in leaf and 5.63% in root), which was larger than that of nine *Ocimum* field-grown plants (Table 1; Omoto et al. 1997).

3.2 Callus Culture

Nine *Ocimum* callus cultures produced a high content of RA as major phenolic constituents under both dark and light conditions (Table 2). The maximum RA content was observed in *O. basilicum* 'Sweet' callus cultured on Gamborg

Table 1. Rosmarinic acid, lithospermic acid and lithospermic acid B contents in *Ocimum* plants. (Omoto et al. 1997)

Material	Part	Contents (% DW)		
		RA	LA	LAB
(In vivo plants)				
O. basilicum "Aromata"	Leaf	2.68	–	–
	Root	3.27	–	–
O. basilicum "Cinnamon"	Leaf	0.29	0.55	0.02
	Root	1.14	Trace	0.01
O. basilicum "Licorice"	Leaf	3.33	–	–
	Root	4.62	–	–
O. basilicum "Purpureum"	Leaf	0.25	0.27	0.02
	Root	0.34	–	0.01
O. basilicum "Purple Ruffles"	Leaf	5.15	–	–
	Root	3.99	–	–
O. basilicum "Sweet"	Leaf	3.98	–	–
O. basilicum "Citriodorum"	Leaf	4.39	–	–
	Root	1.05	–	–
O. basilicum. "Crispum"	Leaf	1.20	1.24	0.05
	Root	1.78	0.03	0.10
O. sanctum	Leaf	0.43	1.01	0.05
	Root	0.91	0.12	0.25
(Shoot cultures)				
O. basilicum "Aromata" (MS)	Leaf	1.06	–	–
	Root	2.18	–	–
O. basilicum "Cinnamon" (MS)	Leaf	4.38	–	–
	Root	3.94	–	–
O. basilicum "Greek" (MS)	Leaf	0.79	0.01	0.10
	Root	0.48	–	0.06
O. basilicum "Purpureum" (MS)	Leaf	4.88	–	–
	Root	4.45	–	–
O. basilicum "Purple Ruffles" (MS)	Leaf	4.65	–	–
	Root	2.78	–	–
O. basilicum "Sweet" (MS)	Leaf	5.01	0.36	–
O. basilicum "Citriodorum" (MS)	Leaf	0.24	0.03	0.09
	Root	0.44	0.01	0.08
O. basilicum. "Crispum" (MS)	Leaf	4.02	–	–
	Root	4.54	–	–
O. basilicum "Minimum" (MS)	Leaf	2.71	–	–
	Root	5.48	–	–
O. sanctum (MS)	Leaf	5.15	–	–
	Root	5.63	–	–

Table 2. Rosmarinic acid, lithospermic acid and lithospermic acid B contents in callus cultures of *Ocimum* plants. (Omoto et al. 1997)

Material (medium, phytohormone, mg/l)		FW[a] (g)	Contents (% DW)		
			RA	LA	LAB
Ocimum basilicum "Cinnamon" (MS, NAA 0.5 + BA, 0.1)	Light	1.17	4.67	–	–
	Dark	2.82	3.21	–	–
O. basilicum "Greek" (MS, NAA 0.5 + 2,4-D 0.5)	Light	2.89	5.75	–	–
	Dark	3.79	5.91	–	–
O. basilicum "Purpureum" (MS, NAA 0.5 + BA, 0.1)	Light	1.88	1.53	–	–
	Dark	2.93	0.43	–	–
O. basilicum "Purple Ruffles" (MS, 2,4-D 1.0 + BA, 0.1)	Dark	0.78	5.02	–	–
O. basilicum "Sweet" (B5, NAA 0.5 + 2,4-D 1.0)	Light	2.56	7.36	–	–
	Dark	1.32	6.78	–	0.62
O. basilicum "Citriodorum" (MS, 2,4-D 1.0)	Light	1.31	0.85	–	–
	Dark	2.80	4.17	–	–
O. basilicum "Minimum" (MS, NAA 0.5 + BA, 0.1)	Light	1.78	6.06	–	–
	Dark	1.29	6.00	–	–
O. sanctum (MS, 2,4-D 1.0)	Light	2.22	2.12	–	–
	Dark	3.22	2.93	–	–

[a] Fresh weight of one callus cultured on 10 ml medium for 8 weeks (inoculation amount: ca. 0.22 g).

B5 (B5) (Gamborg et al. 1968) solid medium supplemented with 0.5 mg/l NAA and 1 mg/l 2,4-D (7.36% in the light and 6.78% in the dark). Only in the 'Sweet' callus was LAB accumulated in the dark, and this content was highest among all investigated in vivo and in vitro *Ocimum* plants (Table 1).

3.3 Adventitious Root Culture of *O. basilicum* 'Sweet'

Adventitious roots of *O. basilicum* 'Sweet' cultured in MS liquid medium containing 1 mg/l NAA showed a linear increase in the biomass (Fig. 4a). At the later period of the culture, week 7, the roots reached the maximum amount (668.5 mg DW/flask). The root cultures produced both RA and LA and, in particular, showed a continuous increment of RA content throughout the culture time (Fig. 4b, RA was the major phenolic produced in the cultures), although LAB was not detected in the cultures. The maximum contents of RA and LA, observed at a later period of culture (6.84% RA at week 8 and 0.52% LA at week 7, DW), were relatively higher than those of the intact plant and in vitro shoots (Table 1). The measured maximum weights of RA and LA obtained in a flask (100 ml volume containing 50 ml medium) in the root cultures reached 55.3 mg RA (at week 8) and 3.4 mg LA (at week 7). *O. basilicum* 'Sweet' adventitious

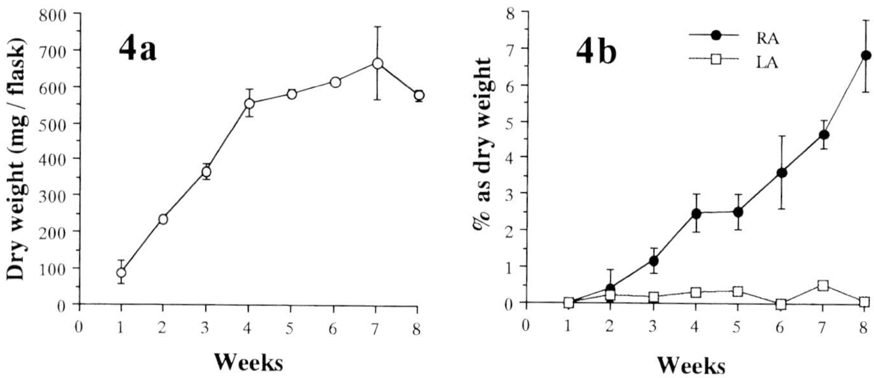

Fig. 4. **a** Growth and **b** phenolics production of *O. basilicum* 'Sweet' adventitious root cultured in MS liquid medium containing 1 mg/l NAA in the dark . *Bars* represent standard errors of three replicates. (Tada et al. 1996a)

root is the first example to succeed in the production of RA and LA in root organ cultures in Lamiaceae (Tada et al. 1996a).

3.4 Transformed Root Culture

3.4.1 O. basilicum *'Sweet'*

Five clones of *O. basilicum* 'Sweet' transformed roots, A-1 and A-2 (induced by *A. rhizogenes* ATCC 15834) and J-1, J-2 and J-3 (by *A. rhizogenes* MAFF 03–01724) were cultured in three hormone-free liquid media of MS, B5 and Woody Plant (WP) (Lloyd and McCown 1980). After inoculation into these media, all clones except one (A-2) started rapid proliferation in the early stage of culture, and particularly in WP medium, the amount of roots increased without an obvious time lag, resulting in a maximum level at week 3 (Fig. 5). The highest amount of roots in all the cultures of A-2 was 702.8 mg (DW) obtained in MS medium at week 6.

RA, LA and LAB contents of these clones are shown in Fig. 6 (Tada et al. 1996b). In particular, high RA contents (over 14% DW) were observed in MS (J-1: 14.1%, at week 8) and B5 (A-2: 14.0%, at week 6) media, in which their levels were almost 3.5-fold higher than that of the intact plant (Table 1). Notably, only the clones J-1, J-2 and J-3 transformed with *A. rhizogenes* MAFF 03–01724 showed a severe decrease in RA content after reaching the maximum levels at week 4 in B5 medium, although in WP medium, RA content of all clones successively decreased in the later culture period. The maximum yield (mg/flask) of RA in all cultures examined was 73.5 mg/flask by J-1 in MS medium at week 5. High LA contents of the hairy roots were observed in MS medium. In particular, the maximum LA production was obtained by J-clones

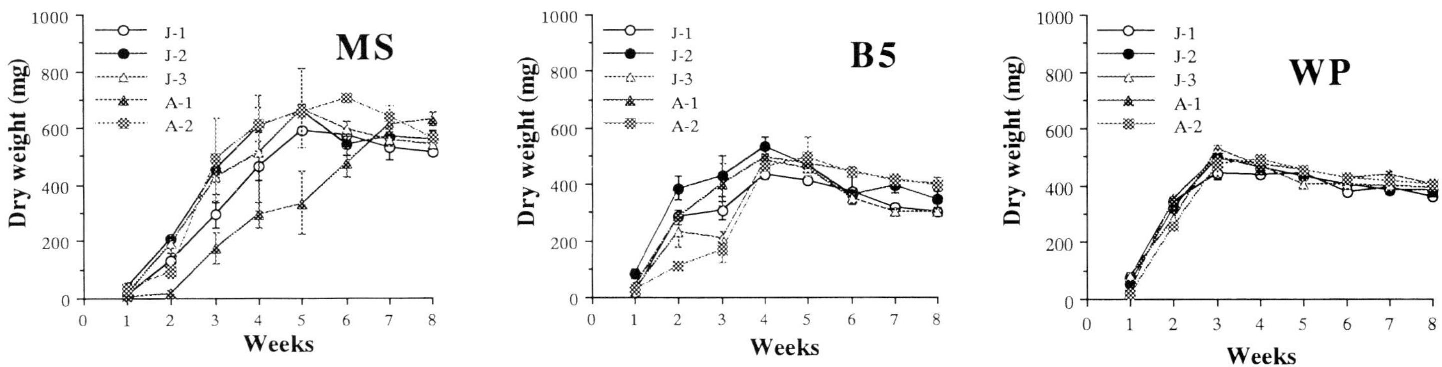

Fig. 5. Growth of *O. basilicum* "Sweet" transformed roots cultured in MS, B5 or WP liquid medium in the dark. (Tada et al. 1996b)

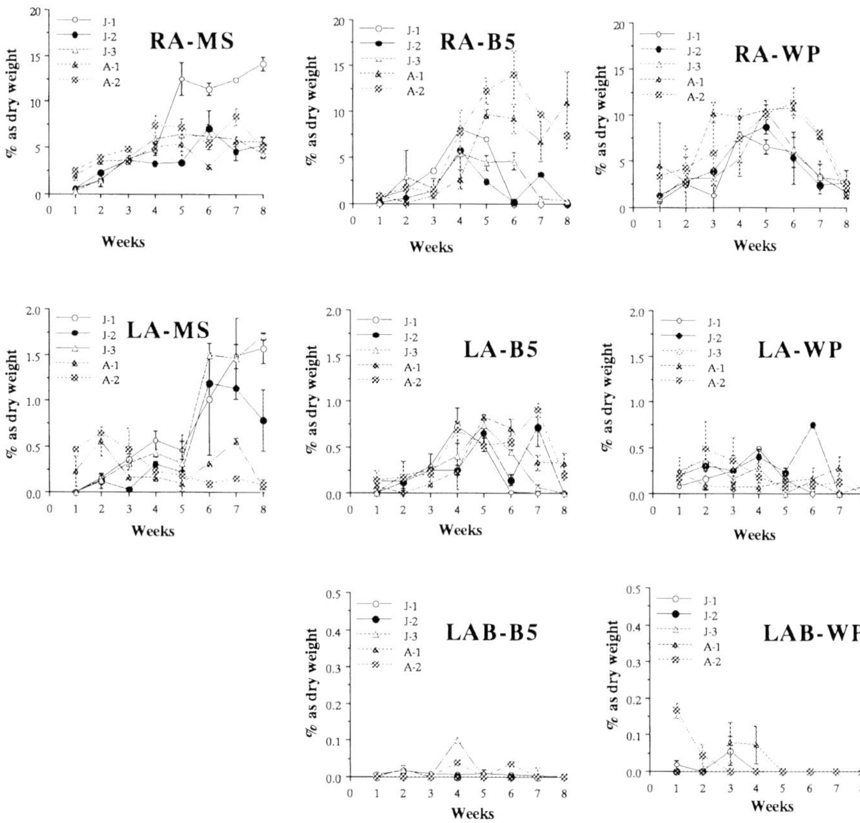

Fig. 6. RA, LA and LAB contents in *O. basilicum* 'Sweet' transformed roots cultured in MS, B5 or WP liquid medium in the dark. (Tada et al. 1996b)

in the later period of culture in MS medium (J-1: 1.57% at week 8, J-2: 1.19% at week 6 and J-3: 1.70% at week 8). Although the content was fairy low (below 0.17%, DW), LAB was certainly produced in some clones when cultured in B5 (A-2, J-1 and J-3) and WP (A-1, A-2 and J-1) media. However, in MS medium, LAB was not detected.

3.4.2 O. basilicum *'Cinnamon'*

Seven clones of the transformed roots of *O. basilicum* 'Cinnamon' induced by the infection with *A. rhizogenes* ATCC 15834 (ObcA-1 and ObcA-2), *A. rhizogenes* MAFF 03–01724 (ObcM-1), *A. rhizogenes* A13/ pBI 121 (ObcA13-1 and ObcA13-2) and *A. rhizogenes* R1000 (ObcR1000-1 and ObcR1000-2) were cultured in four basal liquid media (MS, 1/2 MS, WP and B5) in the dark (Fig. 7). The growth and RA content of respective clones are shown in Fig. 8.

Fig. 7. Transformed root cultures of *O. basilicum* 'Cinnamon'. **a** Induction from stem segment by infection with *A. rhizogenes* ATCC 15834. **b** ObcA-1 cultured on phytohormone-free 1/2 MS solid medium in the dark for 3 weeks. **c** ObcA-1 cultured in phytohormone-free 1/2 MS solid medium in the dark *left* or light *right* for 6 weeks. **d** ObcA-1 cultured in phytohormone-free 1/2 MS liquid medium in the dark (*left*) or light (*right*) for 6 weeks. (Photo by Murakami 1997)

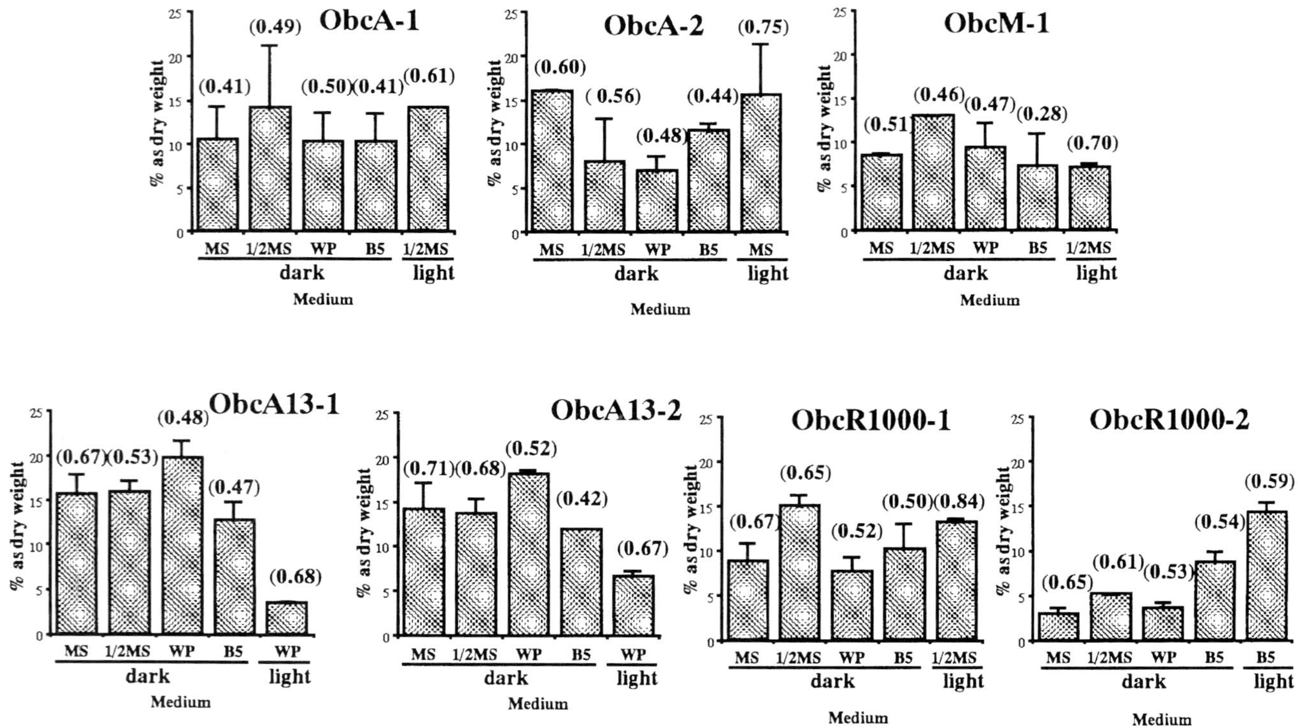

Fig. 8. Growth and RA content of *O. basilicum* 'Cinnamon' transformed roots cultured in various phytohormone-free liquid media for 4 weeks. *Numbers in parentheses* Dry weight/flask; *bars* represent standard errors of three replicates. (Ishimaru et al., unpubl.)

In these clones, LA and LAB were not detected. Among the four media tested in the dark, one medium in which each clone showed the highest RA content was selected and used for culturing the clones under the light condition (growth and RA content are shown in Fig. 8). Although the clones induced by *A. rhizogenes* A13/ pBI 121 (ObcA13-1 and ObcA13-2) showed a larger RA content than those of the other clones in four media in the dark, RA content in these clones (ObcA13-1 and ObcA13-2) was fairly low in the light. On the other hand, ObcR1000-2 showed better RA production in the light compared to that in the dark. Clones ObcA-1 and ObcA-2 induced by *A. rhizogenes* ATCC 15834 showed high RA content in both the dark and light. The maximum RA content (19.7%) was observed in ObcA13-1 cultured in WP medium in the dark.

3.4.3 O. basilicum *'Greek'*

Transformed roots of *O. basilicum* 'Greek' were derived by infection with *A. rhizogenes* ATCC 15834 and one clone (ObgA-1) which showed the best growth was selected for the analysis of phenolics production. ObgA-1 produced a high content of RA in four basal media (MS, 1/2 MS, WP and B5) in the dark and in B5 medium in the light (Fig. 9a). The maximum RA content (15.6%) was observed in B5 medium in the dark. LA and LAB were not detected in ObgA-1.

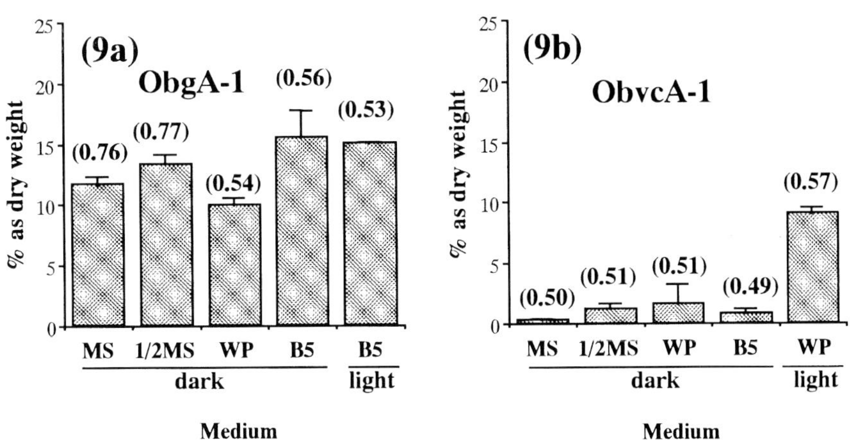

Fig. 9. Growth and RA content of **a** *O. basilicum* 'Greek' (ObgA-1) and **b** *O. basilicum* 'Citriodorum' (ObvcA-1) transformed roots cultured in various phytohormone-free liquid media for 4 weeks. *Numbers in parentheses* Dry weight/flask; *bars* represent standard errors of three replicates. (Ishimaru et al., unpubl.)

3.4.4 O. basilicum *var.* citriodorum

ObvcA-1 clone from the transformed roots of *O. basilicum* 'Citriodorum' derived by infection with *A. rhizogenes* ATCC 15834 showed a low content (below 1.5%) of RA in four basal liquid media in the dark (Fig. 9b). On the other hand, under light, this clone produced a high content (9%) of RA in WP medium. Although RA accumulation in ObvcA-1 seemed to be specifically promoted by illumination, LA or LAB were not detected even in light.

3.4.5 O. sanctum

Growth and RA production of eight clones of the transformed roots of *O. sanctum* induced by infection with *A. rhizogenes* ATCC 15834 (OsA-1 and OsA-2), *A. rhizogenes* MAFF 03–01724 (OsM-1 and OsM-2), *A. rhizogenes* A13/pBI 121 (OsA13-1 and OsA13-2) and *A. rhizogenes* R1000 (OsR1000-1 and OsR1000-2) in four basal liquid media (MS, 1/2 MS, WP and B5) are shown in Fig. 10. Among these clones, OsR1000-2 showed a high content (over 15%) of RA in all media in the dark and in MS medium in the light. Clones OsM-1 and OsM-2, both induced by *A. rhizogenes* MAFF 03–01724, did not show such good production of RA in the dark. OsA-2 produced the maximum content (21.9%) of RA in 1/2 MS medium in the dark, which was the highest observed in all in vivo and in vitro (shoot, callus, suspension cell, adventitious and transformed roots etc.) plant materials which accumulate caffeic acid derivatives (RA, LA and LAB) investigated so far.

4 Shoot Cultures and Volatile Compounds of *Ocimum*

Volatile compounds both in in vivo (field-grown) and in vitro (shoot culture) plants of 11 *Ocimum* varieties were analyzed by the solid phase micro extraction (SPME) method followed by GC/MS detection, in which only the fragrance constituents that naturally come from the materials can be detected. Among 40 volatile compounds detected in the *Ocimum* plants investigated, 26 major constituents (Fig. 11) were selected and their composition (%) in the plants is summarized in Fig. 12 (in vitro plants were analyzed in *O. basilicum* 'Greek', while only in vivo plants were analyzed in *O. basilicum* 'Licorice'). In vivo *Ocimum* varieties were classified into three groups according to their patterns of volatile composition, i.e. the *O. basilicum* 'Citriodorum' and *O. basilicum* 'Aromata' group producing hydrocarbon (5) as a major volatile, the *O. basilicum* 'Crispum', *O. basilicum* 'Licorice', *O. basilicum* 'Purpureum', *O. basilicum* 'Purple Ruffles', *O. basilicum* 'Sweet' and *O. sanctum* group producing ethers (19, 21 and 22) as major volatiles and the *O. basilicum* 'Minimum' and *O. basilicum* 'cinnamon' group producing ester (26) as major volatile.

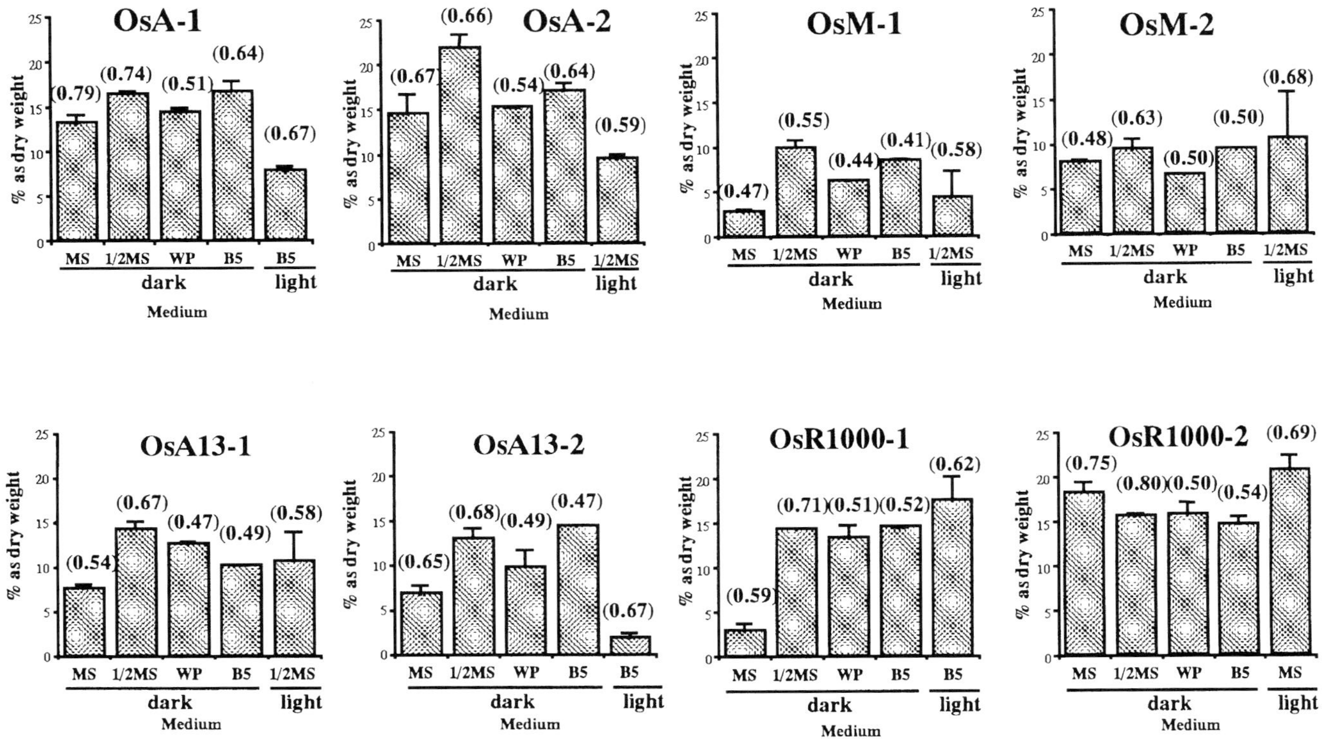

Fig. 10. Growth and RA content of *O. sanctum* transformed roots cultured in various phytohormone-free liquid media for 4 weeks. *Numbers in parentheses* Dry weight/flask; *bars* represent standard errors of three replicates. (Ishimaru et al., unpubl.)

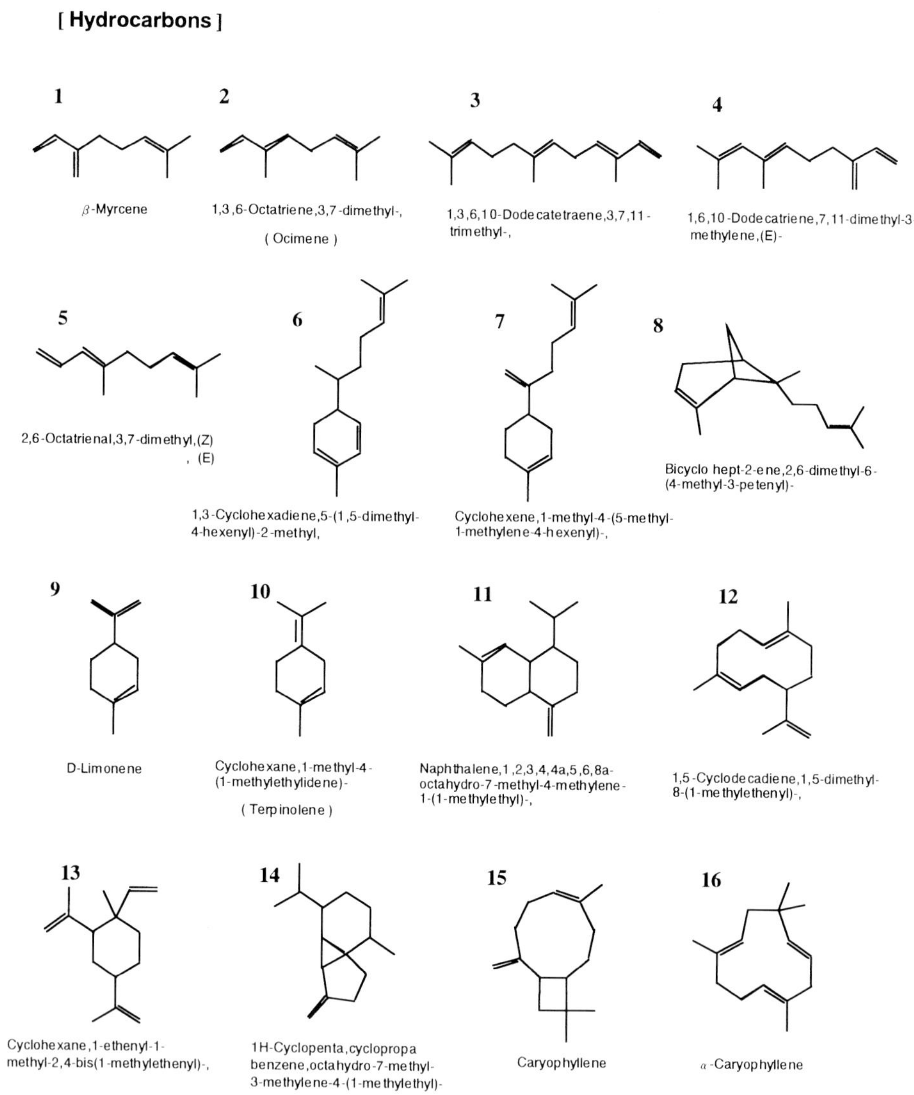

Fig. 11. Major volatile compounds in *Ocimum* varieties. (Ishimaru et al., unpubl.)

In vitro plants were also classified as the hydrocarbon (5 and 8) group; *O. basilicum* 'Citriodorum' and *O. basilicum* 'Purple Ruffles', the alcohol (17) group; *O. basilicum* 'Cinnamon' and *O. basilicum* 'Greek' and the ethers (19 and 20) group; *O. basilicum* ''Crispum', *O. basilicum* 'Aromata', *O. basilicum* 'Minimum', *O. basilicum* 'Purpureum', *O. basilicum* 'Sweet' and *O. sanctum* . In most of the *Ocimum* varieties investigated, the major volatile compounds in vivo were different from those of the in vitro ones. This observation indicated

[Alcohols and Ethers]

17

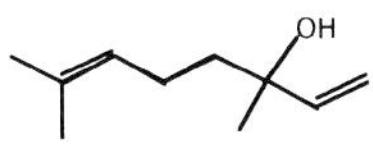

1,6-Octadien-3-ol,3,7-dimethyl-

18

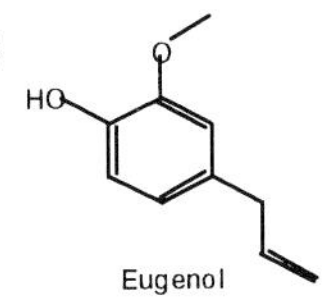

Eugenol
(2-Methoxy-4-allylphenol)

19

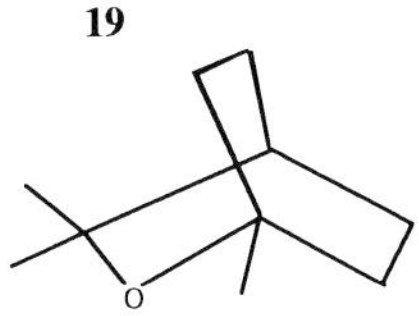

Eucalyptol
(1, 8-Cineol)

20

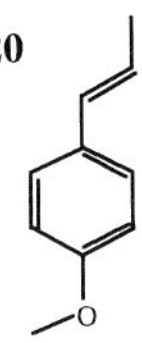

Benzene,1-methoxy-4-(1-propenyl)-
(Anethole)

21

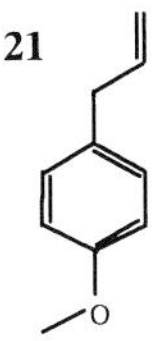

Benzene,1-methoxy-4-(2-propenyl)-
(Estragol)
(Methyl chavicol)

22

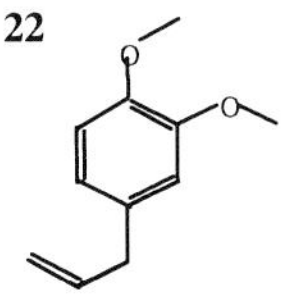

Benzene,1,2-dimethoxy-4-(2-propenyl)-
(Methyl eugenol)
(Eugenol methyl ester)

[Ketones and Ester]

23

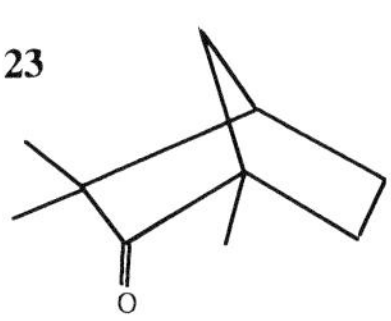

Bicyclo heptan-2-one,1,3,3-trimethyl
(Fenchone)

24

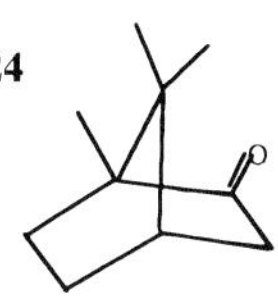

Bicyclo heptan-2-one,1,7,7-trimethyl-,
(Camphor)

25

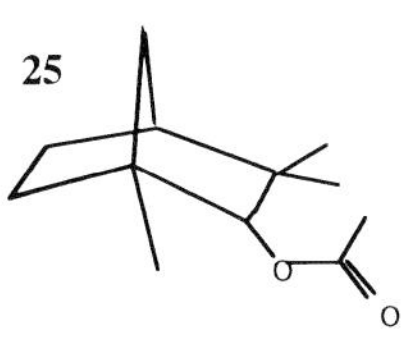

Bicyclo heptan-2-ol,1,3,3-trimethyl,acetate,

26

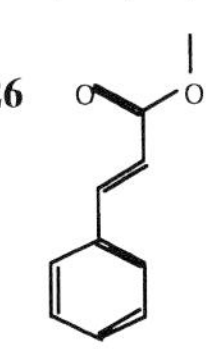

2-Propenoic acid,3-phenyl-,methyl ester
(Methyl cinnamate)

Fig. 11. *Continued*

that volatile constituents of *Ocimum* plants are very variable by changes in culture (cultivation) conditions such as medium, temperature, illumination etc. Methyl chavicol (21) was the major compounds in *O. basilicum* 'Crispum', *O. basilicum* 'Aromata', *O. basilicum* 'Minimum' and *O. sanctum* in vitro shoot cultures. In particular, *O. basilicum* 'Aromata' and *O. basilicum* 'Minimum' produced 21 only in in vitro culture condition.

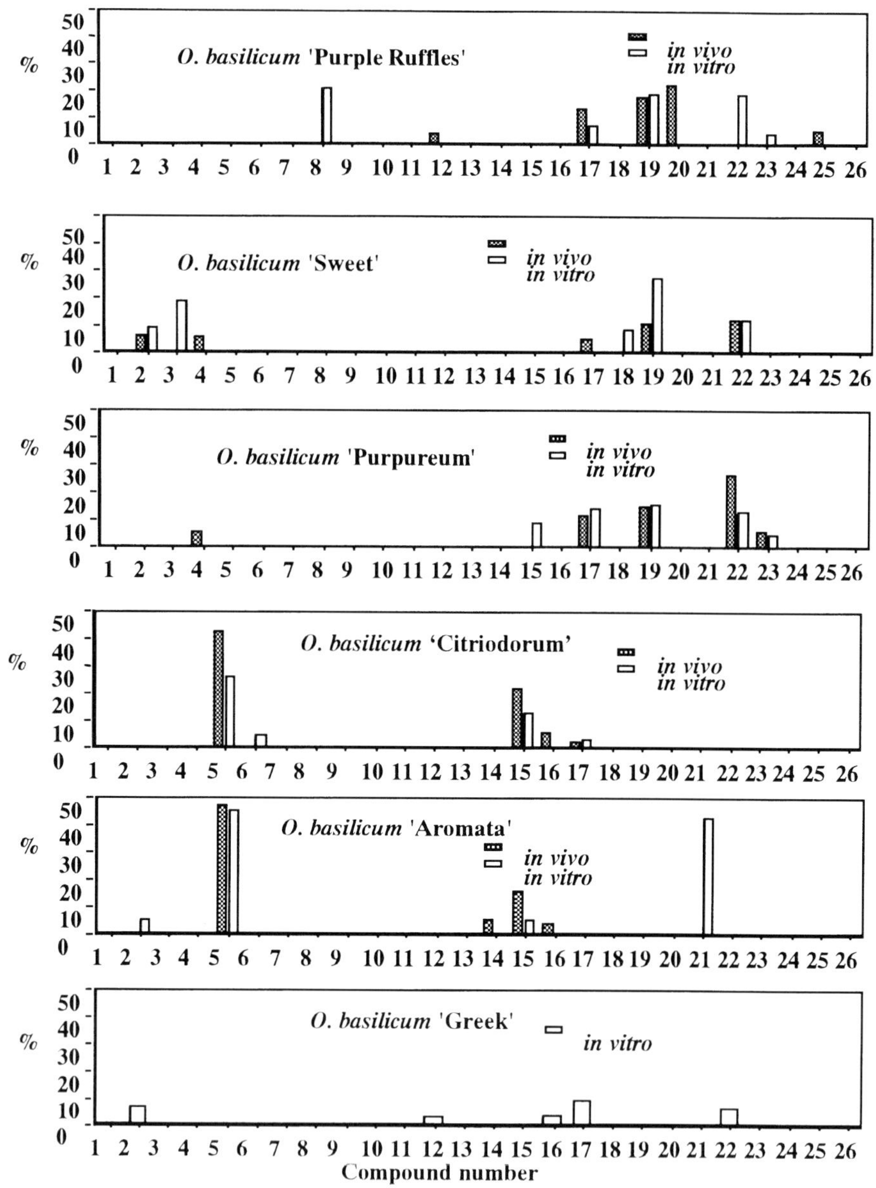

Fig. 12. Composition of volatile compounds (*1—26*) in *Ocimum* varieties. (Ishimaru et al., unpubl.)

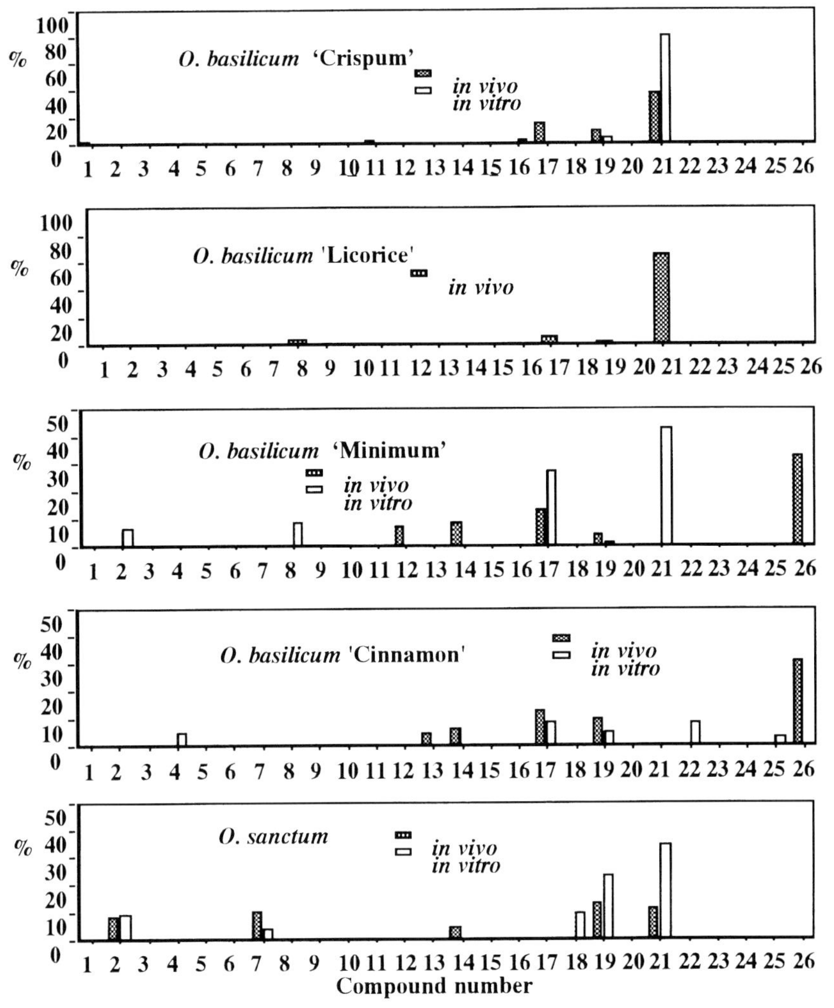

Fig. 12. *Continued*

5 Conclusion and Prospects

Many *Ocimum* varieties are cultivated and used as ingredients for food, cosmetic and medicinal purposes as well as horticultural ornamentals. *Ocimum* plants produce antioxidative, antibacterial and antivirus phenolic constituents such as rosmarinic acid (RA) and other caffeic acid derivatives (lithospermic acid and lithospermic acid B).

Various tissue cultures (shoot, callus, suspension cell, adventitious root and transformed root) of *Ocimum* varieties, as well as in vivo (field-grown) plants, were rich in phenolics, particularly of RA. For example, RA content (21.9%)

observed in transformed roots of *O. sanctum* induced by the infection with *A. rhizogenes* ATCC 15834 (OsA-2) was the highest among all in vivo and in vitro materials investigated so far. Various clones of *Ocimum*-transformed roots showed variations in root growth and RA production. This indicated the importance of the selection of the clones of transformed roots (together with *A. rhizogenes* strains used for the root induction) as well as the optimal culture conditions (medium, culture period etc.) for high production of useful secondary metabolites (RA, LA and LAB). *Ocimum* varieties seem to have a good potential to become one of the most suitable resources for RA supply.

Volatile components in *Ocimum* varieties were also analyzed by SPME-GC/MS method. The composition of volatiles in in vitro shoot cultures was not identical to those of the respective in vivo plant. Some volatile compounds which were not produced under in vivo conditions were observed as major components in the tissue cultures. This observation might show the possibility of *Ocimum* tissue cultures becoming additional material for the supply of a new type of fragrance which has never appeared in nature.

6 Protocol

6.1 Shoot Culture

Seeds of *Ocimum* varieties were surface-sterilized with 2% NaOCl containing Tween 20 for 8 min and germinated aseptically on MS solid medium (solidified with 2.5 g/l gelrite). The axenic plantlets were subcultured on the same medium in the light (16-h photoperiod/day, ca 3000 lux).

6.2 Callus Culture

Leaf segments of the in vitro plants were placed on MS or B5 medium supplemented with various phytohormones (Table 2) and cultured in the dark. After 2 months of culture, callus derived on the segments was cut off, transferred onto the same medium and subcultured in the dark or light.

6.3 Adventitious Root Culture

The roots of the axenic plants in vitro were cut off and transferred into MS liquid medium (50 ml in 100-ml flask) containing 1 mg/l NAA and maintained in the dark on a rotary shaker.

6.4 Transformed Root Culture

A. rhizogenes strains subcultured on YEB agar medium (Vervliet et al. 1975) were inoculated by a needle onto the cut ends of the stems (ca. 1 cm length) prepared from the axenic shoot cultures of *Ocimum* varieties. About 20 days after infection, several transformed roots appeared at the inoculated sites after growing under the light. The roots were cut off and placed on hormone-free MS solid medium containing the antibiotic Claforan (0.5 g/l) and incubated for 1 week at 25 °C in the dark to eliminate the bacteria. Various clones which showed good growth were selected and used for the experiment. The transformation of the hairy roots was proved by the detection of opine using paper electrophoresis (Petit et al. 1983; Tanaka 1990). Clones (ca. 0.2 g fresh wt.) were inoculated into hormone-free liquid media (MS, 1/2 MS, B5 or WP, 50 ml/100-ml flask) and cultured on a rotary shaker (100 rpm) in the dark or light.

6.5 HPLC Analysis

Phenolics (RA, LA and LAB) were analyzed by HPLC. Lyophilized samples (ca. 20 mg) were mashed and extracted with MeOH (2 ml) for 16 h at room temperature. Each extract, after filtration through a Millipore filter (0.5 µm), was subjected to HPLC analysis; column: CAPCELL PAK C18 AG 120 (4.6 × 250 mm), mobile phase; 1 mM tetra-butylammonium (adjusted to pH 2.8 with AcOH)-CH_3CN (4:1 → 1:4, in 25 min), flow rate: 0.65 ml/min, column temperature: 40 °C, detection: 330 nm, Rt (min): RA (15.5), LA (20.7) and LAB (22.5).

6.6 SPME – GC/MS Analysis

Samples (0.3 g fresh wt.) were transferred into a glass vial tube (70 ml volume), which was used for the analysis of SPME (Hisano et al. 1995). The period for the adsorption of volatile compounds was 5 min (at 30 °C). The volatile compounds adsorbed were detected by GC [Hewlett-Packard (HP) 5980 II]/MS (HP 5971 A).

All culture media, MS (containing 30 g/l sucrose), 1/2 MS (containing 30 g/l sucrose), B5 (containing 20 g/l sucrose) and WP (containing 20 g/l sucrose) were adjusted to pH 5.7 before autoclaving at 121 °C for 15 min. All cultures were placed at 25 °C.

Acknowledgements. This work was supported in part by the Ministry of Health and Welfare, Science Research Fund Subsidy granted to the Japan Health Science Foundation.

References

Dalton CC (1983) Chlorophyll production in fed-batch cultures of *Ocimum basilicum* (sweet basil). Plant Sci Lett 32:263–270

Dalton CC (1984) The effect of sugar supply rate on photosynthetic development of *Ocimum basilicum* (sweet basil) cells in continuous culture. J Exp Bot 35:505–516

Everett TH (1981) OCIMUM, the New York botanical garden illustrated encyclopedia of horticulture, vol 7. Garland, New York, 361 pp

Gamborg OL, Miller RA, Ojima K (1968) Nutrient requirements of suspension cultures of soybean root cells. Exp Cell Res 50:151–158

Gerhardt U, Schroter A (1983) Rosmarinsäure–ein natürlich vorkommendes Antioxidans in Gewürzen. Fleischwirtschaft 63:1628–1630

Hikawa M (1998) Basil, science of herb. Yokendo Ltd, Tokyo, pp 164–167

Hisano H, Ishimaru K, Tada H, Ikeda Y (1995) Flavors of heliotrope flowers, analyzed by solid phase microextraction method. Jpn J Food Chem 2:6–8

Kelley CJ, Mahajan JR, Brooks LC, Neubert LA, Breneman WR, Carmack M (1975) Polyphenolic acids of *Lithospermum ruderale* Dougl. ex Lehm. (Boraginaceae). 1. Isolation and structure determination of lithospermic acid. J Org Chem 40:1804–1815

Lee Kuo-H, Kashiwada Y, Nonaka G, Nishioka I, Nishizawa M, Yamagishi T, Bodner AJ, Kikuskie RE, Cheng Yung-C (1992) Tannins and related compounds as anti-HIV agents. In: Chu CK, Cutler HG (eds) Natural products as antiviral agents. Plenum Press, New York, pp 69–90

Lloyd G, McCown B (1980) Commercially-feasible micropropagation of mountain laurel, *Kalmia latifolia*, by use of shoot-tip culture. Int Plant Prop Soc 30:421–427

Matsubara S, Ino M, Murakami K, Kamada M, Ishihara I (1996) Callus formation and plant regeneration of herbs in Perilla family. Sci Rep Fac Agric Okayama Univ 85:23–30

Murashige T, Skoog F (1962) A revised medium for rapid growth and bioassays with tobacco tissue cultures. Physiol Plant 15:473–497

Omoto T, Murakami Y, Shimomura K, Yoshihira K, Mori K, Nakashima T, Tanaka M, Ishimaru K (1997) Caffeic acid derivatives in Lamiaceae and Boraginaceae plants. Jpn J Food Chem 4:11–16

Petit A, David C, Dahl GA, Ellis JG, Guyon P, Casse-Delbart F, Tempe J (1983) Further extension of the opine concept: plasmids in *Agrobacterium rhizogenes* cooperate for opine degradation. Mol Gen Genet 190:204–214

Reschke A (1983) Capillargaschromatographische Bestimmung der rosmarinsäure in Blattgewürzen. Z Lebensm Unters Forsch 176:116–119

Tada H, Murakami Y, Omoto T, Shimomura K, Ishimaru K (1996a) Rosmarinic acid and related phenolics in adventitious root cultures of *Ocimum basilicum* L. Plant Tissue Cult Lett 13:69–71

Tada H, Ikeda Y, Omoto T, Shimomura K, Ishimaru K (1996b) Rosmarinic acid and related phenolics in hairy root cultures of *Ocimum basilicum*, Phytochemistry 42:431–434

Tanaka N (1990) Detection of opines by paper electrophoresis. Plant Tissue Cult Lett 7:45–47

Tanaka T, Morimoto S, Nonaka G, Nishioka I, Yokozawa T, Chung H Y, Oura H (1989) Magnesium and ammonium-potassium lithospermates B, the active principles having a uremia-preventive effect from *Salvia miltiorrhiza*. Chem Pharm Bull 37:340–344

Vervliet G, Holsters M, Teuchy H, Van Montagu M, Schell J (1975) Characterization of different plaque-forming and defective temperate phages in *Agrobacterium* strains. J Gen Virol 26:33–48

Yokozawa T, Chung HY, Lee TW, Oura H, Tanaka T, Nonaka G, Nishioka I (1989) Effect of magnesium lithospermate B on urinary excretion of arachidonate metabolites in rats with renal failure. Chem Pharm Bull 37:2766–2769

Yokozawa T, Lee TW, Chung HY, Oura H, Nonaka G, Nishioka I (1990a) Effect of magnesium lithospermate B on urinary prostaglandins in rats with renal failure. J Nat Prod 53:662–665

Yokozawa T, Lee TW, Chung HY, Oura H, Nonaka G, Nishioka I (1990b) Renal responses to magnesium lithospermate B. J Pharm Pharmacol 42:712–715

10 *Pergularia tomentosa* in Vitro Culture and the Production of Cardenolides and Other Secondary Metabolites

H.M. ELHAG, M.M. EL-OLEMY, and M.S. AL-SAID[1]

1 Introduction

1.1 Distribution, Botany, and Morphology

Pergularia tomentosa L. (*Daemia cordata* Forssk, R. Br. ex Schult) Asclepiadaceae, known locally as “Ghalaka” (Ibn El Bitar 1809; Migahid 1989), is a plant indigenous to Saudi Arabia, Egypt, Abyssinia, North Africa, Iran, Afghanistan, West Pakistan, northern and central Sudan (Engler and Prantl 1891; Maire 1933; Migahid 1989). The plant is one of the abundant plants in Saudi Arabia, where it is found on rocky parts of deserts in various regions.

The vegetative part of this plant is toxic for grazing animals, although it is used by the natives as an antirheumatic and in the treatment of some skin diseases (Vincett 1977). Moreover, many other uses of this plant have also been reported: as a poultice, depilatory, laxative, abortive, anthelmintic, expectorant, for the treatment of asthma and bronchitis, to promote lactation of cows and, when mixed with salt to remove hairs from infected skin areas (Ibn El bitar 1809, El-Antaki 1923; Dalziel 1937; El-Idrissi and Debelmas 1970).

Both earlier and more recent investigations have revealed the important morphological features that may assist in the identification of *P. tomentosa*. The plant is a twining grey, canescent undershrub, with the young branches twining around the old ones. Leaves are rather small, 1.5–3.0 cm long, cordate, petiolate and opposite (Fig. 1). The inflorescence are axillary umbels on peduncles longer than the leaves. The flowers are whitish-yellow with a pedicel much longer than the flowers. Calyx five-parted brownish hirsute; corolla 8 mm broad, with a tube little longer than the calyx and five-parted limb with ciliate margin. The fruit is a pair of spiny follicles, each 3–6 cm long, inflated below, attenuated into a long, hooked beak. The seed is dark brown in colour, ovate, 0.3–1.0 cm long, truncate at the apex, winged and densely velvety pubescent on both sides. The root system consists of a main woody tap root with smaller rootlets and branches (Al-Said et al. 1988a; Migahid 1989).

[1]Department of Pharmacognosy, College of Pharmacy, King Saud University, P.O. Box 2457, Riyadh 11451, Saudi Arabia

Biotechnology in Agriculture and Forestry, Vol. 51
Medicinal and Aromatic Plants XII (ed. by T. Nagata and Y. Ebizuka)
© Springer-Verlag Berlin Heidelberg 2002

Fig. 1. *P. tomentosa* plants grown in the medicinal plant garden of the College of Pharmacy, Riyadh

1.2 Chemical Nature of *P. tomentosa* Constituents

P. tomentosa was extensively studied primarily regarding its cardenolide constituents. Mahran et al. (1967) reported the presence of uzarigenin glycoside, choline and β-sitosterol, whereas El-Idrissi and Debelmas (1970) identified coroglaucigenin in the plant. Al-Said and Hifnawy (1985) reported on the phytochemical and biological screening of *P. tomentosa* growing in Saudi Arabia and indicated the presence of cardenolides in the roots, but not in the aerial parts. Al-Said et al. (1988b) reported the isolation and identification of calactin (**1**) and the new cytotoxic cardiac glycoside ghalakinoside (**2**) from the roots. Hifnawy et al. (1990) identified a new cardiac glycoside, named pergularotoside (**3**),from the roots of the title plant. Gohar et al. (2000), working with aerial parts, reported the isolation of uzarigenin (**4**), coroglaucigenin (**5**)and desglucouzarin (**6**), in addition to β-sitosterol glucoside and the three main glycosides calactin (**1**), ghalakinoside (**2**), and pergularotoside (**3**) previously reported in the roots. (Structure 1).

1.3 Biological Effects of *P. tomentosa*

Preliminary pharmacological screening was undertaken on the aqueous decoction of the root and the chloroform and methanol extracts of the leaves

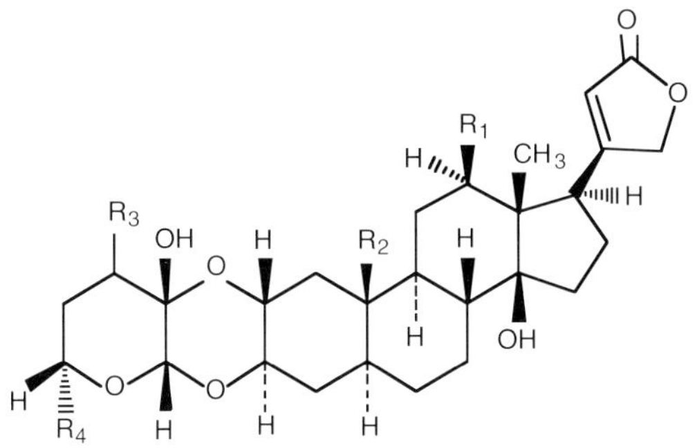

	R_1	R_2	R_3	R_4
Calactin (**1**)	H	CHO	α-H, β-OH	CH_3
Ghalakinoside (**2**)	OH	CH_2OH	α-OH, β-H	CH_2OH
Pergularotoside (**3**)	OH	CHO	α-OH, β-H	CH_2OH

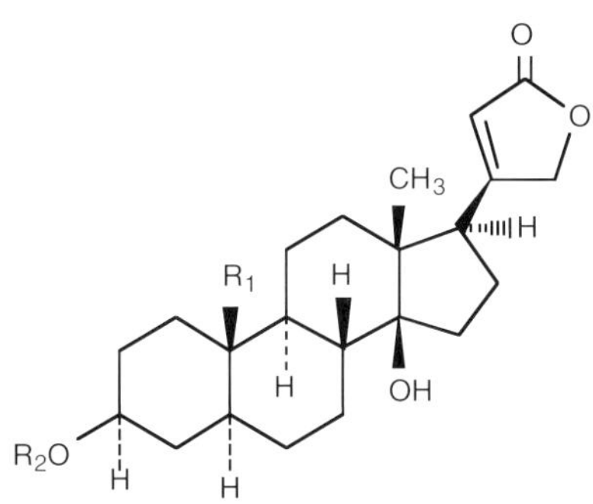

	R_1	R_2
Uzarigenin (**4**)	CH_3	H
Coroglaucigenin (**5**)	CH_2OH	H
Desglucouzarin (**6**)	CH_3	Glucose

Structure 1.

(Al-Said and Hifnawy 1985). The root extract produced a significant positive ionotropic and chronotropic effect on the isolated rabbit heart as well as a significant rise in the blood pressure of anaesthetized rabbits. It also caused contractions of the smooth muscle of guinea-pig ileum, an effect which was not blocked by atropine or antihistamines, suggesting spasmogenic activity (Al-Said and Hifnawy 1985). On the other hand, chloroform and methanolc extracts of the leaves failed to produce any significant effects on myocardium but did cause contraction of the smooth muscle of guinea-pig ileum, which was blocked by atropine, suggesting a parasympathomimetic effect.

Al-Yahya et al. (1985) reported that the chloroform extract of the aerial parts of this plant possessed significant cytotoxic activity. Moreover, one of the main cardenolide constituents of *P. tomentosa*, ghalakinoside, was shown to exhibit cytotoxic activity (ED_{50} = 4 μg/ml) against 9-KB cells (Al-Said et al. 1988b). Several biochemical effects of this glycoside were also studied on certain blood parameters, especially those related to cancer diseases (Al-Said et al. 1989). Sprague-Dawley male rats (200–250 g) were used. Ghalakinoside was found to cause a transient insignificant increase in plasma uric acid with single low doses (1–5 mg/kg) and a significant decrease with single larger doses (20 mg/kg). There was a significant decrease in serum calcium (Ca^{2+}) at 10 mg/kg glycoside. Serum Ca^{2+} is elevated in certain cancers with osteolytic activity (Barker 1976; Varley et al. 1980). γ -Glutamyl transferase (GGT) was found very sensitive to ghalakinoside, causing a noticeable decrease at 1 mg/kg, with a very strong inhibition at higher doses. This suggests that ghalakinoside causes a selective decrease in amino acid transport and uptake, resulting in an interference with de novo protein synthesis (Wunner et al. 1966). An increase in GGT activity has been reported in certain neoplastic diseases (Barker 1976; Varley et al. 1980; Tietz 1982). Ghalakinoside also showed significant inhibitory effects on the level of serum alkaline phosphatase (ALP) and lactic dehydrogenase (LDH) at doses of 10 mg/kg. These enzymes also show moderate to significant elevations in several types of cancer diseases (Varley et al. 1980; Tietz 1982). On the other hand, ghalakinoside was found to increase the level of serum transaminases (SGOT and SGPT) at a dose of 10 mg/kg. Ghalakinoside showed no effect on serum cholesterol, glucose, total proteins and albumin levels of the experimental animals (Al-Said et al. 1989).

In conclusion, the inhibitory effects of a single dose of ghalakinoside on levels of uric acid, Ca^{2+}, GGT, ALP and LDH are in favor of its antitumor potential, since their levels are increased in a variety of cancerous states. Chronic dosing of ghalakinoside could provide more significant responses to these parameters. It should be noted that the family Asclepiadaceae is known to contain cytotoxic cardenolides of the structural type related to ghalakinoside, calactin and pergularotoside of *P. tomentosa*.

2 In Vitro Culture Studies

To the best of our knowledge, in vitro culture studies have not been carried out on this particular plant or on other members of the family Asclepiadaceae. However, there are many reports on the tissue culture studies of other cardenolide-producing plant species, e.g. *Digitalis Lanata* (Ohlsson et al. 1983; Luckner and Dietrich 1987), *Digitalis obscura* (Gavidia and Perez-Bermudez 1997) and *Strophanthus divaricatus* (Kawaguchi et al. 1993).

The root of *P. tomentosa* is the main organ where the highest concentration of cardenolides has been found. Therefore it is the organ of choice for culture studies. However, roots from field-grown plants are difficult to manipulate in vitro, and at best would be slow growing. Therefore, we resorted to explants from sterile axenic germinated seedlings, where we attempted callus initiation from the roots, stems and leaves.

2.1 Establishment of Tissue Cultures

2.1.1 Callus Induction from Axenic Seelings

Seeds were surface-sterilized and grown on Heller-support filter paper dipped in 1/2 MS liquid medium. Axenic seedlings usually emerge after 1 week. One-month-old axenic seedlings were used for callus initiation using leaf, stem and root explants. The explants were cultured on MSB5 basal agar solidified medium inorganic salts [Murashige and Skoog (1962) in combination with Gamborg B5 vitamins (Gamborg et al. 1968)] supplemented with 0.1 mg/l kinetin (K) and auxins 2,4-D (2,4-dichlorophenoxyacetic acid) or NAA (α-naphthaleneacetic acid) at a concentration of 1.0 or 3.0 mg/l. The root explants produced more callus in treatments with 3.0 mg/l 2,4-D and NAA than leaf and stem explants. Callus induced from leaf, stem and root explants differentiated into roots on the medium containing 1.0 or 3.0 mg/l NAA. Such roots emerged from the peripheries of the callus clumps and resembled hairy roots in appearance. None of the callus from other treatments differentiated into roots. Upon transfer to MS liquid medium devoid of NAA, the roots developed and continued to grow. The growth and other characteristics of these axenic roots were studied in comparison to the hairy roots.

2.1.2 Initiation of Suspension Cultures

P. tomentosa suspension cultures could easily be initiated by transferring pieces of friable loose callus into MS or MSB5 liquid medium supplemented with 1.0 mg/l 2,4-D and 0.1 mg/l kinetin (1.0 g callus/100 ml liquid medium). A fine suspension was obtained by the second subculture. Growth measurements

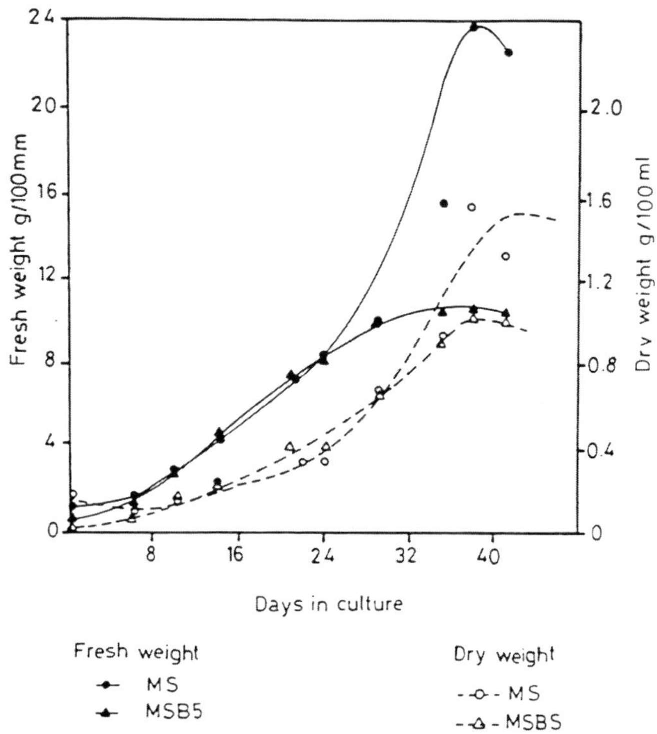

Fig. 2. Growth of hormonal autotrophic *P. tomentosa* suspension cultures on MS and MSB5 liquid media

indicated that the cultures attained maximal growth in 3 weeks, after which the cell suspension went into the stationary phase.

2.1.3 *Hormonal Autotrophic* P. tomentosa *Suspension Cultures*

When root callus was transferred to MS or MSB5 liquid medium devoid of growth regulators, it formed a cell suspension and continued to grow, although with a longer lag phase and an extended growth cycle. However, by the third generation, the growth of this hormonal autotrophic culture was faster, achieving maximal growth in 4 weeks. A comparison of growth on MS and MSB5 liquid medium proved that MS medium was a better medium for the growth of this hormonal autotrophic culture than MSB5 medium (Fig. 2).

The extracts of callus as well as suspension cultures (medium and cells) of *P. tomentosa* were tested for cardenolides.

2.1.4 *Biotransformation Studies with* P. tomentosa *Suspension Cultures*

Hormonal autotrophic *P. tomentosa* cell suspension cultures grown on basal MSB5 liquid medium, were able to biotransform the cardenolide aglycone digitoxigenin (El-Olemy et al. 1994) into the 5β-hydroxy derivative (Major) as well as the 1β-hydroxy derivative and the 5-epimer, uzarigenin (**4**) (Scheme 1).

Other biotransformation experiments performed with *P. tomentosa* cultures included a number of cardenolides: gitoxigenin, digoxigenin, digitoxin, gitoxin and digoxin, the latter three compounds being glycosides. *P. tomentosa* is known to contain a type of cytotoxic cardiac glycosides, common to the Asclepiadaceae (Ohlsson et al. 1983; Al-Said et al. 1988b; Hifnawy et al. 1990; Kawaguchi et al. 1993; Gavidia and Perez- Bermudez 1997; Gohar et al. 2000). The results indicated no or minor transformation for digoxigenin, digoxin and gitoxin, but the highest biotransformation was observed with gitoxigenin, followed by digitoxin (Table 1). Three main compounds, as evidenced by thin layer chromatography (TLC), were obtained as biotransformation products of gitoxigenin and two as conversion products of digitoxin. It is expected that transformation of gitoxigenin might be related to that reported for digitoxigenin, i.e. 5β-hydroxylation, 1β-hydroxylation and 5-epimerization (El-Olemy et al. 1994).

2.1.5 *Formation of Extracellular Polysaccharides (ECP) in* P. tomentosa *Suspension Cultures*

It was observed that *P. tomentosa* suspension cultures release an extracellular material that is responsible for the high viscosity of the medium. The ECP of

Table 1. TLC analysis of the biotransformation of gitoxigenin and digitoxin by *P. tomentosa* and *A. visnaga* cell suspension cultures

Compound	R_f values of biotransformation products[a]	
	P. tomentosa	*A. visnaga*
Gitoxigenin ($R_f = 0.52$)	0.06	0.06
	0.44 (major)	0.44
	–	0.50 (major)
	0.75	–
Digitoxin ($R_f = 0.19$)	–	0.22
	0.31	0.31 (major)
	–	0.33
	0.44 (major)	0.44

[a] TLC system: silica gel GF_{254}/EtOAc-MeOH (10:1); detection with Kedde's reagent.

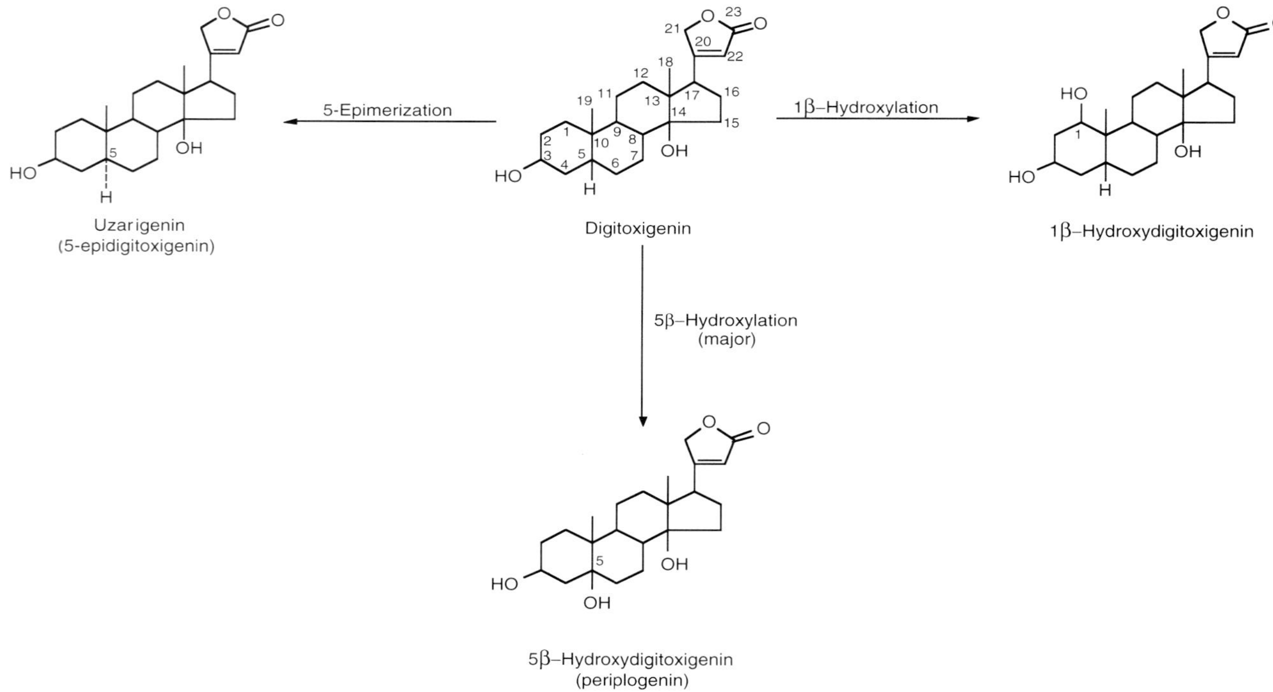

Scheme 1. Biotransformation of digitoxigenin by *P. tomentosa* and *A. visnaga* cell suspensions. (El-Olemy et al. 1994)

the medium of *P. tomentosa* suspension cultures were subjected to acid hydrolysis. HPLC analysis of the hydrolytic products of the ECP showed primarily two neutral sugars: galactose (t_R 6.6 min., major) and xylose (t_R 7.4 min, minor), together with a third unknown peak. Thus the ECP of *P. tomentosa* suspension culture is primarily a xylogalactan. Further work is needed to determine the nature of the third minor hydrolytic product and type of linkages between monosaccharide components.

There are a few reports on the release of extracellular polysaccharides from cell suspension cultures. These include tuberose (*Polianthus tuberosa)* and *Nicotiana plumbaginifolia* cultures (Sims and Bacic 1995; Honda et al. 1996). The ECP of tuberose was proven to be composed of a glucuronomannan backbone, with branches of arabinose and galactose(Honda et al. 1996).

It has been reported that the ECP production by plant tissue cultures is linked to the degree of dedifferentiation. Thus, 2,4-D was found to increase ECP production in *P. tuberosa* suspension cultures, as compared to NAA, due to the higher degree of dedifferentiation caused by 2,4-D (Honda et al 1996). However, *P. tomentosa* suspension cultures produced ECP on 2,4-D or NAA-containing medium as well as on the medium devoid of growth regulators. This suggests that cultures of *P. tomentosa* are capable of synthesizing these auxins de novo.

2.2 Establishment of Transformed Cultures

2.2.1 Induction of Hairy Roots in P. tomentosa *Sterile Seedlings*

Induction of hairy roots of *P. tomentosa* was attempted by infection of sterile seedlings (by injuring the seedling and inoculation of the injured sections) with one of three different strains of *Agrobacterium rhizogenes* (A4, 15834 and 01724). Leaf discs and stem sections were also inoculated with *A. rhizogenes.* Two weeks after inoculation, hairy roots started to form only on the stems of the injured seedlings, and solely with one of the *A. rhizogenes* strains (strain 15834). No hairy root formation was observed with the other two strains. Leaf and stem sections did not form hairy roots, even with the strain 15834. The hairy roots were separated and grown on MS liquid basal medium containing the antibiotic cefotaxime at a concentration of 1.5 g/l to help eliminate the *Agrobacterium.*

Transfer of the hairy roots onto liquid MS medium containing the antibiotic was necessary for several generations. By the third subculture, the antibiotic concentration was reduced to 0.75 g/l and the hairy roots were grown on both liquid or solid MS medium. By the eighth subculture, the bacteria were completely eliminated.

Determination of the growth index of *P. tomentosa* hairy roots using three different initial cell densities (Table 2) proved that at a lower initial cell density the growth index reached almost 98 times the initial fresh weight in 4 weeks.

Table 2. Growth of hairy roots of *Pergularia tomentosa* on antibiotic-free liquid MS medium after 4 weeks

Initial fresh weight (g)	Mean final fresh weight (g)	Mean dry weight	G.I.[a]
0.2 g/50 ml	19.78 ± 2.06	0.82 ± 0.06	97.80
1.0 g/100 ml	43.36 ± 4.26	1.49 ± 0.09	42.40
2.0/100 ml	41.44 ± 3.88	1.52 ± 0.06	19.72

[a] Growth Index

$$\left(\text{G.I.} = \frac{\text{Final Fresh Weight} - \text{Initial Fresh Weight}}{\text{Initial Fresh Weight}} \right).$$

Table 3. Effect of GA_3 on growth of hairy roots of *P. tomentosa* on MS or B5 liquid medium

Medium	GA_3 (mg/l)	Mean fresh weight[a]	Mean dry weight[b]
MS	0	22.93	0.97
	0.01	10.66	0.98
	0.10	5.50	0.68
	1.0	3.26	0.48
B5	0	5.41	0.78
	0.01	5.05	0.84
	0.1	3.10	0.78
	1.0	6.21	0.90

[a] The initial weight of inoculum was 0.5 g/50 ml; mean of three determinations.
[b] Dry weights were pooled for each triplicate.

At higher initial cell density (1.0 g/100 ml or 2.0 g/100 ml), the growth index was much lower, indicating an overload of the capacity of the liquid culture. The growth was also compared on MS, B5 and MSB5 liquid basal media. MSB5 medium resulted in the highest growth (data not shown). The effect of GA_3 on growth and cardenolides was studied on MS or B5 medium, each in combination with GA_3 at (0.0, 0.01, 0.1 and 1.0 mg/l). GA_3 resulted in a reduction of the growth of hairy roots with increasing concentrations of GA_3 in MS or B5 medium (Table 3). The hairy roots grew much more vigorously on MS as compared to B5 medium; moreover, the cultures acquired a reddish-yellow colour on the latter. The difference in growth on MS and B5 medium could be attributed to the difference in their nitrogen composition, i.e. the NH_4:NO_3 ratio.

2.2.2 *Development of Transgenic Callus of* P. tomentosa

When *P. tomentosa* hairy roots were cultured on hormone-free solid MS or MSB5 medium, callus formation was observed. The proliferation of roots and callus continued in a recurrent fashion; when the callus was separated and cultured, it also produced hairy roots. This callus was thus termed transgenic callus. Upon transfer of this transgenic callus culture to the light, it developed pockets of an anthocyanin-like purple pigment. This pigment production was not restricted to the transgenic callus, but was also produced by the hairy roots when grown in the light.

2.2.3 Growth and Purple Pigment Production in Transgenic Callus of P. tomentosa

When grown on media containing the auxins 2,4-D or NAA (1.0–3.0 mg/l), the transgenic callus continued to grow, especially on the NAA-containing medium. The intensity of the pigmented cells also increased with the medium content of NAA (Table 4), while growth was minimal on 2,4-D containing medium, in contrast to the normal callus. The transgenic callus continued to produce the purple pigment for several subcultures only in NAA-containing medium in the light (Fig. 3).

Attempts to select the pigmented cells and to grow them on a similar medium in the light did not improve their growth or pigment production. Instead, no growth occurred and the pigmented cells eventually died. It thus appears, for *P. tomentosa* transgenic callus, that once the signal for anthocyanin synthesis is turned on, the signal for cell division and growth is turned off; only the non-pigmented cells are capable of carrying out the cell division process. This is different from what has been reported for *Aralia cordata* (Sakamoto et al. 1996) and *Glehnia littoralis* (Miura et al. 1998), where it was possible to select the pigmented cells to obtain high anthocyanin-producing cell lines.

For isolation of the purple pigment, the pigmented transgenic callus or hairy roots were harvested at the end of each subculture. The non-pigmented cells were subcultured on the appropriate medium (NAA, 1.0 or 3.0 mg/l) and incubated in the light at 24 °C for growth and further pigment production.

2.2.4 Induction of Crown Gall Tumors on P. tomentosa *Seedlings*

Stems of axenic seedlings (1 month old) were infected with *Agrobacterium tumefaciens* (strains 20305, 30200 and IAM 13575). The resulting greenish crown galls (Fig. 4) were separated from the stems and grown on agar solidified basal MSB5 medium devoid of growth regulators but containing the antibiotic cefotaxime. After eight subcultures, the bacteria were completely eliminated. The crown galls transformed into undifferentiated callus mass without producing shooty teratomas on the basal medium alone or on medium supplemented with growth regulators. The growth of the tissues was slow at the beginning on either media, but attained good growth by the tenth subculture on MSB5 basal medium. When grown in the light, the crown gall undifferentiated callus turned green.

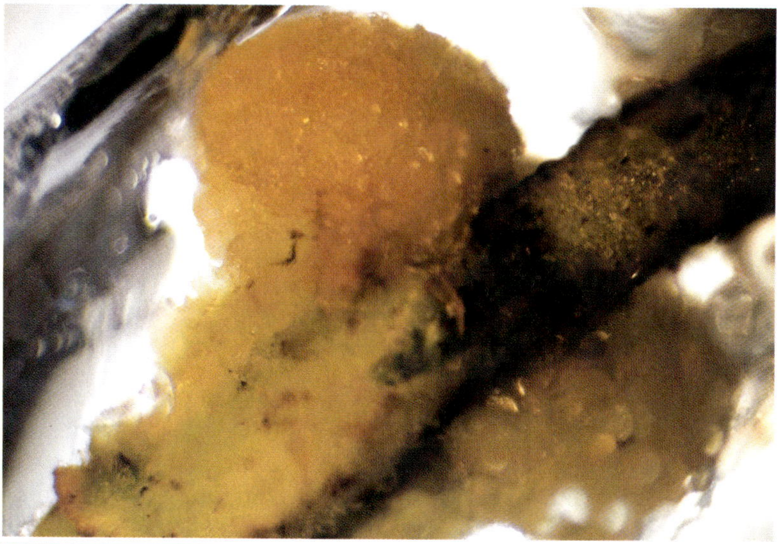

4a

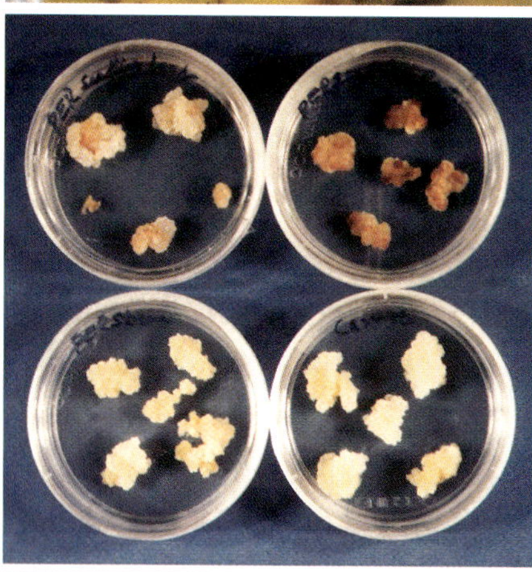

3a

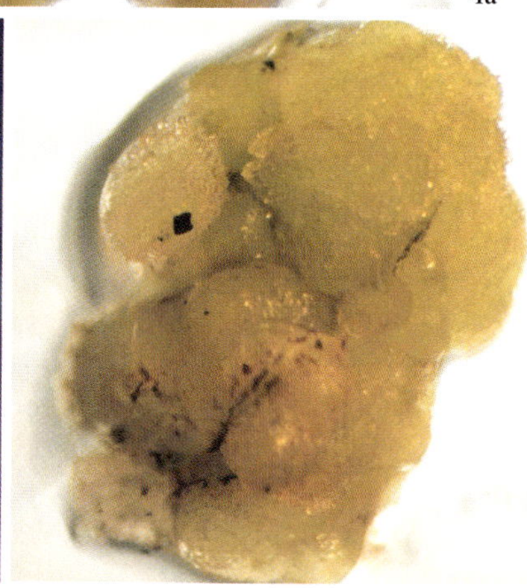

4b

3b

Fig. 3. *P. tomentosa* normal and transgenic callus showing purple pigment production: **a** Normal callus grown on NAA in the light (*top plates*) and in the dark (*lower plates*). **b** Transgenic callus with hairy roots grown (*left* to *right*) on NAA (3.0, 1.0) and 2,4-D (1.0 mg/l) in the light

Fig. 4. Crown galls of *P. tomentosa*: **a** Photomicrograph of *P. tomentosa* crown gall formed on the shoot. **b** Photomicrograph of *P. tomentosa* crown gall separated from the shoot and cultured on MS medium containing antibiotic

Table 4. Effect of 2,4-D and NAA on growth and purple pigment production in *Pergularia tomentosa* normal and transgenic callus

2,4-D or NAA conc. (mg/l)	Normal Callus				Transgenic Callus			
	Initial fresh[a] wt. (g)	Final fresh wt.[a] (g)	G.I.[b]	Pigment production[c]	Initial fresh wt.[a] (g)	Final fresh wt.[a] (g)	G.I.[b]	Pigment production
2,4-D								
1.0	2.44 ± 0.04	8.75 ± 2.8	2.6	–	2.48 ± 0.17	4.01 ± 0.3	0.62	–
3.0	2.65 ± 0.1	9.85 ± 5.4	2.7	–	nd	nd	nd	nd
NAA								
1.0	2.4 ± 0.1	8.22 ± 2.3	2.4	+	2.55 ± 0.1	12.96 ± 3.0	4.08	++
3.0	2.5 ± 0.13	6.58 ± 2.3	1.6	+	2.39 ± 0.11	14.80 ± 0.11	5.19	+++

[a] Mean ± SE.

[b] G.I., Growth Index = $\frac{\text{Final Fresh Weight} - \text{Initial Fresh Weight}}{\text{Initial Fresh Weight}}$.

The harvested tissues were subjected to extraction and analysis similar to the hairy roots and transgenic callus. Cardenolides were not detected in the crown gall tissues.

2.2.5 Detection of Opines in Hairy Roots, Transgenic Callus and Crown Gall Tissues of Pergularia tomentosa

The aqueous extracts of *P. tomentosa* hairy roots, transgenic callus and the undifferentiated crown gall tissues were tested for opines by paper chromatography against authentic agropine and mannopine according to the reported methods (Kamada et al. 1986; Elhag et al. 1996), in comparison with the aqueous extracts of axenic roots and normal callus. Opines were detected in the hairy roots, transgenic callus and crown gall tissues, but not in the normal tissues (axenic roots and normal callus). This confirms the transgenic nature of the hairy roots and the transgenic callus. More evidence of their transgenic nature is their ability to grow in the absence of exogenous growth regulators (Table 2), since hairy roots are reported to produce their own endogenous auxins and cytokinins (Biondi et al. 1995)

2.2.6 Extraction and Analysis of the Purple Pigment of P. tomentosa *Transgenic Callus*

The purple transgenic fresh callus tissues (100g), grown on solid MSB5 medium, supplemented with 1.0mg/l NAA and 0.1mg/l kinetin were macerated in 300ml MeOH containing 0.1% HC1 and kept at 4°C (in the refrigerator) for 24h. This extract was filtered and the cells were extracted with a further 100ml of the same acidified MeOH. The combined methanol extract was evaporated on a rotary evaporator at 35°C in vacuo. The intensely purple residue was dissolved in H_2O and extracted with ether to remove lipids and other impurities. The aqueous solution was evaporated under vacuum on a rotary evaporator, with the help of EtOH. The residue was then dissolved in a small volume of MeOH containing 0.1% trifluoroacetic acid (TFA) and chromatographed through a column of Sephadex LH-20, eluting with the same solvent and collecting the purple eluate as one fraction. This was concentrated on a rotary evaporator in vacuo and subjected to TLC on cellulose plates/butanol-HOAc-H_2O(4:1:5), paper chromatography using butanol-HOAc-H_2O(4:1:5) and HPLC analysis.

2.2.7 HPLC Separation of the Purple Pigment of P. tomentosa *Cultures*

1. Qualitative HPLC analysis: column: C-18 reversed phase analytical column 300 × 3.9mm; solvent system: isocratic using 0.1% TFA in 19% aqueous

acetonitrile (CH_3CN); flow rate: 3 ml/min; volume injected: 20 µl. Detector: UV at λ 520 nm.

2. Preparative HPLC separation: column: C-18 reversed phase preparative column (300 × 7.8 mm); solvent system: 0.1% TFA in 19% aqueous CH_3CN; flow rate: 6 ml/min; volume injected: 100–150 µl. Detector: UV at λ 520 nm.

TLC and paper chromatographic analysis showed several purple spots that turned greenish-blue upon exposure to NH_3 vapour. This colour change lends support to the nature of anthocyanins which are red in acid and turn bluish in alkali. HPLC analysis using reversed C-18 phase with the solvent system 19% acetonitrile in 0.1% aqueous TFA, and UV detection at 520 nm, corresponding to λ_{max} of anthocyanin, revealed the presence of several anthocyanin compounds.

Using preparative HPLC for the isolation of the main anthocyanin components, six fractions were collected (Table 5). Fractions II–VI were concentrated in vacuo using the rotary evaporator to isolate the individual compounds, while fraction I, containing three anthocyanins, would be fractionated further by preparative HPLC into individual components. Each would be checked for purity by HPLC and then subjected to identification and structural elucidation using various spectral methods.

It has been mentioned that the type of auxin affected the production of this purple pigment. NAA enhanced pigment formation, in contrast to 2,4-D which completely inhibited pigment formation (Table 4, Fig. 3). This is in accord with what has been reported for anthocyanin pigments of *Callistephus chinensis* (Rau and Forkmann 1986) and *Aralia cordata* (Sakamoto et al. 1993, 1996).

Anthocyanin pigments are important as suitable natural food colourings, which are non-toxic compared to synthetic food colourings (Sakamoto et al. 1996). The safety of such pigments as food colourings has been confirmed by the lack of mutagenicity as tested biologically by short-term systems for detecting carcinogens (Maron and Ames 1983). Moreover, they possess antioxidant properties and thus can act as anti-aging and cancer-preventing agents (Robak and Ceryglewski 1988).

Table 5. Preparative HPLC fractionation of the purple pigment extract of *P. tomentosa* transgenic callus

Fraction	Retention time (min)
I[a]	2.14, 2.30 and 2.74
II	3.23
III	3.97
IV	4.58
V	5.24
VI	6.66

[a] Mixture of three compounds a, b and c, respectively.

3 Cardenolide Production by *P. tomentosa* Callus and Suspension Cultures

3.1 Extraction of *P. tomentosa* Cultures for Cardenolides

The EtOAc extracts of media and the MeOH extracts of the tissues of *P. tomentosa* cultures (normal as well as hormonal autotrophic suspension cultures), were tested for the presence of cardenolides using TLC analysis on both normal phase (system I) and reversed phase silica (system II). Only a trace of one Kedde's positive spot (pink) was detected in the media and not in the cells (R_f 0.54 in system I and R_f 0.46 in system II). This compound did not correspond with any of the known cardenolides of *P. tomentosa*: ghalakinoside, pergularoside and calactin (Al-Said et al. 1989; Hifnawy et al. 1990) or the newly identified *P. tomentosa* cardenolides of the aerial parts, desglucouzarin, coroglaucigenin and uzarigenin (Gohar et al. 2000), as tested by both TLC systems.

In addition, the MeOH extracts of *P. tomentosa* hairy roots, transgenic callus, normal callus, axenic roots and crown gall tissues were also tested for cardenolides using TLC System I (Table 6). The results indicate that the transgenic callus/hairy root mixed culture grown on solid medium produces the two main glycosides of *P. tomentosa*, ghalakinoside and pergularotoside (based on TLC evidence), together with an unknown Kedde's positive high R_f compound. Supplementation of the solid medium with NAA not only increased the cardenolide production, but also, in addition, several other cardenolides of *P. tomentosa* were detectable by TLC, namely desglucouzarin and calactin, in addition to another unknown high R_f Kedde's positive compound. The axenic roots showed the high R_f (less polar) unknown cardenolides, which were also common components of the hairy roots. However, axenic roots did not produce the main cardiac glycosides of *P. tomentosa*. Crown gall tissues, contrary to what was expected, did not produce any cardenolides, as has been previously mentioned. This could be attributed to their undifferentiated nature, since cardenolide production has been shown to be linked to some form of differentiation.

The production of cardenolides in organized tissues, i.e. hairy roots, in contrast to undifferentiated normal callus or crown gall tissues, is consistent with previous reports with cardenolide-producing plants. It has been shown that higher cardenolide synthesis in *Digitalis lanata* was associated with some form of organogenesis or differentiation (Luckner and Dietrich 1987). Since solid callus cultures, under various conditions, are not suitable for mass production, the transgenic callus/hairy roots culture in liquid medium might be the answer. It has been reported that *Datura* hairy roots grown in liquid cultures provided a suitable system for the production of hyoscyamine (Rhodes et al. 1987).

Table 6. TLC analysis of the ethyl acetate extract of *P. tomentosa* cultures for cardenolides[a]

No.	Culture type	R_f values	Corresponding reference compounds	R_f values
1.	Transgenic callus/hairy roots (solid medium + NAA)	0.045	Ghalakinoside	0.045
		0.12	Pergularotoside	0.12
		0.18	Desglucouzarin	0.18
		0.68	Calactin	0.69
		0.82	Unknown	
		0.92	Unknown	
2.	Transgenic callus/hairy roots (solid MS basal medium)	0.045	Ghalakinoside	0.045
		0.12	Pergularotoside	
		0.92	Unknown	0.12
3.	Axenic roots	0.82	Unknown	
		0.92	Unknown	
4.	Crown gall undifferentiated tissues	No Kedde's positive spots		
5.	Normal callus cultures [medium + 2,4-D (1.0 mg/l) and K (0.1 mg/l)]	No Kedde's positive spots		
6.	Hormonal autotrophic cell suspension cultures			
	Tissues	No Kedde's positive spots	Unknown	
	Medium	0.54	Unknown	

[a] TLC system: silica gel/$CHCl_3$-MeOH (9:1).

Alternatively, it has been mentioned that only a trace of a Kedde's positive compound (i.e. a cardenolide) was detected in the medium of both normal and hormonal autotrophic cell suspension cultures. This compound is releasable in the medium, probably indicating its high polarity. The mere detection of this compound in the cultures, though an unknown compound, points out that such cell suspension cultures are capable of synthesizing the cardenolide ring system. This opens the way for the use of various elicitors for the induction of cardenolide synthesis (Fujita and Tabata 1987; Hagendoorn et al. 1996), as well as manipulation of the medium regarding sucrose content, macro and microelements, pH, and various growth regulators (auxins, cytokinins and gibberellic acid), with the aim of optimizing conditions for cardenolide production. Differentiation, organogenesis or embryogenesis would also be attempted, since higher cardenolide synthesis in *Digitalis lanata* has been shown to be associated with some form of differentiation (Luckner and Dietrich 1987).

4 Summary and Conclusion

Callus and suspension cultures of *P. tomentosa* were induced from various explants (leaf, stem, root) of sterile axenic seedlings. The callus grows well on MSB5 medium containing 1.0 mg/l 2,4-D and 0.1 mg/l kinetin in the dark. Fine suspension cultures were established in liquid medium of the same composition or in medium devoid of growth regulators for hormonal autotrophic cultures. Suspension cultures of root callus origin released a highly viscous compound into the medium. HPLC analysis of the hydrolytic products of this extracellular polysaccharide (ECP) indicated that it is a xylogalactan (i.e. primarily composed of galactose and xylose).

Transformed hairy roots and crown galls were induced on stems of axenic seedlings with *A. rhizogenes* and *A. tumefaciens* respectively. Transformation was confirmed by detection of opines in comparison with normal tissues. The hairy roots spontaneously produced transgenic callus on basal MS medium. The production of hairy roots and callus continued in a recurrent fashion. When the callus was transferred to the light, it developed a purple pigment on the periphery of the callus clumps. The pigment was intensified in medium containing NAA (1.0 or 3.0 mg/l). The newly initiated callus from axenic seedlings also produces anthocyanin-like pigments when grown in the light on MSB5 medium supplemented with NAA, but not 2,4-D. This quality was, however, lost after few generations. Extraction and analysis of the pigment from the transgenic callus lend strong support to the anthocyanin nature of the pigment which turned greenish-blue upon exposure to NH_3 vapour and was easily detectable in HPLC at λ_{max} corresponding to anthocyanins. These anthocyanins have been separated into individual components by preparative HPLC.

Extraction and analysis of various *P. tomentosa* cultures for cardenolides were carried out. Only a trace of an unknown Kedde's positive compound (pink) was detected in the media, but not in the cells of suspension cultures or static callus cultures. This compound is different from the known cardenolides of *P. tomentosa*. On the other hand, the hairy roots and transgenic callus tissues did produce the main cardenolides of *P. tomentosa*. NAA supplementation enhanced the productivity of cardenolides by this system both qualitatively and quantitatively. However, the undifferentiated crown gall tissues were devoid of any cardenolides.

5 Protocol

5.1 Establishment of Tissue Cultures

The best method of establishing successful aseptic cultures of *P. tomentosa* is by using explants from axenic seedlings. Axenic seedlings usually emerge after

1 week. Callus can be initiated from leaf, stem and root explants on MSB5 medium supplemented with 3% sucrose, 0.1 mg/l kinetin, 0.8% agar and 1.0–3.0 mg/l 2,4-D or NAA. However, explants grown on NAA-supplemented medium produce numerous roots.

Rapidly growing suspension cultures were obtained on MSB5 liquid medium containing 1.0 mg/l 2,4-D in combination with 0.1 mg/l kinetin. The growth curve followed the typical sigmoidal shape in 3 weeks. Hormonal autotrophic suspension cultures on basal MS or MSB5 liquid medium were also obtained from solid- or liquid-grown cultures, though with an extended growth cycle at the first and second subcultures.

5.2 Biotransformation Studies

Hormonal autotrophic suspension cultures were used for the biotransformation of the cardenolides: digitoxigenin, gitoxigenin, digoxigenin, digitoxin, gitoxin and digoxin. Each compound was dissolved in 2 ml alcohol (96%) and added to cells growing in the exponential or logarithmic stage of growth. Cells were harvested after 7–10 days. The media and freeze-dried tissues were separately extracted with EtOAc and analyzed by TLC (Table 1) using the system: silica gel GF_{254}/EtOAc-MeOH (10:1).

5.3 Establishment of Transformed Cultures

Hairy roots and crown galls were induced on axenic seedlings by infection with different strains of *Agrobacterium rhizogenes* and *A. tumefaciens* respectively. The formed hairy roots and crown galls were separated and grown on solidified basal medium containing the antibiotic cefotaxime at 1.5 g/l in the beginning and reduced to 0.75 g/l by the fifth subculture, until the tissues were free of the bacteria. The hairy roots spontaneously produced callus on MS or MSB5 basal medium. The proliferation of roots and callus continued in a recurrent fashion. This callus was thus termed transgenic callus. When transferred to the light, it produced purple pigments on solid medium containing NAA. Extraction and HPLC analysis (see Sects. 2.2.6 and 2.2.7) revealed the presence of several anthocyanin compounds. The effect of GA_3 on growth of the hairy roots and transgenic callus was studied on MS and B5 media, each in combination with 0.0, 0.01 and 1.0 mg/l GA_3. The crown gall tissues, on the other hand, formed undifferentiated callus on MSB5 medium that were slow growing in the beginning, but attained good growth by the tenth subculture.

Both types of transformed tissues were tested for opines by established procedures (Kamada et al. 1986; Elhag et al. 1996) and then analyzed for cardenolide production.

5.4 Extraction and Analysis of Cultures for Cardenolides

Fresh tissues (solid cultures) or filtered cells (using a Buchner funnel) from suspension cultures (10 g), were extracted with MeOH (three times each with 50 ml) with the aid of a magnetic stirrer at room temperature. The combined filtered methanol extract was concentrated on a rotary evaporator at 40 °C. The residue was dissolved in a few drops of $CHCl_3$-MeOH (1:1) and used for TLC analysis.

The medium from suspension cultures was extracted three times with an equal volume of EtOAc. The EtOAc extract was evaporated in a rotary evaporator at 40 °C. The residue was dissolved in $CHCl_3$-MeOH (1:1) and used for TLC analysis.

5.5 Thin-Layer Chromatography of *P. tomentosa* Cardenolides

The following TLC systems were found useful for separation of cardenolides of *P. tomentosa*. They were used to screen extracts of callus and cell cultures (media and cells) for production of cardenolides:

- System I: silica gel GF_{254}/chloroform:methanol (9:1)
- System II: C18 reversed phase silica/water:acetonitrile (6:4)
- Detection was done with Kedde's reagent.

Acknowledgements. The authors extend their thanks to King Abdelaziz City for Science and Technology (KACST), Riyadh, Saudi Arabia, for financial support (Project No. AT–15–39) and to Dr. K. Shimomura (Tsukuba Medicinal Plant Research Station, National Institute of Health Sciences, Japan), for providing the *Agrobacterium rhizogenes* strains and authentic opines.

References

Al-Said MS, Hifnawy MS (1985) Phytochemical and biological screening of *Pergularia tomentosa* L. Proc Saudi Biol Soc 8:155–161

Al-Said MS, Kadertargan HU, Hifnaway MS (1988a) Pharmacognostical and preliminary phytochemical investigation of the fruit of *Pergularia tomentosa* L. Int J Crude Drug Res 26:9–16

Al-Said MS, Hifnawy MS, McPhail AT, McPhail DR (1988b) Ghalakinosaide, a cytotoxic cardiac glycoside of *Pergularia tomentosa*. Phytochemistry 27:3245

Al-Said MS, Abu-Jayyab A, Hifnawy MS (1989) Biochemical studies on ghalakinoside, a possible antitumor agent from *Pergulaira tomentosa*. J Ethnopharmacol 27:235–240

Al-Yahya MA, Mossa JS, Al-Meshal IA, Antoun MD, McCloud TG, Cassady JM, Jacobsen LB, McLaughlin JL (1985) Phytochemical and biological studies on Saudi medicinal plants, part 9. Antitumor testing. Int J Crude Drug Res 4:45–66

Barker PF (1976) The regulation of intracellular calcium. In: Duncan CJ (ed) Calcium in biological systems. Cambridge University Press, Cambridge, pp 67–88

Biondi S, Mengoli M, Bagni N (1995) Auxin and ethylene in hairy root cultures of *Hyoscyamus Muticus*. In: Terzi M, Cella R, Falavigna A (eds) Current issues in plant molecular and cellular biology. Kluwer, London, pp 487–492

Dalziel JM (1937) The useful plants of West Africa. The Crown Agents for the Colonies, London
El-Antaki D (1923) Tazkaret Oli Al-Albab. The Azharia Press, Cairo, pp 225
Elhag H, Gohar A, El-Olemy MM, Mossa JS (1996) Transgenic plants from Ri-transformed callus of *Datura innoxia*. Alkaloid content of normal and transformed tissues. Arab Gulf J Sci Res 14:455–462
El-Idrissi T, Debelmas AM (1970) Moroccan asclepiad. *Pergularia tomentosa*. Preliminary research. Plant Med Phytother 4:176–181
El-Olemy MM, Elhag H, El-Domiaty M, Abdel Sattar E, Al-Azizi MM, Al-Said MS (1994) Biotransformation of digitoxigenin by cell suspension cultures of *Pergularia tomentosa* and *Ammi visnaga*. Saudi Pharm J 2:76–83
Engler A, Prantl K (1891) Die Natürlichen Pflanzenfamilien, vol 4, 1st edn. Wilhelm Engelmann, Leipzig, pp 257–258
Fujita Y, Tabata M (1987) Secondary metabolites from plant cells. Pharmaceutical applications and progress on commercial production. In: Green CE, Somers DA, Hackett WP, Biesboer DD (eds) Plant tissue and cell cultures. Liss, New York, pp 165–185
Gamborg OL, Miller RA, Ojima K(1968) Nutrient requirements of suspension cultures of soybean root cells. Exp Cell Res 50:151–158
Gavidia I, Perez-Bermudez P (1997) *Digitalis lanata* tissue cultures. Phytochemistry 22:2447–2450
Gohar AA, El-Olemy MM, Abdel Sattar E, Al-Said MS, Niwa M (2000) Cardenolides and β-sitosterol glucoside from *Pergularia tomentosa*. Nat. Prod Sci 6:142–146
Hagendoorn MJM, Van der Plas LHW, Van Walraven HS (1996) *Petunia hybrida*: in vitro culture and the production of anthocyanins and other secondary metabolites. In: Bajaj YPS (ed) Biotechnology in agriculture and forestry, vol 37. Springer, Berlin Heidelberg New York, pp 284–303
Hifnawy MS, Al-Said MS, El-Hawary SS (1990) Pergularotoside, a new cardiac glycoside from *Pergularia tomentosa* L. Al-Azhar J Nat Prod 5:1–8
Honda Y, Inaoka H, Takel A, Sugimura Y, Otsuji K (1996) Extracellular polysaccharides produced by tuberose callus. Phytochemistry 41:15127–1521
Ibn-El Bitar A (1809) Mofradat Al-Adwia Wa Al-Agezia. The Azharia Press, Cairo, pp 151
Kamada H, Okamura N, Satake M, Hardad H, Shimomura K (1986) Alkaloid production by hairy roots cultures of *Atropa belladonna*. Plant Cell Rep 5:239–242
Kawaguchi K, Asaka I, Hirotani M, Furuya T, Katsuki S. (1993) Cardenolides in the regenerated plants obtained from *Strophanthus divaricatus* calli. Phytochemistry 34:1317–1321
Luckner M, Dietrich B (1987) Biosynthesis of cardenolides in cell cultures of *Digitalis lanata*. The results of a new strategy. In: Green CE, Somers DA, Hackett WP, Biesboer DD (eds) Plant tissue and cell culture. Liss, New York, pp 187–197
Mahran JH, Saber AH, Khairy NM (1967) Phytochemical investigation of *Pergularia tomentosa* plants growing in Egypt. UAR J Pharm Sci 8:101
Maire R (1933) Etude sur la Flore et végétation du Sahara central. Mem Soc Hist Nat Afr Nord 3:172–175
Maron DM, Ames BN (1983) Revised methods for *Salmonella* mutagenicity test. Mutat Res 113: 173–215
Migahid MA (1989) Flora of Saudi Arabia, vol 2, 3rd edn. King Saud University Publications, Riyadh, pp 84–87
Miura H, Kitmaura Y, Ikanaga T, Mizobe K, Shimuzu T, Nakamura M, Kato Y, Yamada T, Maitani T, Goda Y (1998) Anthocyanin production of *Glehnia littoralis* callus cultures. Phytochemistry 48 :279–283
Murashige T, Skoog F (1962) A revised medium for rapid growth and bioassays with tobacco tissue cultures. Physiol Plant 15:473–497
Ohlsson AB, Bjork L, Gatenbeck S (1983) Effect of light on cardenolide production by *Digitalis lanata* tissue cultures. Phytochemistry 22:2447–2450
Rau D, Forkmann G (1986) Anthocyanin synthesis, in tissue cultures of *Callistephus chinensis* (China aster). Plant Cell Rep 5:435–435
Rhodes MJC, Robins RJ, Hamill JD, Parr AJ, Walton NJ (1987) Secondary product formation using *Agrobacterium rhizogenes* – transformed hairy root cultures. Newslett IAPTC 53:2–15

Robak J, Ceryglewski RJ (1988) Flavonoids are scavengers of superoxide anions. Biochem Pharmacol 37:837

Sakamoto K, Lida K, Sawamura K, Hajiro K, Yoshihisa A, Yoshikawa T, Furuya T (1993) Effects of nutrients on anthocyanin production in cultured cells of *Aralia cordata*. Phytochemistry 33: 357–360

Sakamoto K, Asada Y, Furuya T (1996) *Aralia cordata* Thumb: in vitro culture and production of anthocyanins. In: Bajaj YPS (ed) Biotechnology in agriculture and forestry, vol 37. Springer, Berlin Heidelberg New York pp 38–63

Sims IM, Bacic A (1995) Extracellular polysaccharides from suspension cultures of *Nicotiana plumbaginifolia*. Phytochemistry 38:1398–1405

Tietz NW (1982) Fundamentals of clinical chemistry. WB Saunders, Philadelphia, pp 603–672

Varley H, Gowenlock AH, Bell M (1980) Practical clinical biochemistry, vol 1, 5th edn. William Heineman Medical Books Ltd, London, pp 714–999

Vincett BAL (1977) Wild flowers of central Saudi Arabia., PIME Editric, Milano, Italy, 103 pp

Wunner WH, Bell J, Munro HN (1966) The effect of feeding with a tryptophan-free amino acid mixture on rat liver polysomes and ribosomal ribonucleic acid. Biochem J 10:417–429

11 Indolopyridoquinazoline, Furoquinoline, Canthinone and Protoberberine-Type Alkaloids from *Phellodendron amurense* Callus Tissues

AKIRA IKUTA[1]

1 Introduction

1.1 Distribution and Importance of the Plant

Phellodendron amurense Rupr. (amur cork tree, Fig. 1) is a northern species belonging to the family Rutaceae and is found in Japan, Korea, northern China, Manchuria, Ussuri and Amur. It is a dioecious tree with rather thick branches reaching more than 20 m high in the mountains, whose flowers bloom greenish in May–July. The fruit is nearly globose, is about 1 cm across and turns black when ripe. Another three species are found in Japan: *P. amurense* Rupr. var. *sachalinense* F. Schmidt., *P. amurense* Rupr. var. *japonicum* (Maxim.) Ohwi, *P. amurense* var. *lavallei* (Dode) Sprague, and also *P. chinensis* Schneid. in China (Ohwi 1984). Cortex *Phellodendri* is the dry stem bark of *P. amurense* Rupr. or *P. chinensis* Schneid., and is officially listed in the Japanese and Chinese Pharmacopoeia (Tang and Eisenbrand 1992). It is used as a crude drug in Japan, China and Korea: The bark and fruit have been used as a remedy for fever and the fruit has also been used as an expectorant. A decoction or powder is used for stomach relief and an anodyne, and "daraniske" made from the extract is a famous Japanese medicine. It is a remedy for kidney trouble, stomach, and bladder problems. It is also a strong antibiotic to treat bacillary dysentery, typhus and typhoid. The bark of another species *P. chinensis* is a better grade than that of the northern species. *P. wilsonii* Hay. and Kaneh. is used as an endemic; a decoction of the inner bark and root is taken for stomach relief and a remedy for kidney trouble. Furthermore, it is used in Formosa as an eyewash to treat red or inflamed eyes (Perry and Metzger 1980).

Some pharmaceutically interesting compounds from the plant have been extensively studied in the past.

[1]Research Institute for Science and Technology, Science University of Tokyo, 2669 Yamazaki, Noda, Chiba 278-8510, Japan

Biotechnology in Agriculture and Forestry, Vol. 51
Medicinal and Aromatic Plants XII (ed. by T. Nagata and Y. Ebizuka)
© Springer-Verlag Berlin Heidelberg 2002

Fig. 1. ***Phellodendron amurense***

1.2 Alkaloid/Secondary Metabolites

1.2.1 Phellodendron amurense Rhur

The major chemical constituents of its bark are the isoquinoline alkaloids, berberine, palmatine, jatrorrhizine, phellodendorine (Tomita and Nakano 1957) and magnoflorine (Tomita and Kunitomo 1960), and recently, berberine has been found in the leaves (Yamaguchi et al. 1998). Cadicine (an aliphatic quaternary ammonium) has been isolated from the bark (Tomita and Kunitomo 1960; Fig. 2). The indole alkaloid canthin-6-one has also been found in the roots of young (1-year-old) trees and their aerial parts, and the benzylisoquinoline alkaloid, polyberbine, was isolated along with a large amount of berberine from the fruits (Ikuta and Nakamura 1995; Ikuta and Urabe 1999). In addition to the alkaloids, limonoids and sitosterol were isolated from the

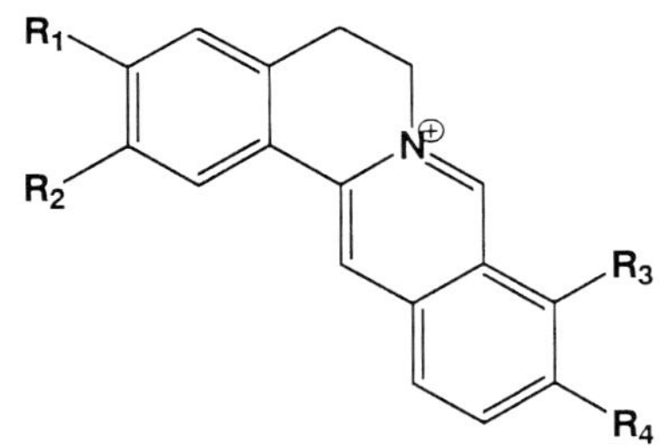

	R_1	R_2	R_3	R_4
(1) Berberine	-O-CH_2-O-		OCH_3	OCH_3
(2) Palmatine	OCH_3	OCH_3	OCH_3	OCH_3
(3) Jatrorrhizine	OH	OCH_3	OCH_3	OCH_3

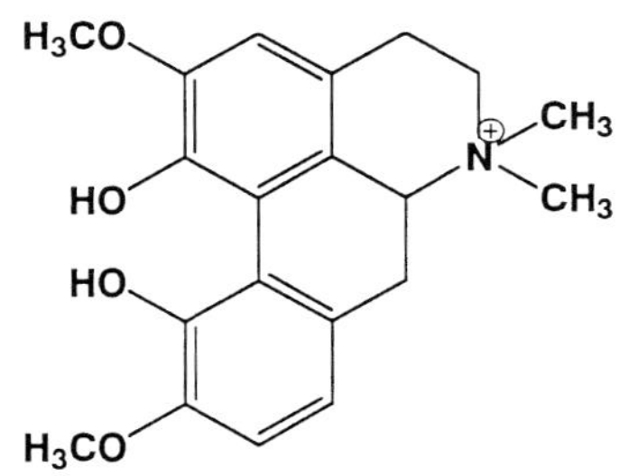

(4) Magnoflorine

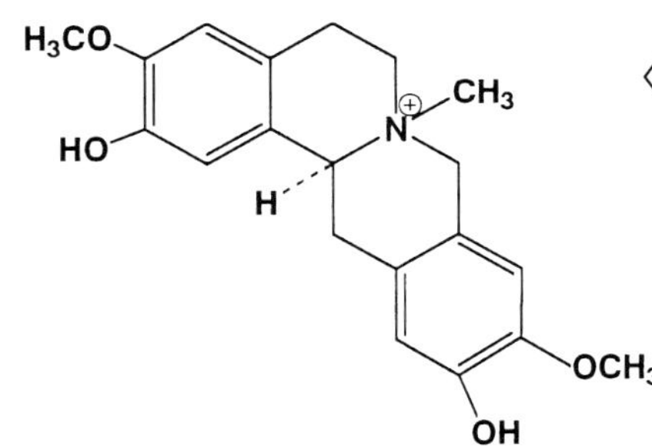

(5) Phellodendrine

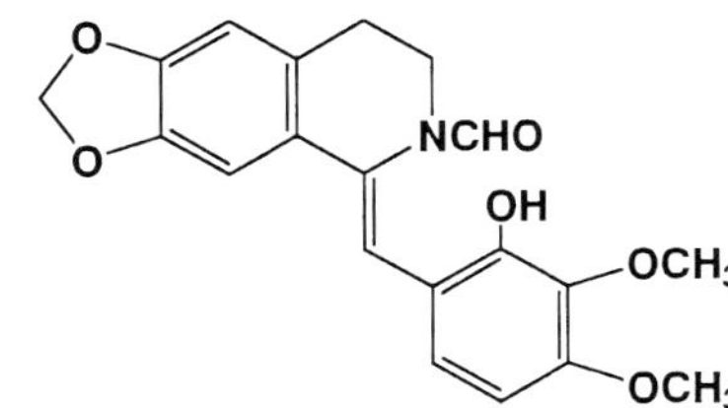

(13) Polyberbine

Fig. 2. Alkaloids from *Phellodendron amurense*

unsaponifiable matter of *P. amurense* (Tsukamoto et al. 1958; Nishioka 1958) and also limonin and obacunone from the bark shortened the sleeping time induced by α-chloralose and urethane in mice (Wada et al. 1990). Two new limonoids, kihadalactones A and B, and the protolimonoids have been reported in the fruit along with seven tirucallane-type triterpenoids (niloticin, dihydroniloticin, niloticin acetate, piscidinol A, hispidol B, bourjotinolone A and hispidone) (Kishi et al. 1992). Limonoids and their glycosides are further reported in seeds and bark. Seeds: limonin, obakunone, limonin 17-β-D-glucopyranoside and obakunonne 17-β-D-glucopyranoside. Bark: limonin, obakunone and nomilin (Miyake et al. 1992). Flavonoid glycosides, phellamurin and amurensin have been isolated from fresh leaves (Hasagawa and Shirato 1953). The flavone aglycone and the glucosides, phellodendroside, hyperin, phelloside, 2,3-dihydrophelloside, noricariside, phellatin and phellavin have also been reported in leaves (Tang and Eisenbrand 1992). On the other hand, 3-*O*-feruloylquinic acid (**I**) and syringaresinol di-*O*-β-glucopyranoside (**II**), (±)-5,5′-dimethoxyariciresinol (**III**), and *N*-methylhigenamine-7-*O*-β-glucopyranoside have been isolated from the dried bark together with eight other phenolic compounds (Ida et al. 1994). Furthermore, the phenolic components **I** and **II** exhibited comparable antioxidant activity to that of α-tocopherol, while component **III** exhibited higher activity than that of α-tocopherol. However, the alkaloid components such as *N*-methylhigenamine-7-*O*-β-glucopyranoside and berberine chloride did not show any antioxidant activity (Hino et al. 1995).

1.2.2 Phellodendron chinense Schneid

Canthin-6-one was isolated from the root bark (Sheridan and Bhandari 1992). Furthermore, *N*-methylflindersine and melianon, a new *N*-isobutylamide, along with two known amides have been isolated from the fruit of *P. chinense* as antifeeding constituents against *Reticulitermes spertus* (Su et al. 1990b, 1994). In addition, new tirucallane-type triterpenoids, niloticin, niloticin-acetate and dihydroniloticin (Gray et al. 1988), phellocechin, friedelin and 4,10-dimethylene-7-isopropyl-5 (E)-cyclodecenol have also been reported in the fruit (Su et al. 1990a). Also, six new flavonoid glycosides were isolated from *P. japonicum* Maxim. together with eight known compounds (Miyaichi et al. 1994).

The production of some pharmaceutically interesting berberine using plant cell culture has been extensively studied. Recently, berberine production by the two-stage culture method using the in vitro culture of *P. amurense* has been reported (Choi et al. 1966) and the details of the in vitro culture and micropropagation have recently been described (Park and Choi 1999). In this chapter, I have focused on the production of the indolopyridoquinazoline, furoquinoline and canthinone-type alkaloids from *P. amurense* callus tissues.

2 In Vitro Cultures (Table 1)

2.1 Establishment of Callus and Suspension Cultures

The stems and fruit of *P. amurense* Rupr. (Rutaceae) were collected in April 1992 and in October 1996, respectively, at the Medicinal Plant Garden of the Science University of Tokyo.

The callus tissues from stems of *P. amurense* were established in April 1992. The MS (Murashige and Skoog 1962) medium (minus glycine) containing 1–3 mg/l 2,4-dichlorophenoxyacetic acid (2,4-D) and 0.1 mg/l kinetin (Kin) as plant growth regulators were used for induction of the callus tissues. The callus tissues were subcultured every 5–6 weeks on fresh MS medium containing 2,4-D (1 mg/l) and Kin (0.1 mg/l) at 25 ± 1 °C in the dark. The cell suspension cultures were generated from 30- to 40-day-old solid cultures (agar medium) grown at 25 ± 1 °C on MS medium containing 1.0 mg/l NAA (1-naphthalene-acetic acid) and 0.1 mg/l BA (6-benzylaminopurine). Approximately 8 g of callus was transferred to 500-ml Erlenmeyer flasks containing 100 ml of liquid MS media and cultured every 4 weeks on a rotary shaker (120 rpm/min). Ten-ml aliquots of the cell suspension were added to Erlenmeyer flasks (500 ml) containing 100 ml of MS with 1.0 mg/l NAA and 0.1 mg/l BA for the time course and elicitor experiments.

2.2 Alkaloid Extraction and Identification (Ikuta et al. 1998a,b)

2.2.1 Extraction and Isolation of Alkaloids

The *P. amurense* callus tissues were subcultured at 5-week intervals for 7–8 months on solid MS medium containing 2,4-D (1 mg/l) with Kin (0.1 mg/l) as plant growth regulators, and each callus tissue was harvested at 5-week intervals, then preserved in a glass bottle together with an adequate amount of MeOH for alkaloid investigations. The stored callus tissues, 4.5 kg fresh wt. and 130 g dry wt. (subcultured on agar medium during 1993–1994 at 5-week intervals from establishment of the callus tissues), for the first investigation, and 8.5 kg fresh wt. and 220 g dry wt. (subcultured during 1995–1996 from establishment of callus tissues), for the second investigation, were individually extracted with cold MeOH and EtOAc, respectively, in a Waring blender and both extracts were combined. The extracts were concentrated under reduced pressure and the residue was extracted using $CHCl_3$ and H_2O repeatedly to obtain the soluble organic fraction. The $CHCl_3$ solutions were evaporated to dryness and the extracts chromatographed on a column of silica gel by gradient elution using $CHCl_3$ with increasing proportions of MeOH. The obtained crude alkaloid mixtures were repeatedly purified by column chromatography on a silica gel column eluted with hexane-EtOAc to give the following alkaloids.

Table 1. Summary of in vitro studies on *Phellodendron amurense*

Purpose of study	Medium[a]	Results/remarks	References
Berberine and palmatine production by callus tissues	MS: 2 mg/l 2,4-D, 0.3 mg/l Kin		Kuroda and Ikekawa (1975)
High vitamin B6 production	LS: 10^{-5} M NAA, 10^{-6} M BA		Yamada et al. (1980)
Mass propagation of mature tree	WPM: 0.23–0.64 mg/l BA or 0.02 mg/l IBA	Shoot multiplication, in vitro rooting	Kunio (1986)[b]
Mass propagation of mature tree		Shoot multiplication, in vitro rooting, soil acclimatization	Kim et al. (1992)[b]
Berberine production by cell culture		Immobilization, berberine secretion	Park et al. (1992)[b]
Berberine production by cell culture	LS: 1.0 mg/l IAA, 7% sucrose	Two-stage culture High level of nitrate (80 mM) and phosphate (8.98 mM) for berberine production Low level of nitrate (40 mM) and phosphate (2.25 mM) for cell growth	Choi et al. (1996)
Indolopyridoquinazolines, furoquinolines, canthin-6-one, berberine and palmatine production	MS: 1 mg/l 2,4-D, 0.1 mg/l Kin		Ikuta et al. (1998a)
New indolopyridoqunzoline alkaloids	MS: 1 mg/l 2,4-D, 0.1 mg/l Kin		Ikuta et al. (1998b)

[a] MS: Murashige and Skoog (1962); LS: Linsmaier and Skoog (1965).
[b] Cited from Park and Choi (1999).

2.2.2 Identification of Alkaloids

The extracts were chromatographed on a silica gel column to give three indole alkaloids (**5**–**9**), three furoquinoline alkaloids (**10**–**12**), and small amounts of known isoquinoline alkaloids, berberine (**1**), palmatine (**2**), and magnoflorine (**4**). Alkaloids **5** and **6** were identified by direct comparison with authentic canthin-6-one and rutaecarpine, respectively. These compounds were the first to be reported from this plant source. Alkaloids **7**, **8** and **9** were extensively analyzed by NMRs including ^{1}H-^{1}H COSY, HOHAHA, ^{1}H-^{13}C heteronuclear correlated 2D-NMR spectra (HMBC) and HMQC experiments. From these data, the structure of **7** was established to be 7,8-dehydrorutaecarpine by comparing the spectral data with those published. This compound has been reported as a synthetic product from rutaecarpine (Mohrle et al. 1980), but this is its first isolation from a natural source. Furthermore, **8** was established to be 7,8-dihydroxyrutaecarpine. This is the first report from a natural source. Alkaloid **9** was identified as 7-hydroxyrutaecarpine reported previously from *Tetradium glabrifolium* (*Evodia meliafolia*) (Wu et al. 1995) and it is the second report from natural source. Three known furoquinoline-type alkaloids, dictamnine (**10**), γ-fagarine (**11**) and skimmianine (**12**) were further isolated and identified by ^{1}H-NMR, ^{13}C-NMR, and mass spectrometry, and comparison of their spectral data with those already published (Vrkoc and Sedmenra 1972; Pussetet et al. 1991). The alkaloids produced from the callus tissue are shown in Fig. 2. The canthinone and indolopyridoquinazoline-type alkaloids (**5**–**9**) and the three furoquinoline-type alkaloids (**10**–**12**), along with a small amount of the known protoberberine-type alkaloids, berberine (**1**), palmatine (**2**), and the aporphine-type, magnoflorine (**4**), have been produced from the callus tissues. However, jatrorrhizine (**3**), produced from the original plant, was not detected, this being somewhat surprising given its occurrence in ranunculaceous, berberidaceous and menispermaceous callus tissues (Ikuta 1988). Callus tissues subcultured 4 years after their establishment from the callus were next re-checked for the reproducibility of the alkaloid production. The same alkaloids (**1**–**2** and **4**–**10**) obtained from the previously cultured callus tissue (2 years after establishment of the callus) were again produced. That is, the biosynthetic ability for the production of the alkaloids by the callus was retained for 5 years following callus establishment. The alkaloids (**5**, **6**, **7**, **8**, **9** and **10**) have not been previously reported as constituents of the original plant, *P. amurense*. Next, it was of interest to compare the biosynthetic capability between the callus and the plant, and to re-investigate whether the canthinone, indolopyridoquinazoline and furoquinoline-type alkaloids were present or not in the original plant.

2.2.3 Investigation of the Original Plant

In order to compare the alkaloid constituents with the original plant, the constituents of the fine roots (500 g fresh wt.) and aerial parts (350 g fresh wt.) of

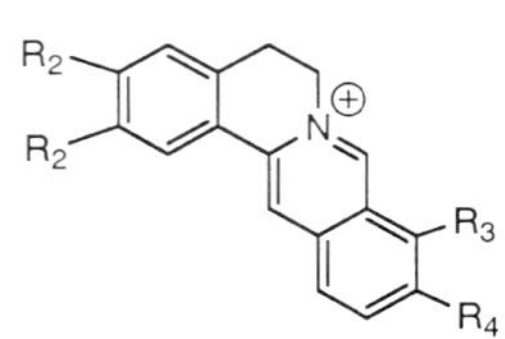

	R1	R2	R3	R4
(1) Berberine	-O-CH2-O-		OCH3	OCH3
(2) Palmatine	OCH3	OCH3	OCH3	OCH3

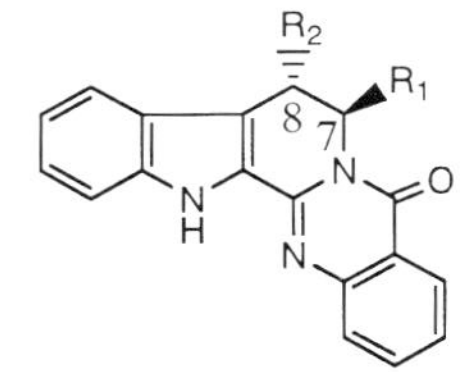

	R1	R2
(6) Rutaecarpine	H	H
(7) 7,8-Dehydrorutaecarpine	Δ	
(8) 7-Hydroxyrutaecarpine	OH	H
(9) 7,8-Dihydroxyrutaecarpine	OH	OH

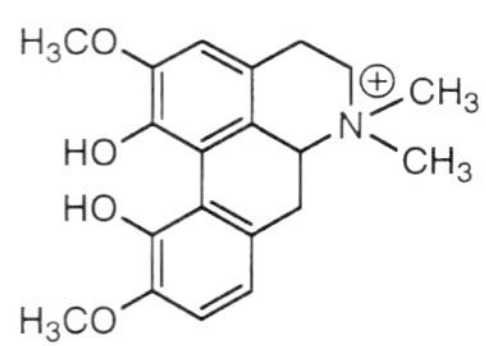

(4) Magnoflorine

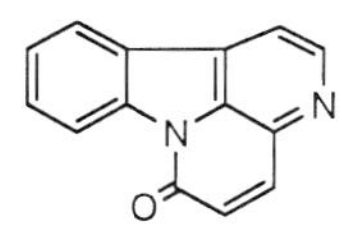

(5) Canthin-6-one

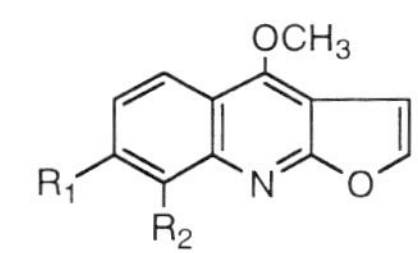

	R1	R2
(10) Dictamnine	H	H
(11) γ-Fagarine	H	OCH3
(12) Skimmianine	OCH3	OCH3

Fig. 3. Alkaloids from *Phellodendron amurense* callus tissues

a young intact plant (*P. amurense*, 1 year from germination) collected from a cultivated field in October 1994, and the ripe fruits (2.2 kg fresh wt., 412 g dry wt.) in October 1996 were investigated.

1. Fine root and aerial part: the fine root and aerial part were repeatedly extracted with cold and hot MeOH as described above, which afforded a small amount of canthin-6-one (**5**; 0.04% dry wt.), and larger amounts of berberine (**1**), palmatine (**2**), jatrorrhizine (**3**) and magnoflorine (**4**) (Ikuta and Nakamura 1995). However, the indolopyridoquinazoline-type alkaloids, rutaecarpine (**6**) and 7,8-dehydrorutaecarpine (**7**), 7-hydroxyrutaecarpine (**8**) and 7,8-dihydroxyrutaecarpine (**9**), and the furoqunoline-type alkaloids, dictamnine (**10**), γ-fagarine (**11**) and skimmianine (**12**), which were produced as the major alkaloid products in the callus tissues, were not detected.
2. Fruits: in addition, the ripe fruits of *P. amurense* were repeatedly extracted with cold and hot MeOH in the same manner. The total extracts were concentrated under reduced pressure and the residue was submitted to chromatography on a column of silica gel to afford the crude alkaloid fractions. The crude fractions were also identified using TLC analysis by comparison with standard samples (**1**, **2**, **4**, **5** and **6**). Alkaloids **6**, **7** and **8** were repeatedly purified by column chromatography on a silica gel and identified by comparison with their spectral data. As a result of this investigation, the isoquinoline alkaloids, berberine (**1**), palmatine (**2**), jatrorrrizine (4) and magnoflorine (**5**), and also rutaecarpine (**6**), 7-hydroxyrutaecarpine (**8**), and the benzylisoquinoline alkaloid, polyberbine (**13**) were identified by TLC analysis and by a comparison of the spectral data (^{1}H, ^{13}C-NMR and MS) (Ikuta and Urabe 1999). However, as before, furoquinoline (skimmianine, dictamnine and γ-fagarine) and canthin-6-one alkaloids were not detected in the fruits.

3 Chemotaxonomy

As shown in Table 2, the callus tissues produced the indolopyridoquinazoline, canthinone, and furoquinoline-type alkaloids as the main alkaloids, and whereas the protoberberine-type alkaloid, berberine is produced in small amounts, it is nevertheless the main product from the original plant. The furoquinoline-type alkaloids widespread in Rutaceae have not yet been detected in the original plant, and the furoquinoline-type alkaloids have been produced as the main alkaloid from rutaceous callus tissues *Toddalia asiatica*. *Evodia rutaecarpa* callus have further produced the furoquinoline-type alkaloids instead of producing the indolopyridoquinazoline-type alkaloid such as evodiamine and rutaecarpine isolated from the original plant as the main alkaloids (Ikuta et al. 1999). The furoquinoline-type alkaloids have also been

Table 2. Comparison of alkaloids in callus tissue and plants of *Phellodendron amurense*

Type of alkaloids	Callus	Plant			
		Stem	Root	Fruit	Bark[a]
Protoberberine type					
Berberine	+	+	+	+	+
Palmatine	+	+	+	+	+
Jatrorrhizine	–	+	+	–	+
Aporphine type					
Magnoflorine	+	+	+	+	+
Furoquinoline type					
Dictamnine	+	–	–	–	–
γ-Fagarine	+	–	–	–	–
Skimmianine	+	–	–	–	–
Canthinone type					
Canthin-6-one	+	+	+	–	–
Indolopyridoquinazoline type					
Rutaecarpine	+	–	–	+	–
7,8-Dihydrorutaecarpine	+	–	–	–	–

–, Absent; +, present.
[a] Commercial crude drug.

reported as the main alkaloids in *Fagara zanthxyloides* callus tissues (Couillerot et al. 1994).

It is interesting from a chemotaxonomic point of view that the indolopyridoquinazoline and furoquinoline-type alkaloids are produced as the main alkaloids from *P. amurense* callus tissues. Canthin-6-one (**5**) was found exclusively in the families Rutaceae and Simaroubaceae of the Rutales (Ohmoto and Koike 1989). However, the distribution of the indolopyridoquinazoline alkaloids, having the more complex structure, such as rutaecarpine (**6**), appears to be restricted to the genera *Euodia* and *Zanthoxylum* of the Rutaceae (Mester 1983).

4 Speculative Biogenesis of the Indolopyridoquinazoline, Furoquinoline, Canthinone and Protoberberine-Type Alkaloids in Callus Tissues (Ikuta et al. 1998a)

As shown in Table 2, the callus tissues produced the indolopyridoquinazoline, furoquinoline and canthinone-type alkaloids as their main alkaloids along with a small amount of the protoberberine-type alkaloids. Therefore, it can be presumed that in the callus tissue, the biosynthetic pathway to the indole alkaloids from chorismic acid via anthranilic acid is activated more than the isoquinoline-type alkaloid pathway from chorismic acid via prephenic acid in the plant. Therefore, the individual differences in the metabolic ability to form the indole and furoquinoline or isoquinoline-type alkaloids from chorismic acid were probably affected by the growth conditions of the dedifferentiated callus tissues and its plant, respectively (Scheme 1), although the reasons are still unknown. The obtained results coincide with the metabolic distance from anthranilic acid being the branch point intermediate in the three types of alkaloid biosynthetic pathways; The furoquinoline-type alkaloids were biosynthesized from anthranilic acid and acetic acid (Matsuo et al. 1966; Grundon 1988) and the indolopyridoquinazoline-type alkaloids were biosynthesized from tryptophan plus anthranilic acid (Yamazaki and Ikuta 1966; Yamazaki et al. 1967; Takagi et al. 1979). Furthermore, canthin-6-one was biosynthesized from tryptophan via anthranilic acid (Crespi-Perellino et al. 1986). This is actually interesting from a signal transduction perspective, and from chemotaxonomic and phylogenetic points of view.

5 Isolation and Characterization of cDNA Clones Encoding Aromatic Amino Acid Decarboxylase During Alkaloid Biosynthesis from *P. amurense* (Suzuki et al. 1999)

We intended to clarify the differences in the expression ability between the TDC (tryptophan decarboxylase) and the TyDC (tyrosine decarboxylase) known as the rate-limiting step encoding the key enzyme in the indole and protoberberine-type alkaloid pathways of the callus tissue in order to compare the alkaloid biosynthetic ability (Scheme 1). Two cDNA clones [the PA, (*P. amurense*) TDC 1806 and PATyDC 1763 base pairs] for tryptophan/tyrosine decarboxylase (designated as PATDC/PATyDC) were isolated by nested genomic PCR (polymerase chain reaction) on a cDNA library from the *P. amurense* cell and by the plaque hybridization method. Furthermore, the two clones were heterologously expressed in the transformed *Escherichia coli* enzyme activities. Each enzyme showed marginally higher substrate specificity for the L-tryptophan or the L-tyrosine/DOPA, but did not accept the L-tyrosine/DOPA or the L-tryptophan as substrates, respectively. The tryptophan and the tyrosine decarboxylase cDNA clones encoded a putative protein of 492 amino acids with a calculated molecular mass of 54,888 daltons and 523 amino acids with a calculated molecular mass of 58,492 daltons, respectively. The tryptophan decarboxylase sequence of PATDC showed a high degree of similarity with the decarboxylase of *Camptotheca acuminata* (82%), *Catharanththus roseus* (79%), *Papaver somniferum* (68%) *and Petroselinum crispum* (67%), *P. amurense* [TyDC (PATyDC) 65%] and *Bos taurus* (56%). The tyrosine decarboxylase sequence of PATyDC also showed a significant similarity to the decarboxylase of *P. crispum* (74%), *P. somniferum* (78%), *C. roseus* (67%), *C.*

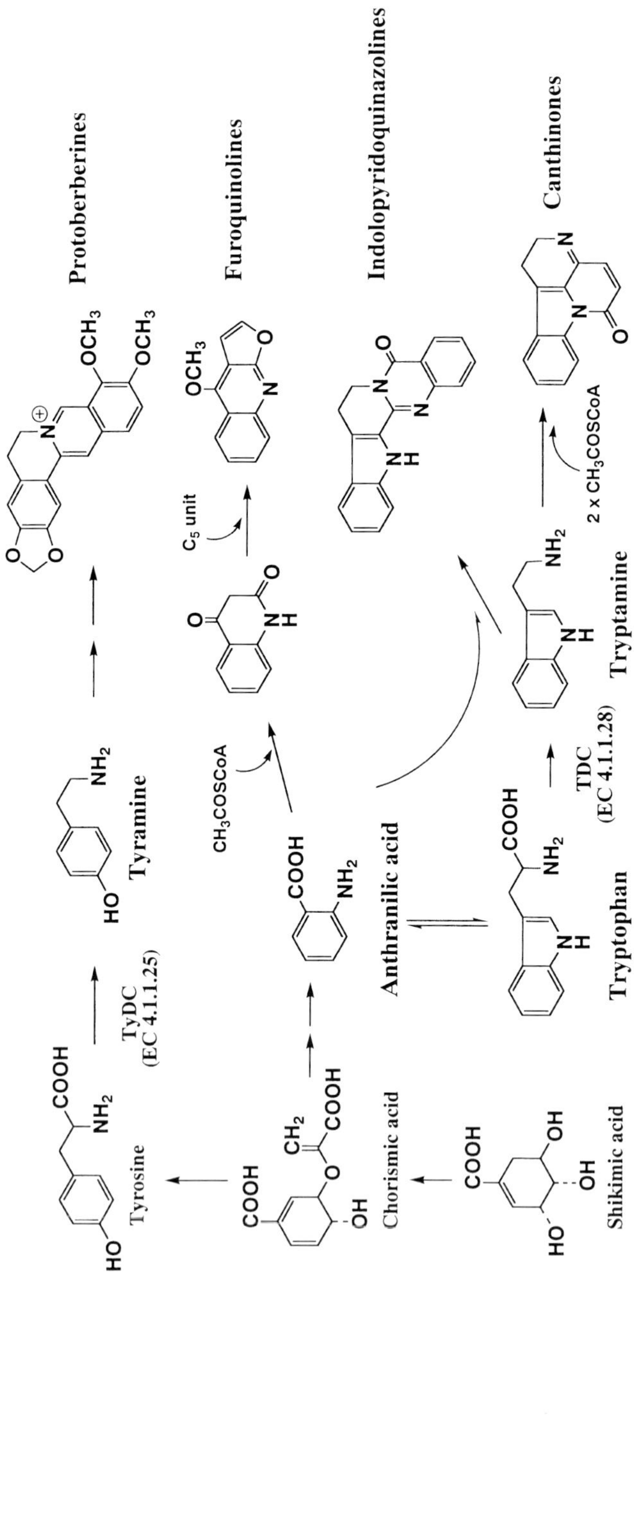

Scheme 1. Biosynthetic pathway of chorismic acid metabolism in callus tissue of *Phellodendron amurense*

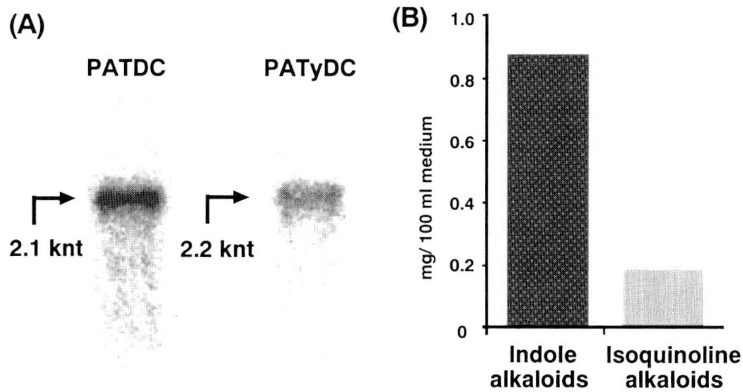

Fig. 4. (**A**) Northern analysis of mRNA and (**B**) comparison of alkaloid contents isolated from callus tissue of *P. amurense*

acuminata (67%) and *B. tautus* (51%), suggesting a possible evolutionary link between these amino acid decarboxylases. The expressed amount from the PATDC was compared to that of the PATyDC by a Northern analysis of mRNA from the callus tissues of *P. amurense*. The expression amount of PATDC was about 3.5 times higher than that of PATyDC (Fig. 4). The result of the expression closely agreed with that of the productivity of the indolopyridoquinazoline and protoberberine-type alkaloids from the callus tissues of *P. amurense*. It is suggested that the differences in the expression from both decarboxylases determine the productivity of the different types of alkaloids.

6 Time Course Study

A time course study was also undertaken in order to investigate the changes in the composition of the various alkaloids over a 40-day period using cell suspension cultures with NAA and BA. Ten-ml aliquots of the cell suspension were added to Erlenmeyer flasks (500 ml) containing 100 ml of MS with hormones (1.0 mg l^{-1} NAA and 0.1 mg l^{-1} BA). After inoculation, the flasks were placed onto a rotary shaker (125 rpm/min) at 25±1 °C, and allowed to culture for 4, 8, 12, 16, 20, 24, 28, 32, 36, and 40 days (Fig. 4). Cells were harvested at each timepoint by suction filtration and extracted by sonication in MeOH (10 ml) for 30 min at room temperature, then concentrated in vacuo. The callus MeOH extract and the media were extracted separately with $CHCl_3$ and the amount of alkaloids in each fraction was determined by HPLC. In the suspension cultures, alkaloid productivity was different from the above-mentioned static cultures. The indolopyridoquinazoline-type alkaloids, rutaecarpine (**6**) and dehydrorutaecarpine (**7**), were produced in small amounts, nevertheless, minor indolopyridoquinazoline-type alkaloids (**8**) and (**9**) were produced under solid culture conditions in relatively large amounts instead of the above two alkaloids (**6** and **7**). In the time course (Fig. 5), the quantities of the alkaloids were determined in both the tissue and the medium by HPLC, focusing on the furoquinoline-type (dictamnine and skimmianine), the indolopyridoquinazoline-type (**8** and **9**) alkaloids, and canthin-6-one (**5**). The furoquinoline-type [dictamnine (**10**) and skimmianine (**12**)] and indolopyridoquinazoline-type (**8** and **9**) alkaloids were produced in the early to later stage of the culture period (40 days). The former were produced in concentrations which ranged from 0.5 to 3.4 mg/g (0.05–0.3% dry wt.) as the main alkaloid products, with the latter ranging from 0.2 to 0.7 mg/g (0.02–0.07% dry wt.). The furoquinoline-type alkaloid pathway, which was not active in the intact plant (*P. amurense*) produced the main alkaloids throughout the culturing period in the suspension cultures. On the other hand, canthin-6-one (**5**) was produced in a small amount (0.1 mg/g, 0.01% dry wt.) during the latter half of the culturing period. Berberine (**1**), the main alkaloid of the intact plant was produced only in trace amounts throughout the culturing period. However, the furoquinoline-type alkaloid pathway, which was not active in the intact plant (*P. amurense*), produced the main alkaloids throughout the culturing period in the suspension culture (Fig. 5).

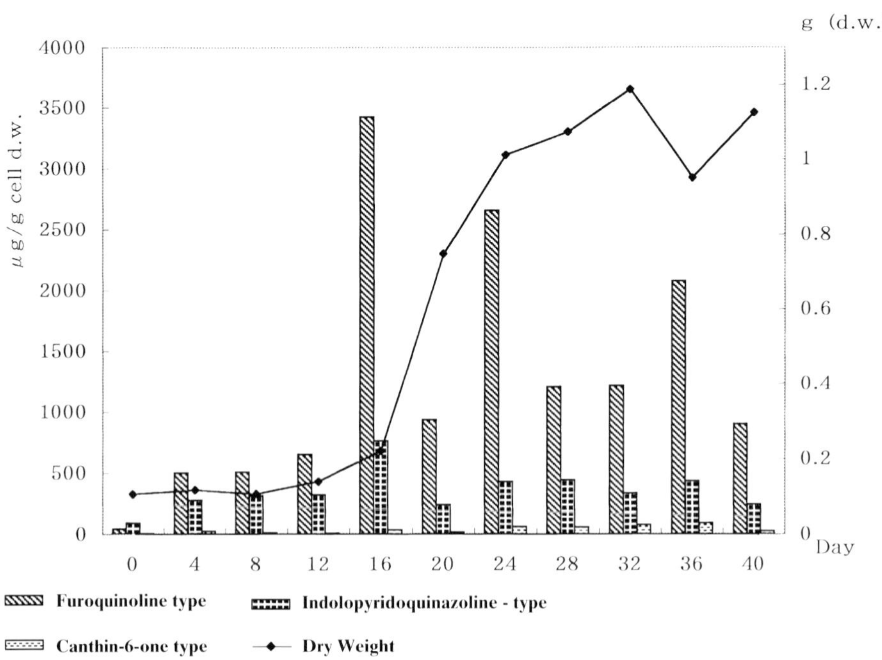

Fig. 5. Time course of alkaloids and dry mass accumulation in NB cell suspension culture *Phellodendron amurense*

7 Productivity of Alkaloids from *P. amurense* Callus Tissues

The pharmacological study of these indolopyridoquinazoline-type alkaloids reported them to possess antileukemia properties, and uterotonic activity, and furthermore, to potentiate the uterine-contracting action of acetylcholine, serotonine, oxytocine and PGF (Bergman 1983). The pharmacological properties of the canthin-6-one alkaloids are also known to include antibiotic and cytotoxic properties, CNS activities and a strong inhibitory effect on cAMP phosphodiesterase (Roberts 1991). Accordingly, it is of interest to investigate the productivity of the alkaloids berberine (**1**), canthin-6-one (**5**) and 7,8-dihydroxyrutaecarpine (**8**). We studied the effects of plant hormones on the alkaloid productivity (Fig. 6) and the elicitor-induced accumulation of these alkaloids in *P. amurense* suspension cultures (Fig. 7). Furthermore, the accumulation was dependent on medium composition, and the MS medium was more effective than the Woody plant medium (WPM; Figs. 8, 9). When co-cultured with the biotic (YE: yeast extract) and abiotic ($CuCl_2$) elicitors, the callus responded to the addition of the former with accumulation of these alkaloids. A time-course experiment with YE in the MS medium (final concentration adjusted to 0.5 and 1.0 mg/l, respectively) resulted in an increase at the higher concentration (1.0 mg/l) in the accumulation of these alkaloids up to a fivefold amount at 72–88 h (Figs. 7, 8). The use of elicitors such as YE could be a method to increase these alkaloid accumulations in the *P. amurense* callus tissues (Ikuta et al. 1996). Further work is now in progress in an attempt to enhance their production by the elicitation method.

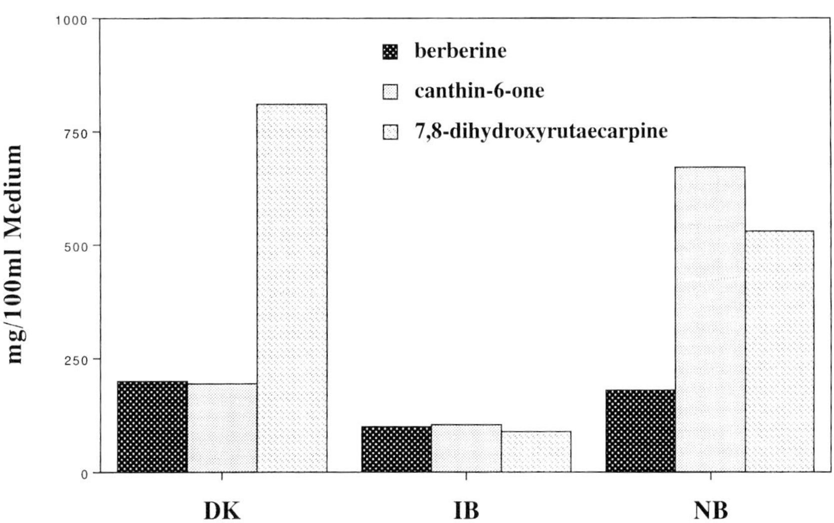

Fig. 6. Productivity of alkaloids by several hormones

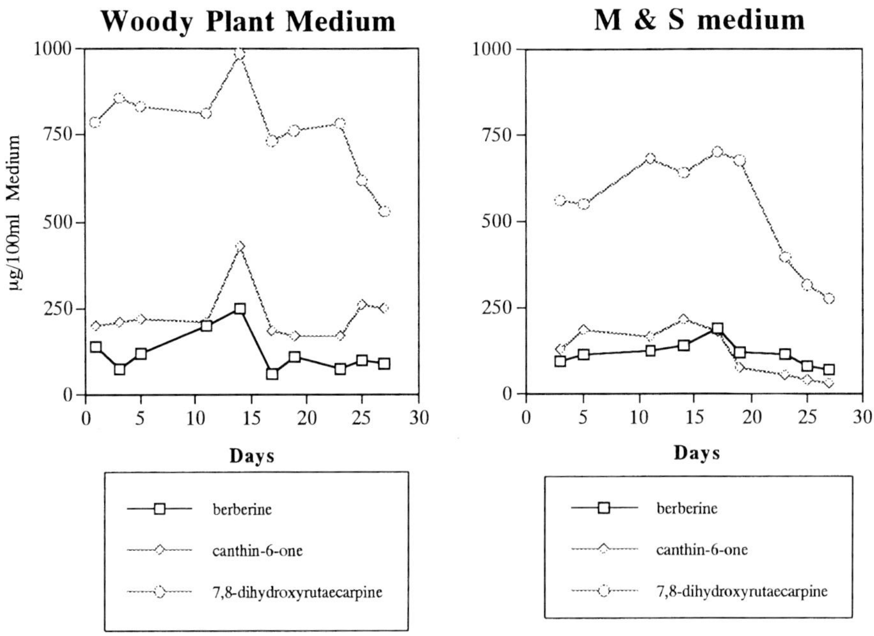

Fig. 7. Time course studies of alkaloid production in cell suspension culture of *Phellodendron amurense* in WP and MS media (1.0 mg/l 2,4-D, 0.1 mg/l Kin)

8 Summary

Callus tissue from the stems of *Phellodendron amurense* (Rutaceae) produced the indolopyridoquinazoline-type alkaloids, rutaecarpine, 7-hydroxy-ruteacarpine, and 7, **8**-dehydrorutaecarpine and 7,8-dihydroxyrutaecarpine (which are new from a natural source), canthin-6-one and three furoquinoline-type alkaloids, dictamnine, γ-fagarine and skimmianine, as the main alkaloid components, along with a small amount of two protoberberne-type alkaloids (berberine and palmatine) and magnoflorine. The indolopyridoquinazoline and furoquinoline-type alkaloids were isolated for the first time from the *P. amurense* plant source. The productivity of these alkaloids was investigated by the elicitor experiment and the use of an elicitor such as YE could be a method to increase these alkaloid accumulations in *P. amurense* callus tissues. Recently, the indolopyridoquinazoline-type alkaloids have been reported to have anti-noiceptive (evodiamine and rutaecarpine; Matsuda et al. 1997, 1998), vasodilatory (dehydroevodianmine, evodiamine and rutaecarpine; Chiou et al. 1996) and antiplatelet activities (rutaecarpine; Sheu et al. 1996, 1998) and 1-hydroxy-rutaecarpine (Sheen et al. 1996). The furoquinoline-type alkaloids showed a further significant inhibition on platelet aggregation induced by arachidonic

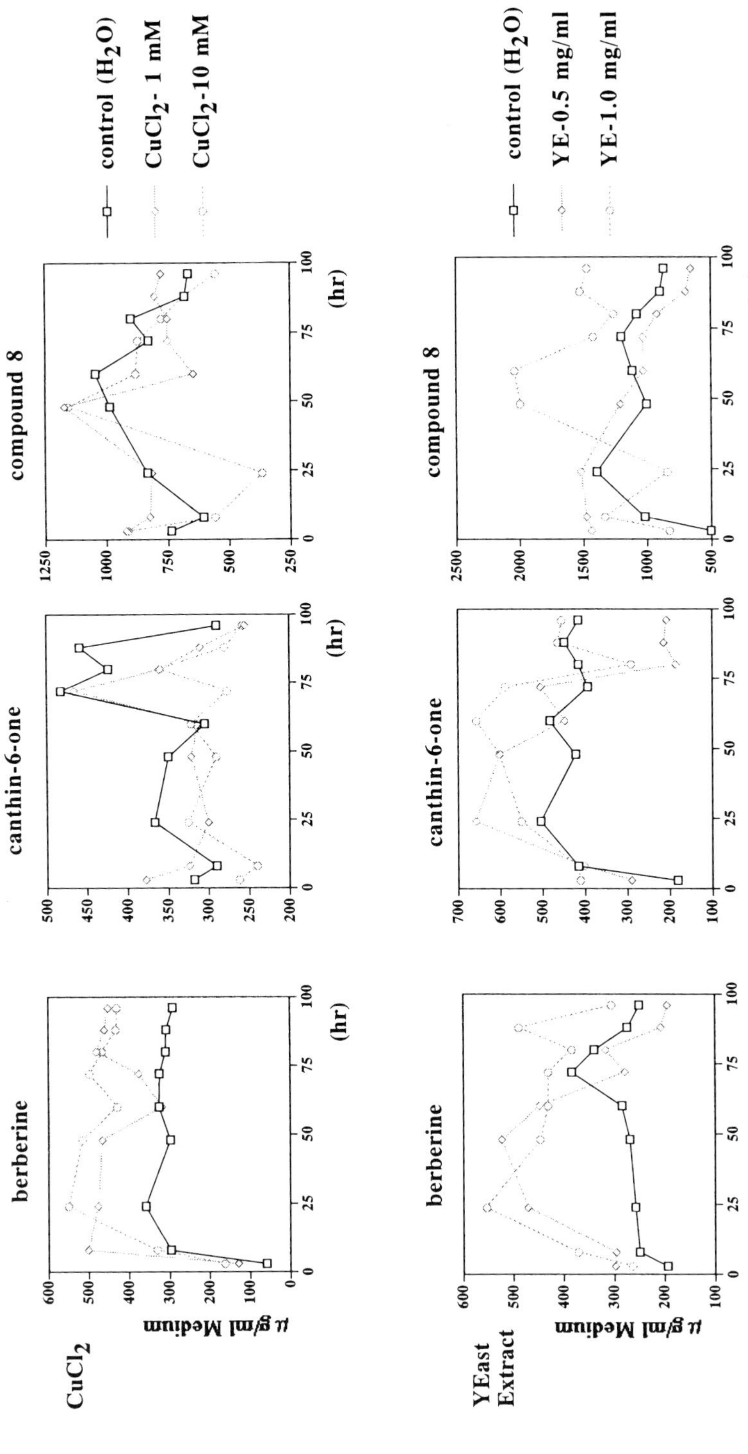

Fig. 8. Effect of elicitor in WP medium (compound 8 = 7,8-dihydroxyrutaecarpine)

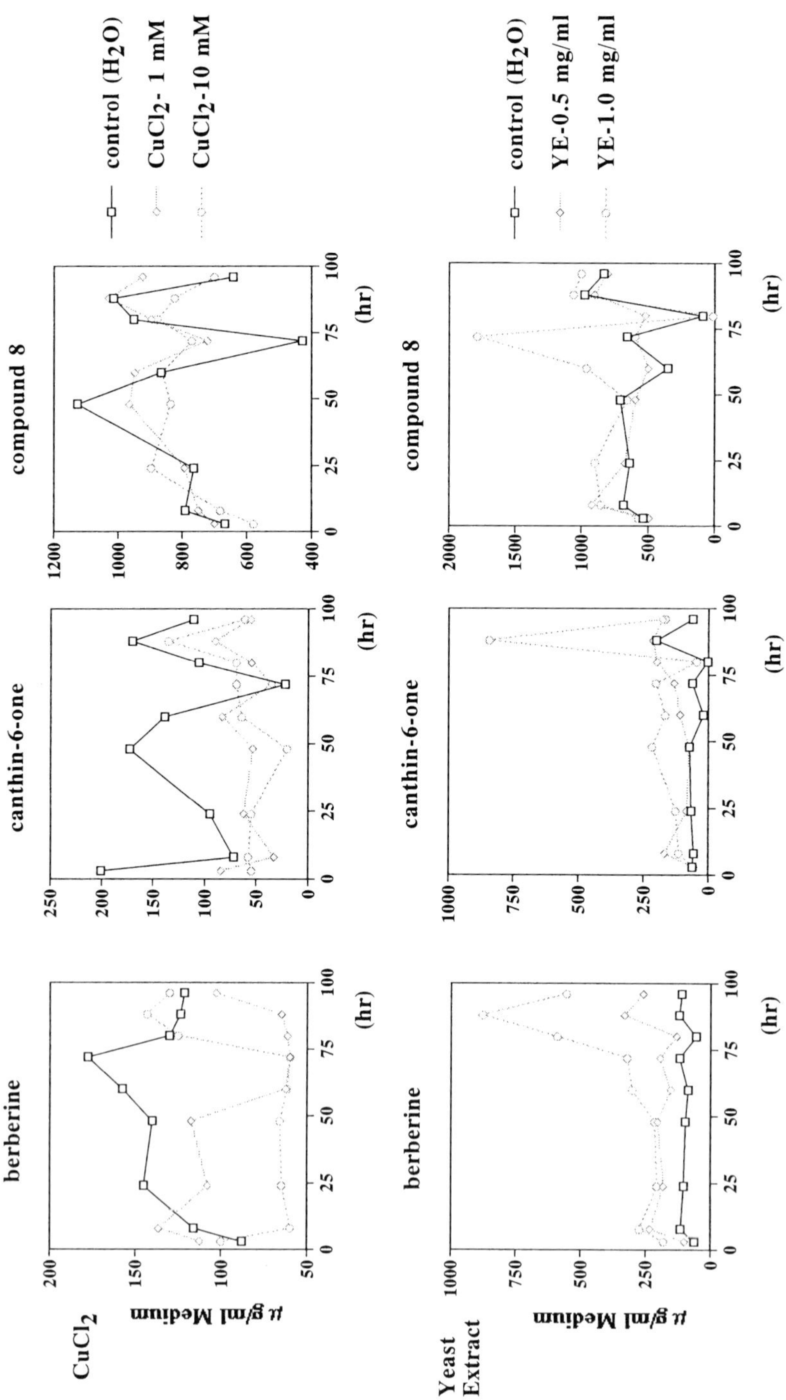

Fig. 9. Effect of elicitor in MS medium (compound 8 = 7,8-dihydroxyrutaecarpine)

acid and collagen (I.S. Chen et al. 1994; J.T. Chen et al. 2000). Cheng et al. (1994) also suggested that the skimmianine and related furoquinoline alkaloids are an antagonist of the 5-hydroxytryptamine receptors in animals. These callus tissues are an interesting new source of these alkaloids, especially for the indolopyridoquinazoline and furoquinoline-type alkaloids which have antiplatelet aggregation activity. Pharmacological studies of antiplatelet aggregation activity in these alkaloids is now in progress. They are considered to be of importance in view of our aging society.

9 Protocol

9.1 Alkaloid Extraction and Isolation

9.1.1 Callus Tissues

The *P. amurense* callus tissues were subcultured at 5-week intervals during 7–6 months on static MS medium containing 2,4-D (1 mg/l) with Kin (0.1 mg/l) as plant growth regulators. Each callus tissue was harvested at 5-week intervals and preserved in a glass bottle together with adequate MeOH for alkaloid investigation. The total stored amounts of callus tissues, 4.5 kg fresh wt., 130 g dry wt. (subcultured during 1993–1994 from establishment of callus tissue), for the first investigation, and 8.5 kg fresh wt., 220 g dry wt. (subcultured during 1995–1996 from establishment of callus tissue), for a second investigation, were individually extracted with cold MeOH and EtOAc in a Waring blender. The extracts were concentrated under reduced pressure and the residue was separated repeatedly between $CHCl_3$ (2 l) and H_2O (0.3 l) to obtain the organic soluble fraction. The $CHCl_3$-soluble fractions were evaporated to dryness and the extracts chromatographed on a column of silica gel by gradient elution using $CHCl_3$ with increasing proportions of MeOH. The crude alkaloid mixtures so obtained were purified repeatedly by column chromatography (CC) on a silica gel column eluted with hexane-EtOAc to give the alkaloids. These alkaloids were determined by the spectral data (^1H, ^{13}C NMR, MS, IR and UV).

9.1.2 Fruits

Fruits (2.2 kg, 412 g dry wt.) of *P. amurense* were collected in October 1996, as described above, and repeatedly extracted with cold and hot MeOH (3–4 l). The total extracts were concentrated under reduced pressure and the residue submitted to chromatography on a column of silica gel to afford the crude alkaloid fractions. Further, the crude fractions were developed by using solvent systems with standard samples for TLC analysis. (a; C_6H_6:EtOAc:CH_3CN =

7:3:0.5, b; $MeOH-H_2O-NH_4OH$ = 8:1:0.5, c; C_6H_6:EtOAc:*n*-propyl-OH:CH_3OH :NH_2Et = 8:4:2:1:1).

9.2 Time Course Study

P. amurense cell suspension cultures were generated from 30- to 40-day-old static cultures (agar medium) grown at 25±1 °C on MS containing 1.0 mg/l NAA (1-naphthaleneacetic acid) and 0.1 mg/l BA (6-benzylamino purine). Approximately 8 g of callus was transferred to 500-ml Erlenmeyer flasks containing 100 ml of liquid MS media and cultured every 4 weeks on a rotary shaker (120 rpm/min). Ten-ml aliquots of the cell suspension were added to Erlenmeyer flasks (500 ml) containing 100 ml of MS with hormones (1.0 mg/l NAA and 0.1 mg/l BA). After inoculation, the flasks were placed onto a rotary shaker (125 rpm/min) at 25 ± 1 °C and allowed to culture for 4, 8, 12, 16, 20, 24, 28, 32, 36, and 40 days. Cells were harvested at each time-point by suction filtration, extracted by sonication in MeOH (10 ml) for 30 min at room temperature and concentrated in vacuo. The callus MeOH extract and the media were extracted separately with $CHCl_3$ and the amount of alkaloids in each determined by HPLC (Waters 600; HPLC column CAPCELL PAK C18 AG120, 5 mm, 4.6 × 250 (SHISEIDO); flow rate 0.5 ml/min; detection UV 310 nm).

References

Bergman J (1983) The quinazolinocarboline alkaloids. In: Mansk RHF (ed) The alkaloids, vol XXI. Academic Press, London, pp 29–54

Berinzaghi B, Muruzabal A, Labriola R, Deulofeu V (1945) Studies on argentine plants. VII. The structure of γ-fagarine. J Org Chem 10:181–183

Chen IS, Wu SJ, Lin YC, Tsai IL, Seki H, Ko FN, Teng CM (1994) Dimeric 2-quinolone alkaloid and antiplatelet aggregation constituents of *Zanthoxylum simulans*. Phytochemistry 36:237–239

Chen KS, Chang YL, Teng CM, Chen CF, Wu YC (2000) Furoquinolines with anti-platelet aggregation activity from leaves of *Melicope confusa*. Planta Med 66:80–81

Cheng JT, Chang TK, Chen IS (1994) Skimmianine and related furoquinolines function as antagonists of 5-hydroxytryptamine receptors in animals. J Auton Pharmacol 14:365–374

Chiou WF, Liao JF, Chen CF (1996) Comparative study on the vasodilatory effects of three quinazoline alkaloids isolated from *Evodia rutaecarpa* J Nat Prod 59:374–378

Choi MS, Shin DI, Park YG (1966) Berberine production by callus suspension cultures of cork tree (*Phellodendron amurense* Rupr.). Korean J Pharmacogn 27(1):32–36

Couillerot E, Caron C, Comoe L, Audran JC, Molinatti P, Zeches M, Le Men-Olivier L, Jardillier JC, Chenieux JC (1994) Benzophenanthridine and furoquinoline accumulation in cell suspension cultures of *Fagara zanthoxyloides*. Phytochemistry 37:425–428

Crespi-Perellino N, Guicciardi A, Malyszko G, Minghetti A (1986) Biosynthetic relationship between indole alkaloids produced by cell cultures of *Ailanthus altissima*. J Nat Prod 49:814–822

Gray AI, Bandari P, Waterman PG (1988) New protolimonoids from the fruits of *Phellodendron chinense*. Phytochemistry 27:1805–1808

Grundon MF (1988) Quinoline alkaloids related to anthranilic acid. In: Brossi A (ed) The alkaloids, vol 32. Academic Press, London, pp 341–438

Hasegawa M, Shirato T (1952) Two new flavonoid glycosides from the leaves of *Phellodendron amurense* Rupr. J Am Chem Soc 75:5507–5511

Hino K, Yamaguch S, Ida Y, Satoh Y, Maoka T, Itoh Y (1995) Antioxidative activities of constituents in *Phellodendron amurense* bark. Igaku to Seibutugaku 131:59–62

Ida Y, Satoh Y, Ohtsuka M, Nagasao M, Shoji I (1994) Phenolic constitutents of *Phellodendron amurense* bark. Phytochemistry 35:209–215

Ikuta A (1988) Isoquinoline. In: Constabel F, Vasil IK (eds) Cell culture and somatic cell genetics of plants, vol 5. Phytochemicals in plant cell cultures. Academic Press, New York, pp 289–314

Ikuta A, Nakamura T (1995) Canthin-6-one from the roots of *Phellodendron amurense*. Planta Med 61:58l-582

Ikuta A, Urabe H (1999) 7-Hydroxyrutaecarpine from fruit of *Phellodendron amurense*. Nat Med 53:333

Ikuta A, Nakamura T, Urabe H (1996) Indole alkaloids from *Phellodendron amurense* callus tissues and their regulation. In: Xu Z-H, Chen Z-H (eds) Plant biotechnology for sustainable development of agriculture. China Forestry Publishing House, pp 263–267

Ikuta A, Nakamura T, Urabe, H (1998a) Indolopyridoquinazoline furoquinoline and canthinone type alkaloids from *Phellodendron amurense* callus tissues. Phytochemistry 48:285–292

Ikuta A, Urabe H, Nakamura T (1998b) A new indolopyridoquinazoline-type alkaloid from *Phellodendron amurense* callus tissues. J Nat Prod 61:1021–1041

Ikuta A, Tani T, Si S (1999) Study of rutaceous plant tissue cultures. Constituents of production of alkaloids from *Evodia* and *Toddalia* callus tissues.120th General Meet Japanese Pharmaceutical Society of Japan Tokushima, Part II., 160 pp (31po 014–003)

Kim JH, Goo GH, Choi MS, Park YG (1992) Micropropagation and soil adjustment of cork tree (*Phellodendron amurense* Rupr.) through in vitro culture. Korean J Plant Tissue Cult 9(1): 37–42

Kishi K, Yoshikawa K, Arihara S (1992) Limonoids and protolimonoids from the fruits of *Phellodendron amurense*. Phytochemistry 31:1335–1338

Kunio A (1986) Plantlet formation by intrapetiolar bud culture of *Phellodendron amurense* Rupr. For Tree Breed 138:24–26

Kuroda H, Ikekawa T (1975) Berberine and palmatine JP 50013519 6

Linsmaier E, Skoog F (1965) Organic growth factor requirement of tobacco tissue cultures. Physiol Plant 18:100–127

Lloyd G, McCown B (1980) Commercially feasible micropropagation of mountain laurel, *Kamia latifolia* by use of shoot-tip culture. Proc Int Plant Prop Soc 30:421–427

Matsuda H, Wu JX, Tanaka T, Iinuma M, Kubo M (1997) Antinociceptive activities of 70% methanol extract of *evodiae fructus* (fruit of *Evodia rutaecarpa* var. *bodinieri*) and its alkaloidal components. Bio Pharm Bull 20:243–248

Matsuda H, Yoshikawa M, Ko S, Iinuma M, Kubo M (1998) Antinociceptive and antiinflammatory activities of evodiamine and rutaecarpine. Nat Med 52:203–208

Matsuo M, Yamazaki M, Kashida Y (1966) Biosynthesis of skimmianine. Biochem Biophys Res Commun 23:679–682

Mester I (1983) Structural diversity and distribution of alkaloids in the rutales. In: Waterman PG, Grundon MF (eds) Chemistry and chemical taxonomy of the Rutales. Academic Press, London, pp 31–96

Miyaichi Y, Segi H, Tomimori T (1994) Studies on the constituents of the leaves of *Phellodendron japonicum* Maxim. Yakugaku Zasshi 114:186–199

Miyake M, Inaba N, Ayano S, Ozaki Y, Maeda H, Ifuku Y, Hasegawa S (1992) Limonoids in *Phellodendron amurense* (Kihada). Yakugaku Zasshi 112:343–347

Möhrle H, Kamper C, Schmid R (1980) Eine neue Synthese von Rutaecarpin. Arch Pharm 313:990–995

Murashige T, Skoog F (1962) A revised medium for rapid growth and bioassays with tobacco tissue cultures. Physiol Plant 15:473–497

Nishioka I (1958) Studies on the sterols from *Phellodendron amurense* Rupr. (2). Isolation of 7-dehydrostigmasterol. Yakugaku Zasshi 78:1432–1434

Ohmoto T, Koike K (1989) Chapter 3, canthin-6-one alkaloids. In: Bross A (ed) The alkaloids, vol 36. Academic Press, San Diego, pp 135–170

Ohwi J (1984) A combined, much revised, and extended translation by the author of this flora of Japan (1953) and flora or Japan-Pteridophyta (1957). In: Meter FG, Walker EH (eds) Flora of Japan. Smithsonian Institute, Washington, pp 583–584

Park YG, Choi MS (1999) *Phellodendron amurense* (cork tree): in vitro culture, micropropagation, and the production of berberine. In: Bajaj YPS (ed) Biotechnology in agriculture and forestry, vol 43. Medicinal and aromatic plants XI. Springer, Berlin Heidelberg New York, pp 337–349

Park JC, Kim LS, Choi YW (1992) Production of protoberberine-type alkaloid from cell cultures. II Culture methods for effective berberine production. Korean J Plant Tissue Cult 19(6): 323–329

Perry LM, Metzger J (1980) Medicinal plants of East and Southeast Asia: attributed properties and uses. The MIT Press, Cambridge, MA pp 367

Pusset J, Lopez JL, Pais M, Neirabeyeh MA, Veillon JM (1991) Isolation and 2D NMR studies of alkaloids from *Comptonella sessilifoliola*. Planta Med 57:153–155

Roberts MF (1991) The quinazolinocarboline alkaloids. In vitro culture and the formation of secondary metabolites. In: Bajaj YSP (ed) Biotechnology in agriculture and forestry, vol 15. Medicinal and aromatic plants Ill. Springer, Berlin Heidelberg New York, pp 39–57

Sheen WS, Tsai IL, Teng CM, Ko FN, Chen IS (1996) Indolopylydoquinazoline alkaloids with antiplatelet aggregation activity from *Zanthoxylum integrifolium*. Planta Med 62:175–176

Sheu JR, Hung WC, Lee YM, Yen MH (1996) Mechanism of inhibition of platelet aggregation by rutaecarpine, an alkaloid isolated from *Evodia rutaecarpa*. Eur J Pharmacol 318:469–475

Sheu JR, Kan YC, Hung WC (1998) The antiplatelet activity of Rutaecarpine, an alkaloid isolated from *Evodia rutaecarpa*, is mediated through inhibition of phospholipase C. Thromb Res 92:53–64

Sheridan H, Bhandari P (1992) Canthin-6-one from the root bark of *Phellodendron chinese*. Planta Med 58:299

Su R, Kim M, Kawaguchi H, Yamamoto T, Goto K, Taga T, Miwa Y, Kozuka M, Takahashi S (1990a) Triterpenoids from the Fruits of *Phellodendron chinense* SCHNEID.: the stereostructure of niloticin. Chem Pharm Bull 38:1616–1619

Su R, Kim M, Yamamoto T, Takahashi S (1990b) Antifeeding constituents of *Phellodendron chinense* fruit against *Reticulitermes speratus*. J Pestic Sci 15:567–572

Su R, Kim M, Nakajima S, Liu M (1994) Amides from the fruits of *Phellodendron chinense*. Acta Bot Sin 36:817–820

Suzuki H, Kurihara S, Yoshikawa T, Ikuta A (1999) Isolation and characterization of cDNA clones encoding aromatic amino acid decarboxylase in alkaloid biosynthesis from *Phellodendron amurense* cell culture. 17th General Meet Japanese Society for Plant Cell and Molecular Biology, 40 pp (1Bp-06)

Takagi S, Akiyama T, Kinoshita T, Sankawa U, Shibata S (1979) Minor basic constituents of Evodia Fruits. Shoyakugaku Zasshi (Jpn J Pharmacog) 33:30–34

Tang W, Eisenbrand G (1992) *Phellodendron amurense* Rupr. In: Chinese drugs of plant origin, chemistry, pharmacology, and use in traditional and modern medicine. Springer Berlin Heidelberg New York, pp 759–761

Tomita M, Kunitomo J (1960) Studies on the alkaloids of rutaceous plants. Yakugaku Zasshi 80:880–884, 885–887, 1238–1244, 1300–1301

Tomita M, Nakano T (1957) Studies on the alkaloids of Rutaceous plants. I. Alkaloids of *Phellodendron amurense* Prur. (1). Chem Pharm Bull 5:10–12

Tsukamoto T, Nishioka I, Mihashi K, Miyahara H (1958) Studies on the sterols from *Phellodendron amurense* Rupr. I. Yakugaku Zasshi 78:1099–1101

Vrkoc J, Sedmera P (1972) Extractives of *Chloroxylon swietenia*. Phytochemistry 11:2647–2648

Yamaguchi S, Kawamura T, Noro Y, Tanaka T (1998) Berberine contents in *Phellodendron* leaves. Nat Med (Tokyo) 52:452–454

Yamazaki M, Ikuta A (1966) Biosynthesis of *Evodia* alkaloids. Tetrahedron Lett 3221–3224

Yamazaki M, Ikuta A, Mori T, Kawana T (1967) Biosynthesis of *Evodia* alkaloids II. The participation of C1-unit to the formation of indoloquinazoline alkaloids. Tetrahedron Lett 3317–3320

Wada K, Yagi M, Matsumura A, Sasaki K, Sakata M, Haga M (1990) Isolation of limonin and obacunone from *Phellodendri* cortex shorten the sleeping time induced in mice by α-chloralose-urethane. Chem Pharm Bull 38:2332–2334

Wu TS, Yeh JH, Chen KT, Lin LC, Chen CF (1995) 7-hydroxyrutaecarpine from *Tetradium glabrifolium* and *Tetradium rutaecarpum*. Hetreocycles 41:1071–1076

Yamada Y, Watanabe K (1980) Selection of high vitamin B6 producing strains in cultured green cells. Agric Biol Chem 44(11):2683–2687

12 In Vitro Culture and the Production of Secondary Metabolites by *Sempervivum* spp. (Houseleek)

É. Kertész-Dobos, Á. László-Bencsik, and B. Dános[1]

1 Introduction

1.1 Distribution and Importance of *Sempervivum*

The genus *Sempervivum* (Crassulaceae) contains approximately 80 species and several subspecies (Parnell 1988). The name is indicative of their evergreen, sempervirent nature (semper = always, vivum = living) (Csapody 1982). Houseleeks (*Sempervivum* spp.) grow mainly on sunny, barren hillsides, mostly at 1000–2000 m (Codignola et al. 1990). They are favourite plants in rock-gardens, because they grow on walls and roofing tiles (Aurnhammer 1979).

One of the most important species, *Sempervivum tectorum* L. (common houseleek; Fig. 1) is native to the Alps, but it can be found sporadically as far as the Pyrenees and the northern regions of the Balkan Peninsula, in central Anatolia (Oztan and Arslan 1993). It grows well under extreme conditions (Kolb et al. 1983; Kolb and Schwarz 1984), usually in calcareous soil (Roessner and Popp 1986). It is a cosmopolitan species living in dry circumstances (Hegnauer 1964). It can be used on extreme sites (notably in urban environments), if its basic ecological and growth requirements are respected (Kutkova 1990).

One of the most important ecophysiological features of *Sempervivum* – in which it is similar to other members of Crassulaceae family – is nocturnal CO_2 fixation; this physiological adaptation to a dry environment enables tolerance of water deficiency.

This metabolism, known as CAM (crassulacean acid metabolism), is an alternative pathway for C_4-type CO_2 fixation (Street and Cockburn 1972). This route results in the diurnal disconnection of the Calvin cycle and CO_2 fixation. In CAM plants phosphoenolpyruvate (PEP) carboxylation occurs nocturnally – oxaloacetic acid is reduced to malate by malate dehydrogenase, and the malate is accumulated in the vacuole. In this phase the stomata are open, thus CO_2 flows into the cytosol. Without ATP and $NADPH_2$ the direct fixation of CO_2 is impossible in the Calvin cycle, because in the dark the decarboxylation of malate stops rapidly. Under illumination, malate is returned to the cytosol and decarboxylated, the CO_2 being donated into the Calvin cycle, so that

[1]Research Institute for Medicinal Plants, Lupaszigeti str 4, 2011 Budakalász, Hungary

Biotechnology in Agriculture and Forestry, Vol. 51
Medicinal and Aromatic Plants XII (ed. by T. Nagata and Y. Ebizuka)
© Springer-Verlag Berlin Heidelberg 2002

Fig. 1. Perennial *Sempervivum tectorum* L. with young sprout-rosettes suitable for vegetative propagation

light-dependent CO_2 fixation can occur under closed conditions. Desiccation of these plants is, consequently, prevented when exposure to daylight causes warming and dryness (Kelly and Latzko 1982).

Thomas and Andre (1987) examined the effect of water stress on hourly and daily patterns of *Sempervivum* growth. Water stress reduced gross photosynthesis and modified its division between the reduction of CO_2 and O_2. Schuber and Kluge (1979) examined CO_2 exchange and measured the diurnal rhythms of organic acids in situ in some species of *Sempervivum*. According to measurements by Pilou et al. (1991), PEP carboxylase (CAM key enzyme) activity fluctuated seasonally, with maxima during the summer which were 8- to 30-fold higher than winter values. Daytime temperature seemed to be the most influential factor; light intensity had a small but significant effect. Increasing the temperature increased efflux from the tissue in a nonlinear manner (Friemert et al. 1988). In tissue which had accumulated malate (acidified state), a temperature increase caused reduction of the rate of dark fixation, whereas the rate of CO_2 fixation in the light remained largely unaffected. This result indicates that in the acidified tissue increasing the temperature increases the efflux of malate from the vacuole.

In folk medicine *Sempervivum* is called 'ear-herb', because the juice squeezed from the plant can effectively cure diseases of the ear. It is believed to be effective for treatment of inflammation and skin burns. According to Nicolae et al. (1986), the gel used by them for treatment of burns contains 3–5%

methyl cellulose, 5–8% pectin, and 20–25% propolis extract, and the balance is *Sempervivum tectorum* juice. The gel can also be used for treatment of ear inflammations. The leaf infusion is suggested for treatment of bronchitis and mouth abrasions.

Leaves (diameter 6–10 cm) from fully developed plants are used to prepare the drug 'Sempervivi tectori herba' (Kéry 1993). Because it is not an official pharmacopoeial drug, it has no standard. In 1974 it was proposed that *Sempervivum tectorum* be included in the list of protected plants (Kovács and Priszter 1974); all *Sempervivum* spp. are protected plants in Hungary, two cantons of Switzerland (Nidwalden, Obwalden), Austria, and some provinces of Germany (Dános 1992).

1.2 Conventional Practices for Its Propagation, and for the Production of Medicinal Components

The leaves of this perennial plant are flat and succulent, and form rosettes. Houseleek develops young sprout rosettes on short or long runners (stolons); these form roots and in 3–4 years evolve into flowering plants. The inflorescence forms from the middle of the rosette, and is of ramifying structure. After flowering and seed ripening the flowering rosettes perish (Priszter 1974).

It is recommended that the plant be propagated by means of leaf cuttings or runners, in the conventional manner, because propagation by sowing seeds results in slow development.

Propagation is usually performed with rooted, young daughter rosettes, which develop on the runners, with a runner 2–3 cm in length (Müssel and Kiermeier 1993).

Leaves from 2-year-old plants are cut, squeezed, and the freshly squeezed juice is used. The succulent leaves can be dried by lyophilization, a method which is always successful. A kilogram of the drug can usually be extracted from 7–8 kg fresh leaves; these are of light green color and the original, star-shaped form of the leaf is preserved.

Sempervivum has rarely been investigated phytochemically. Carbohydrates, organic acids, alkaloid, and flavonoids have been isolated from the whole plant; the structures of these are given in Fig. 2. Of the polysaccharides, accumulation of sedoheptulose is characteristic; sucrose and fructose also occur in the leaves of the plants (Nordal and Klevstrand 1951; Frigot 1960). Significant acid metabolites are isocitric (5–9%), citric, and malonic acids (Soderstrom 1962; Kull 1967).

A method for isolation of isocitric acid has been described by Vickery and Wilson (1958). Fumaric acid has also been detected in *Sempervivum* species (Ramstad and Lieberman 1955).

The total alkaloid content of the leaves of *Sempervivum tectorum* has been found to be 0.01–0.03% (Paris and Frigot 1959). Nicotine was identified among the alkaloids by Frigot (1960).

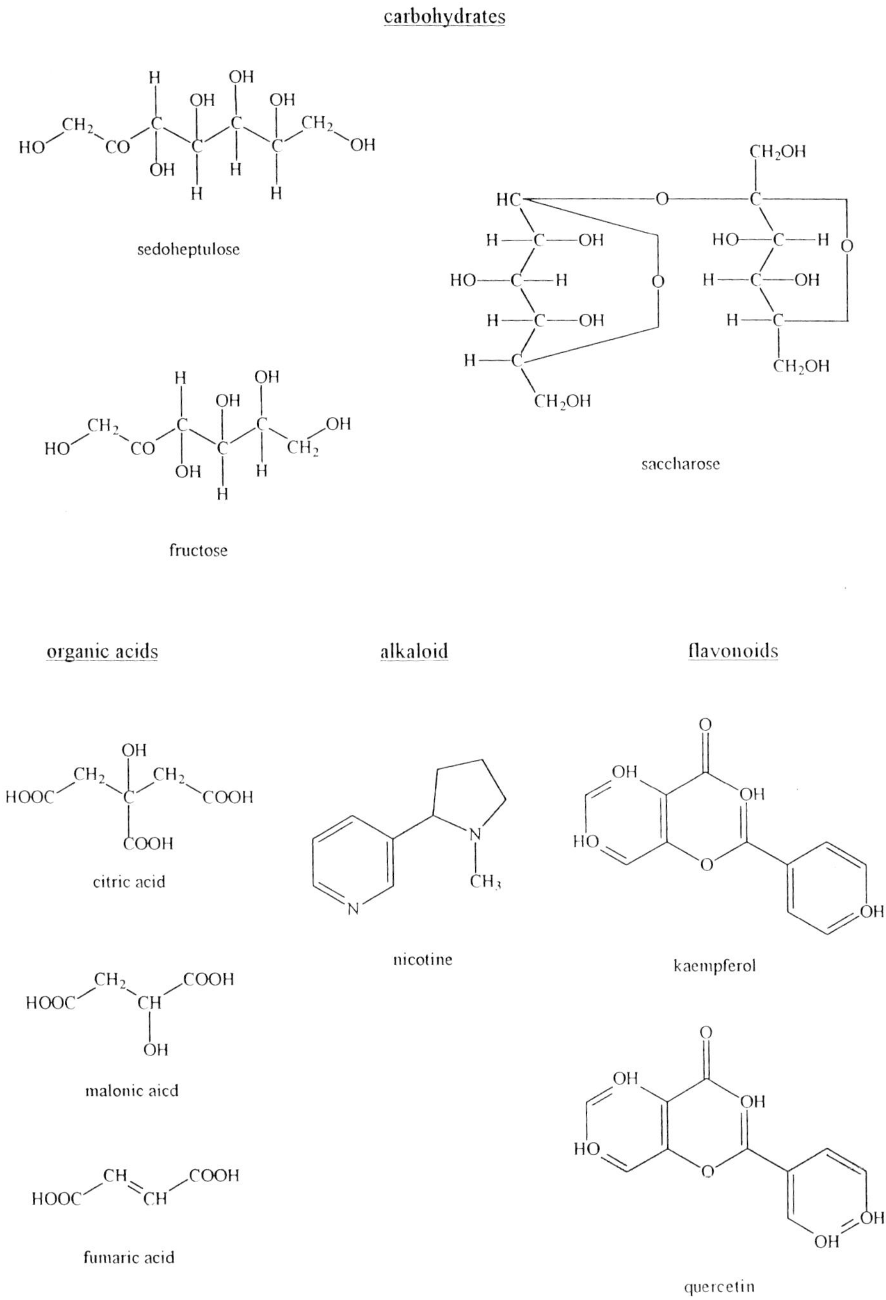

Fig. 2. The chemical structures of compounds isolated from *Sempervivum* spp.

Among the free amino acids asparagine is dominant; caffeic acid and chlorogenic acid are the dominant phenolic acids.

The houseleek also contains approximately 0.7% flavonoids (quercetin, kaempferol, and flavone glycosides), 3–4% tannins (including procyanidines), and significant ascorbic acid and mucilage (Paris and Frigot 1959). Gumenyuk et al. (1982) detected complex phenolic compounds in the flowers of *S. ruthenium* and also isolated and identified five flavonoid glycosides. Flavonoid variation in Eurasian *Sempervivum* was minimal (kaempferol was the principal flavonol); this reflects the morphological uniformity of the genus and the similar ecological preference of the species (Stevens et al. 1996).

Stevens et al. (1995) compared alkaloid and tannin levels in 36 species of Crassulaceae including three *Sempervivum* spp. Proanthocyanidins (condensed tannins) and galloyl esters were found in all the genera, but *Sempervivum* spp. were free from piperidine alkaloids.

Lotti et al. (1991) detected fatty acid compounds in *Sempervivum chrysanthemum*.

Gumenyuk and Borisenko (1979) established that Russian houseleek contained a complex mixture of phenolic substances including quercetin, kaempferol, isorhamnetin, and scutelarein derivatives. The complex had an inflammatory activity.

According to Kéry et al. (1992), phytochemical screening of the plant extract proved the presence of flavonoids (0.7% w/w), polyphenolic compounds (4.2% w/w) and polysaccharides (11.2% w/w). Water-soluble polysaccharides of plant origin are of interest as potential biostimulators with immunomodulating and antitumor activity (Barbakadze et al. 1989). The biologically active polysaccharides of the plants have antitumor activity (Kukhta et al. 1988). Plant extracts containing low molecular mass and antioxidative compounds have been successfully used in phytotherapy since ancient times, because reactive oxygen-containing species are involved in several diseases (Elstner 1987; Blázovics et al. 1993a).

It has been demonstrated that many naturally occurring antioxidants have activity as radical scavengers and lipid peroxidation inhibitors (Fehér and Vereckei 1985; Larson 1988). Stevens et al. (1994) detected no triterpenes in waxes of *Sempervivum*.

Sátory et al. (1988) have reported experiments on lyophilized extracts of *Sempervivum tectorum* L. in which they discovered marked dose-dependent topical antiinflammatory activity which inhibited the edematous response of rabbit ears to Croton oil (28–64%) within 6–24 h.

The pressed juice of *S. tectorum* is a well-known antiphlogistic medicine (Blázovics et al. 1993b).

2 In Vitro Approaches

2.1 Induction of Unorganized Cultures

Experiments on callus induction were conducted by Dobos et al. (1994) with plants of different ages grown from *Sempervivum tectorum* L. seeds which were disinfected with sodium hypochlorite and then raised in vitro.

Plant experiments were performed on culture basal media MS (Murashige and Skoog 1962) and W (White 1943). Sucrose (30 $g l^{-1}$) was added, as a carbon source, to basal medium solidified by addition of 1% Difco Bacto-agar. The cultures were incubated at 26°C under 4000 lux illumination for 16-h days.

Seeds of *Sempervivum tectorum* L. started to germinate in 3–4 days. Initial development of the young plants was extremely slow; their diameter reached 1–1.5 cm by the 30th day.

Plants grown on White medium became yellow and could not be maintained. In contrast, plants grown on MS medium supplemented with thiamine (5 $mg l^{-1}$), pyridoxine (1 $mg l^{-1}$), and myo-inositol (100 $mg l^{-1}$) grew well.

To induce callus production, the authors used 30-day-old sterile plants and pieces (3–5 mm^2) of their leaves to inoculate MS medium supplemented with combinations of different growth regulators – benzylaminopurine (BAP), kinetin (KN), 1-naphthaleneacetic acid (NAA), and indole-3-acetic acid (IAA). BAP (1–2 $mg l^{-1}$) always induced callus production, although to different extents when used in combination with auxin. Callus production failed when the medium was supplemented with kinetin. According to an observation by Dudits (1982), dicotyledons often need cytokinin for the development of dedifferentiated cells.

The first visible callus was observed on the 16th day after inoculation of the medium supplemented with BAP (1 $mg l^{-1}$) and IAA (0.1 $mg l^{-1}$); this combination proved to be the best. Continuous transfer and maintenance of primary calluses were achieved on this medium in 25-day periods.

BAP with different combinations of auxins had different effects on callus production (Table 1; evaluation was 25 days after inoculation).

With a combination of BAP (1–2 $mg l^{-1}$) and IAA (0.1 $mg l^{-1}$) callus weight increased tenfold within 25 days, both for leaf and shoot explants.

Calluses with two different types of morphology could be distinguished — a brownish-yellow soft type and a bright green hard type.

2.2 In Vitro Regeneration

Primary callus development from leaf and shoot explants of common houseleek can be maintained and propagated with different inocula.

Induction of cell differentiation can also be achieved in vitro. The result of the redifferentiation process is plant regeneration.

Table 1. Effect of plant growth regulators added to MS medium on callus production from leaf and shoot explants of *Sempervivum tectorum*

Growth regulator combinations ($mg\,l^{-1}$)			Callus production (g fresh wt.) Explant-type	
BAP	IAA	NAA	Leaf	Shoot
1.0	0.1	–	5.2 ± 0.2	5.3 ± 0.1
1.0	0.2	–	4.5 ± 0.3	4.1 ± 0.1
1.0	–	0.1	3.8 ± 0.2	3.7 ± 0.2
1.0	–	0.2	4.0 ± 0.1	3.8 ± 0.3
2.0	0.1	–	5.0 ± 0.1	5.0 ± 0.2
2.0	0.2	–	3.5 ± 0.4	3.1 ± 0.3
2.0	–	0.1	3.0 ± 0.2	3.0 ± 0.2
2.0	–	0.2	3.8 ± 0.3	3.5 ± 0.4

The 0.5 g callus obtained after the second subculture was used as the initial inoculum.
Data were recorded 25 days after inoculation and are tabulated as means ± SE of results from 12 replicates per medium.

The aim of plant regeneration is to obtain plants from undifferentiated callus cells, i.e., to induce ontogenesis. In the in vitro system for *Sempervivum tectorum* L., one of the main alternative means of ontogenesis, organogenesis, could be reproduced and maintained (Dobos et al. 1994); somatic embryogenesis has not yet been induced by the authors.

Organogenesis occurred as a result of synchronous differentiation of several callus cells; shoot and root meristems were produced.

2.2.1 Organogenesis

In dedifferentiated callus production by *Sempervivum tectorum* L., partly differentiated cells, with different morphogenetic potency, can be observed.

Formation of shoot and root meristematic cells can be regulated and induced by changing the proportions of auxins and cytokinins.

After successful induction from the undifferentiated parenchyma cells, tracheids are first differentiated, as a result thickening their cell-wall (Fig. 3).

Among the tracheids, meristematic cells with large nuclei can be distinguished. From these cells – as a result of prolific cell division – meristemoids are formed in the callus (Fig. 4). In the next phase, depending on the growth-regulator content of the medium, the direction of morphogenesis, i.e., the type of developing organ, will be decided.

Globular cell clusters are formed at certain points before organ development.

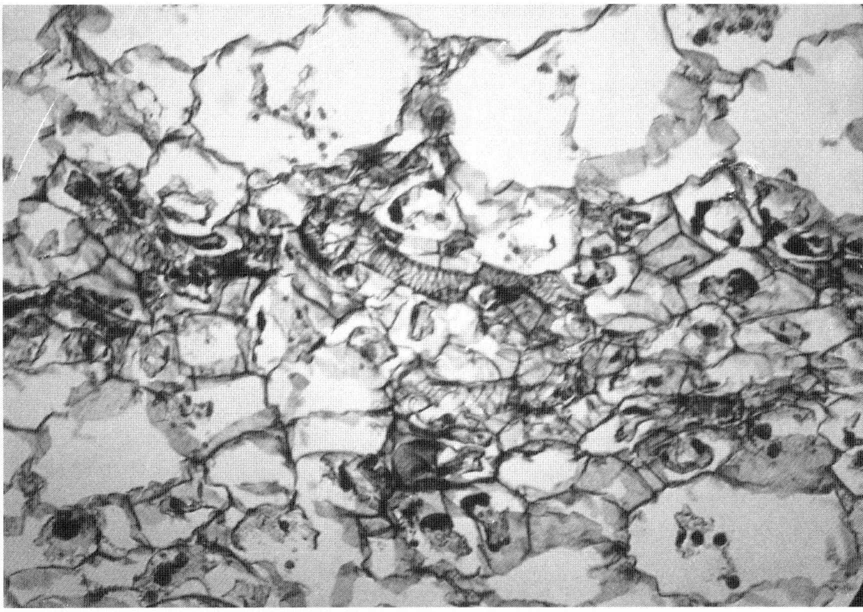

Fig. 3. Differentiation of tracheids from undifferentiated parenchyma cells in callus culture

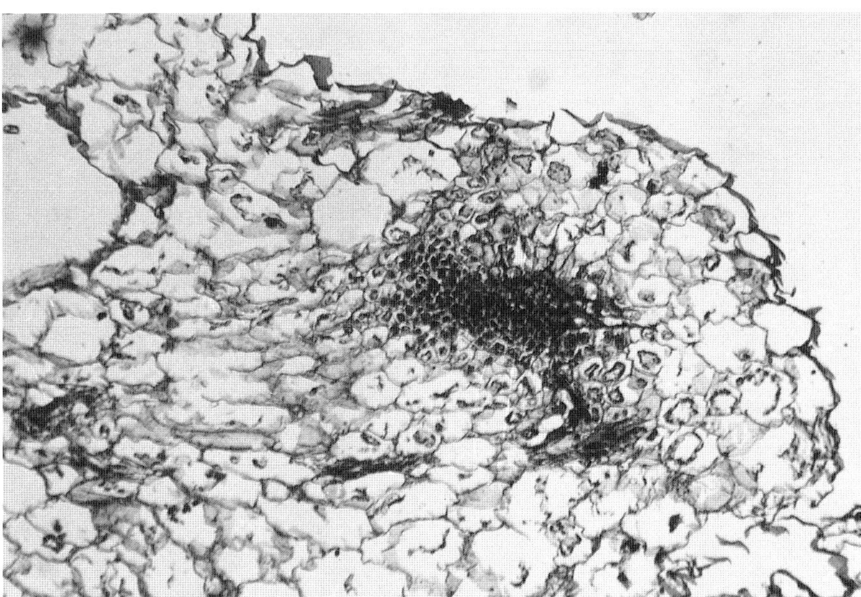

Fig. 4. Effect of BAP ($1\,mg\,l^{-1}$) and IAA ($0.1\,mg\,l^{-1}$) on meristemoid formation

2.2.1.1 Shoot Organogenesis

In shoot differentiation, meristematic tissue can be divided into two parts – tunica and corpus. One-cell-layers of protoderm are formed from the outer cell-layer of the tunica; hair growth follows later. The peripheral meristem develops into leaf primordium, or, near the shoot tip, shoot primordium (Fig. 5). Shoot differentiation was induced on MS medium supplemented with different auxin and cytokinin combinations (Dobos et al. 1994). The grown and undifferentiated callus was used for induction and differentiation. Soft yellow, and hard green calluses, which developed during callus induction and maintenance, differed not only in color and hardness – no differentiation was observed for the soft yellow calluses – whereas in the green calluses shoot organogenesis was observed 5 weeks after inoculation, depending on the plant growth regulators present (Figs. 6, 7).

Optimum shoot organogenesis was induced by IAA and 0.2–0.5 $mg\,l^{-1}$ indole-3-butyric acid (IBA), in combination with 0.5 $mg\,l^{-1}$ BAP. Under these conditions shoot organogenesis was detected in 62.6–78.3% of the cultures.

There was no difference between the morphogenetic potency of calluses originating from leaves and shoots.

When the medium contained KN, shoot organogenesis began in one or two cultures only. In callus cultures of *Sedum telephium* L., another member

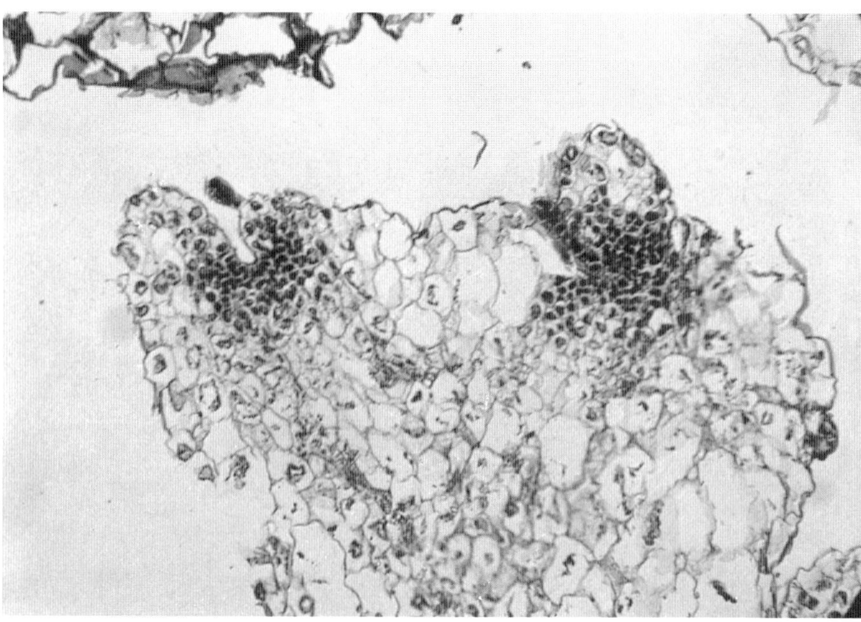

Fig. 5. Development of shoot primordium from the peripheral meristem

Fig. 6. Shoot organogenesis in a callus culture of *S. tectorum* 5 weeks after inoculation

Fig. 7. Regeneration and isolation of fully developed plants from green, hard callus

of the Crassulaceae, Brandao and Salema (1977) induced shoot differentiation in media containing BAP in combination with NAA, whereas KN was ineffective.

In callus cultures in which differentiation was observed, an average of 26.5 shoot buds were detected by Dobos et al. (1994) from 10 g callus mass of shoot origin after 5 weeks on media containing $0.5\,mg\,l^{-1}$ BAP and $0.2\,mg\,l^{-1}$ IAA. A few weeks later, some of the buds had developed into shoots.

2.2.1.2 Root Organogenesis

Histogens constituting root primordium develop from the meristemoids formed from the undifferentiated parenchyma cells of callus cultures.

Root organogenesis has not yet been induced in callus cultures of common houseleek; only the initial cells and meristematic tissues of the shoot tip had any activity.

Rooting plants could be obtained from shoots induced by shoot organogenesis; these shoots were excised and transferred to plant-growth-regulator-free MS medium. In all other cases, shoots developed in the callus cultures without root regeneration.

Rooted plants obtained in vitro were transferred to growth medium comprising a mixture of sand and perlite and adapted in a greenhouse before being planted out of doors.

3 Clonal Propagation Technique

Micropropagation of *Sempervivum tectorum* L. is based on the formation of shoots and buds in the shoot axils developed in callus cultures.

The in vitro propagation of the species is very important because plants grown by classic generative propagation develop extremely slowly. Only a limited number of plants can, however, be produced by clonal propagation using young, small rosettes.

3.1 Materials and Methods

Experiments were performed with plants of different ages grown from *Sempervivum tectorum* L. seeds (Botanical Garden, Research Institute for Medicinal Plants, Budakalász, Hungary) and raised in vitro. Seeds were disinfected in sodium hypochlorite (6% aqueous solution) for 15 min, then washed three times with sterile water. They were then sown on growth regulator-free MS basal medium containing $30\,g\,l^{-1}$ sucrose and $10\,g\,l^{-1}$ Difco Bacto-agar. The pH was adjusted to 5.8 before addition of the agar; the medium was then

Table 2. Effect of hormones on multiple shoot formation

Hormone ($mg\,l^{-1}$)				No. of shoots formed per culture	No. of plantlets formed
BAP	KN	IAA	NAA		
0.5				8.2	–
1.0				9.5	–
2.0				9.8	–
	0.5			8.0	–
	1.0			8.8	–
	2.0			9.0	–
0.5		0.1		10.0	4.1
1.0		0.1		10.5, 12.5[a]	6.8, 11.8[a]
2.0		0.1		10.6, 12.0[a]	6.5, 11.5[a]
0.5			0.1	9.5	3.5
1.0			0.1	9.8	4.0
2.0			0.1	9.8	4.0
	0.5	0.1		11.7, 13.5[a]	6.5, 10.5[a]
	1.0	0.1		11.5	13.8, 10.9[a]
	2.0	0.1		11.0	7.6
	0.5		0.1	10.4	5.6
	1.0		0.1	10.2	5.5
	2.0		0.1	10.1	5.0

[a] These data were obtained by use of different media supplemented with $80\,mg\,l^{-1}$ adenine sulphate in addition to the plant growth regulators.

autoclaved at 120 °C for 15 min. The sterile cultures were incubated at 26 °C under warm white fluorescent lamps giving 6000 lux for 16-h days.

Five weeks after germination, roots were cut from the seedlings and shoot inocula were transferred to MS basal media containing different cytokinins (BAP, KN), auxins (IAA, NAA), gibberellic acid (GA_3), casein hydrolysate, and adenine sulfate (Table 2). Vitamins ($5\,mg\,l^{-1}$ thiamine, $1\,mg\,l^{-1}$ pyridoxine) and myo-inositol ($100\,mg\,l^{-1}$) were also added to every basal media variant.

Shoots developing from axil buds were cut after 6–8 weeks and root growth was induced on MS basal media containing IBA and no growth regulators.

Rooted plants were adapted in a greenhouse, in growth medium comprising a mixture of sand and perlite, before being planted out of doors.

3.2 Results

The rate of germination of disinfected seeds of *Sempervivum tectorum* L. reached 85% by the 10–25th days on growth regulator-free MS medium. Plantlets reached the 0.5–0.6 cm diameter rosette phase within 4 weeks. Removal of the roots of plantlets and shoot induction were conducted during this phase of development (Table 2).

Multiple shoot formation was observed for all freshly excised rosettes on every medium, although the rate of formation varied.

BAP and KN added alone (i.e., not in combination) to MS medium resulted not only in shoot induction, but also in fully developed new shoots and plants.

Addition of IAA or NAA ($0.1\,mg\,l^{-1}$) to the basal medium in combination with BAP (1.0–$2.0\,mg\,l^{-1}$) or KN (0.5–$1\,mg\,l^{-1}$) promoted multiple shoot formation.

In these variants not only shoot induction, but also shoot development was observed, and whole plants were isolated successfully.

The best results (obtained by use of $1\,mg\,l^{-1}$ BAP or KN and $0.1\,mg\,l^{-1}$ IAA) were the development of an average of 6.8–7.8 new plants within 6–7 weeks.

The effect of casein hydrolysate ($100\,mg\,l^{-1}$) and adenine sulphate ($80\,mg\,l^{-1}$) were then determined for the best four media.

Casein hydrolysate proved ineffective, whereas adenine sulfate in combination with KN (0.5–$1\,mg\,l^{-1}$) and IAA ($0.1\,mg\,l^{-1}$) increased the number of new shoots by 60–70%; 56–57% more plants could be isolated on BAP (1–$2\,mg\,l^{-1}$) and IAA ($0.1\,mg\,l^{-1}$) media supplemented with adenine sulphate (Fig. 8).

Sixty to 70% of shoots isolated in the proliferation phase produced roots on growth-regulator-free medium; rooting was observed for 70–75% of shoots on IBA medium within 3 weeks.

Rooted plantlets were adapted, in a mixture of sand and perlite as growth medium, in a greenhouse for 6 weeks, ensuring 90% relative humidity for 14 days, before being transplanted out of doors.

Fig. 8. Shoot multiplication of *S. tectorum* on MS medium plus $1\,mg\,l^{-1}$ BA, $0.1\,mg\,l^{-1}$ IAA, and $80\,mg\,l^{-1}$ adenine sulphate

In vitro plants grown out of doors do not subsequently require special growing conditions.

4 Conclusion and Prospect

As previously described, 'Sempervivi tectori herba' contains isocitric and citric acids, alkaloids, flavonoids, and polysaccharides. The drug is used in folk medicine for curing skin burns, inflammation of the throat, and ear diseases.

Because the plant is protected in several European countries, collecting houseleek growing in the wild is prohibited, so production of *Sempervivum tectorum* L. plants is of great importance.

The disadvantage of generative proliferation is that plants grown from seeds develop very slowly; the disadvantage of vegetative multiplication is that a limited number of young rosettes can be obtained from the runners.

Soft brownish-yellow and hard bright green calluses have been developed in callus cultures of *Sempervivum tectorum* L. of leaf and shoot origin, respectively.

During the redifferentiation process only the hard green type of callus had morphogenetic activity. Shoot organogenesis was induced by KN only sporadically, whereas shoot development could be observed in 62–78% of callus cultures when BAP was combined with IAA.

Although root organogenesis could not be detected in callus cultures, the isolated shoots could be rooted on plant-growth-regulator-free medium.

Callus cultures cannot, however, be used for 100% successful production of large quantities of propagation material.

There is still much uncertainty in the successful induction of morphogenesis, because the causes of regeneration and the molecular changes involved are not completely understood (Ammirato 1983; Flich et al. 1983). We do not know which genes regulate the induction of morphogenesis, nor the endogenous conditions which must be maintained in the cells if this gene is to be activated (Dudits 1990).

The engineering of plant regeneration will be planned and reproduced easily only when conditions necessary for realization of cell totipotency are known on a molecular level.

In contrast to plant regeneration from calluses, the technology of in vitro micropropagation of *Sempervivum tectorum* L. can be applied even under farming conditions. In the propagation phase, six to eight new shoots can be isolated within 6–7 weeks under the action of BAP and KN in combination with IAA.

The efficiency of propagation can be improved by 50–70% by adding the reduced form of nitrogen, adenine sulphate, to the medium.

New shoots can be rooted on media containing IBA, but without the use of growth regulator; after adaptation plants can be placed out of doors.

This method of micropropagation is suitable for producing homogeneous propagation material from *Sempervivum tectorum* L. Shoot cultures maintain their capacity to regenerate during a long period of culture, in contrast to calluses.

The possibility of genetic changes during the propagation of axil-buds and axil-shoots is minimal, so this method is a highly suitable means of preserving genotypes and maintaining an in vitro gene-bank. This is very important in countries where *Sempervivum tectorum* L. is a protected plant.

Besides producing homogeneous propagation material, micropropagation is suitable for the production of standard pharmaceutical products of uniformly high quality for use in phytotherapy.

Different methods and results from in vitro propagation of *Sempervivum tectorum* are summarized in Fig. 9.

5 Protocol

5.1 Plantlet Regeneration by Morphogenesis from Callus Culture

1. Callus induction and transfer to MS medium containing BAP ($1\,mg\,l^{-1}$) and IAA ($0.1\,mg\,l^{-1}$) at 26°C under light.
2. Shoot organogenesis from the green callus on MS medium containing BAP ($0.5\,mg\,l^{-1}$) and IAA ($0.2\,mg\,l^{-1}$) or IBA ($0.5\,mg\,l^{-1}$).
3. Culture regenerated shoots on MS medium supplemented with BAP ($0.5\,mg\,l^{-1}$) and IAA ($0.2\,mg\,l^{-1}$) at 26°C for 5 weeks to obtain plantlets.
4. Rooting of regenerated plantlets on MS medium without hormones.

5.2 Technique for Clonal Propagation by Use of Axil-Buds

1. Culture of sterile rosette on MS medium containing KN ($0.5–1.0\,mg\,l^{-1}$) and IAA ($0.1\,mg\,l^{-1}$), or BAP ($1–2\,mg\,l^{-1}$) and IAA ($0.1\,mg\,l^{-1}$) supplemented with adenine sulphate ($80\,mg\,l^{-1}$) at 26°C for 3 weeks to form multiple shoot complex.
2. Transfer shoots to MS medium without hormones or containing IBA ($0.5\,mg\,l^{-1}$) at 26°C for 3 weeks to obtain rooting plantlets.
3. Adaptation of in vitro-rooted plantlets in a greenhouse in pots containing a mixture of sand and perlite at a relative humidity of 90% for 6 weeks.
4. Transplantation of the acclimatized in vitro-multiplied plants out of doors.

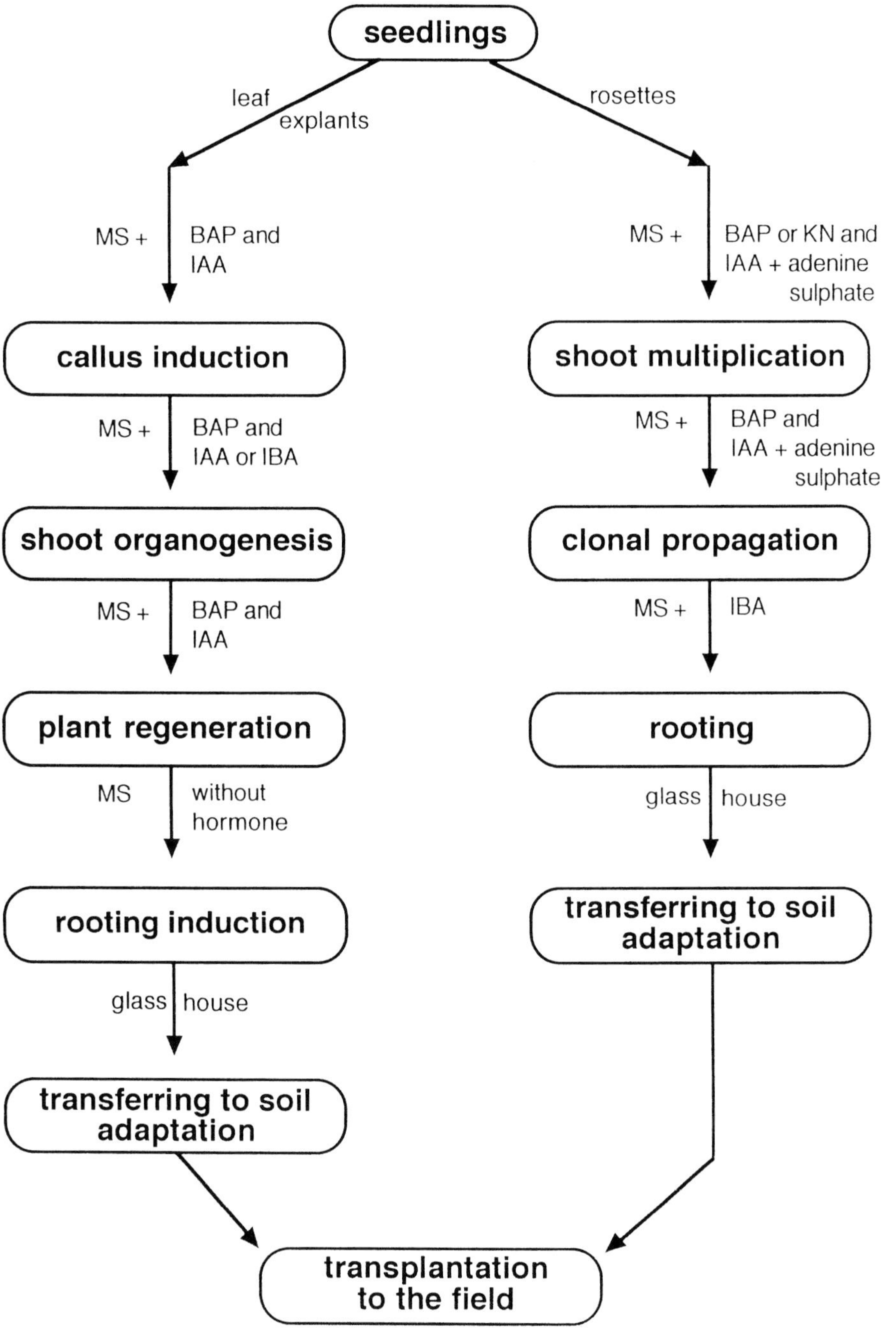

Fig. 9. Possible pathways to the in vitro propagation of *S. tectorum*

References

Ammirato PV (1983) Embryogenesis. In: Evans DA, Sharp WR, Ammirato PV, Yamada Y (eds) Handbook of plant cell culture, vol 1. Macmillan, New York, pp 82–123

Aurnhammer G (1979) *Sempervivum*. In: List PH, Hörhammer L (eds) Hagers Handbuch der pharmazeutischen Praxis, VI. Springer, Berlin Heidelberg New York, p 361

Barbakadze W, Gakhokidze RA, Shengeliya Zh (1989) Preliminary investigation of water soluble polysaccharides from Georgian plants. Chem Nat Compounds 25(3):281–286

Blázovics A, Pronai L, Fehér J, Kéry á, Petri G (1993a) A natural antioxidant extract from *Sempervivum tectorum*. Phytother Res 7(1)95–97

Blázovics A, Fehér J, Fehér E, Kéry á, Petri G. (1993b) Liner protecting and lipid lowering effects of *Sempervivum tectorum* extract in the rat. Phytother Res 7(1):98–100

Brandao J, Salema R (1977) Callus and plantlets development from cultured leaf explants of *Sedum telephium L*. Z Pflanzenphysiol 85:1–8

Codignola A, Fieschi M, Maffei M, Fusconi A (1990) Leaf anatomy and photosynthetic characteristics of succulent alpine plants growing at high elevations. Nordic J Bot 10(1):49–56

Csapody I (1982) Crassulaceae–*Sempervivum tectorum L*. In: Csapody I (ed) Védett növényeink. Gondolat Press, Budapest, pp 93–96

Dános B (1992) *Sempervivum tectorum*. In: Dános B (ed) Gyógynövényismeret. Semmelweis Press, Budapest, p 41

Dobos É, Dános B, László-Bencsik á (1994) Callus induction and shoot regeneration in *Sempervivum tectorum*. Plant Cell Tissue Organ Cult 36:141–143

Dudits D (1982) Kallusztenyészetek. In: Dudits D (ed) Fuzionált sejtek, hibrid növények. Akadémiai Press, Budapest, pp 21–24

Dudits D (1990) A növényregenerálás eredményei és feltételei. In: Dudits D, Heszky L (eds) Növénybiotechnológia. MezŒgazdasági Press, Budapest, pp 116–117

Elstner EF (1987) Metabolism of activated oxygen species. In: Davis DD (ed) Biochemistry of plants, vol 11. Academic Press, New York, pp 253–315

Fehér J, Vereckei A (1985) Szabadgyök-reakciók jelentŒsége az orvostudomány-ban. Biotéka 'Biogal' Pharmaceutical Industry, Medicinal Press, Budapest

Flich CE, Evans DA, Sharp WR (1983) Organogenesis. In: Evans DA, Sharp WR, Ammirato PV, Yamada Y (eds) Handbook of plant cell culture, vol 1. Macmillan, New York, pp 13–81

Friemert V, Heininger D, Kluge M, Ziegler H (1988) Temperature effects on malic acid efflux from the vacuoles and on the carboxylation pathways in Crassulacean acid metabolism plants. Planta 174(4):453–461

Frigot P (1960) Contribution à l'étude chimique de quelques Crassulacées et en particulier de divers. Sedum, Thése (Pharm) Univ Paris

Gumenyuk LA, Borisenko AN (1979) Russian houseleek – new source of biologically active compounds. Farm Zh (Kiev) 3:72–73

Gumenyuk LA, Golovatanya VF, Perova TV (1982) Phytochemical study on *Sempervivum ruthenicum*. Farm Zh (Kiev) 3:61–65

Hegnauer R (1964) *Sempervivum*. In: Hegnauer R (ed) Chemotaxonomie der Pflanzen 3. Birkhauser, Basel, pp 572–584

Kelly JG, Latzko E (1982) Carbon metabolism: the profound effects of illumination on the metabolism of photosynthetic cells. Progress in botany 44. Springer, Berlin Heidelberg New York, pp 103–131

Kéry Á (1993) *Sempervivum tectorum*. In: Bernáth J (ed) Vadon termŒ és termesztett gyógynövények. MezŒgazdasági Press, Budapest, pp 452–453

Kéry Á, Blázovics A, Rozlosnik N (1992) Antioxidative properties of extracts from *Sempervivum tectorum*. Planta Med 58(7):A661–A662

Kolb W, Schwarz T (1984) Properties and costs of substances for extensive covering of flat roofs with vegetation. Gartenamt 33(2):83–90

Kolb W, Schwarz T, Trunk R (1983) Planting roofs having a gravel covering. Zeitschrift für Vegetationstechnik im Landschafts und Sportstättenbau 6(4):143–151
Kovács M, Priszter Sz (1974) A flóra és a vegetáció változása Magyarországon az utolsó száz évben. Bot Közlemények 61(3):185–197
Kukhta EP, Aleksandrova IV, Paukov VN (1988) Polysaccharides from plant tissue cultures I. Properties and partial structure. Khimiya Prirodnykh Soedineniy Tashkent 3:323–346
Kull V (1967) Der Einfluss der Bodentrockenheit auf den Kohlenhydratgehalt einiger Crassulaceen. Planta 72:344
Kutkova T (1990) Herbaceous plants for sites with extreme conditions in landscape architecture. Acta Univ Agric 5(1):67–73
Larson RA (1988) The antioxidants of higher plants. Phytochemistry 27:969–978
Lotti G, Paradossi C, Marchini F (1991) A composition of new seed oils. Agrochimica 35:1–3, 56–68
Müssel H, Kiermeier P (1993) Experience with xeromorphic plants for extensive green cover. Gartenamt 32(6):376–381
Murashige T, Skoog F (1962) A revised medium for rapid growth and bioassays with tobacco tissue cultures. Physiol Plant 15:473–497
Nicolae M, Constantin N, Vectoria N, Cosmin NC (1986) Gel for regeneration of the epithelium. Rom. Ro 88580 (IPC A 17 K-031/715) Appl 112771
Nordal A, Klevstrand R (1951) Studies of the constituents of crassulaceous plants. Acta Chem Scand 5:85 898
Oztan I, Arslan M (1993) Research carried out to determine the species of succulent plants that can be grown in the ecological conditions of Central Anatolia region and the possibilities of their use as ground cover from the landscape architectural point of view. Doga Turk Tarim ve Ormanclik Dergisi 17(2):347–358
Paris R, Frigot P (1959) Sur l'isolement d'une flavanone á partir d'une Crassulacée indigène, le *Sedum altissimum* Poir. C R Hebdomadaires Séances Acad Sci 249:560
Parnell J (1988) Revision of the genus *Jovibarba* and consideration of the *Sempervivum tectorum/S. marmoreum* complex and *S. montanum* subsp. *carpaticum*. Acta Bot Hung 34:1–2 209–224
Pilou-Smith EAH, t'Hart H, Brederode J (1991) Seasonal variation of phosphoenolpyruvate carboxylase specific activity in fifteen species exhibiting facultative or obligate crassulacean acid metabolism. J Plant Physiol 138(5):581–586
Priszter SZ (1974) *Sedum* – és *Sempervivum* vizsgálatok. BotKözlemények 61(3)233–234
Ramstad E, Lieberman H (1955) Studies on the acid metabolism of Crassulaceae. Medd Norsk Farm Selskap 17:214
Roessner H, Popp M (1986) Ionic patterns in some Crassulaceae from Austrian habitats. Flora (Jena) 178(1):1–10
Sátory É, Kéry Á, Petri G (1988) The measurement of the effect of an antiphlogistic agent against otitis. Gyógyszerészet XXXII. 9:486
Schuber M, Kluge M (1979) Crassulacean acid metabolism (CAM) in succulents of Middle Europe: ecophysiological investigations in species of the genus *Sempervivum*. Flora (Jena) 168:1–2 205–216
Soderstrom THR (1962) The isocitric acid content of Crassulacean plants and a few succulent species from other families. Am J Bot 49:850
Stevens JF, t'Hart H, Block A (1994) Epicuticular wax composition of some European *Sedum* species. Phytochemistry 35(2):389–399
Stevens JF, t'Hart H, Van Ham R, Elena ET (1995) Distribution of alkaloids and tannins in the Crassulaceae. Biochem Syst Ecol 23(2):157–165
Stevens JF, t'Hart H, Elema ET, Block A (1996) Flavonoid variation in Eurasian *Sedum* and *Sempervivum*. Phytochemistry 41(2):503–512
Street HE, Cockburn W (1972) The path of carbon in photosynthesis. In: Street HE, Cockburn W (eds) Plant metabolism. Pergamon Press, Oxford, pp 144–151

Thomas DA, Andre M (1987) Oxygen and carbon dioxide exchanges in Crassulacean acid metabolism plants: I. Effect of water stress on hourly and daily patterns. Plant Physiol Biochem (Fr) 25(2):85–93

Vickery HB, Wilson DG (1958) Crassulaceae. In: Hegnauer R (ed) Chemotaxonomie der Pflanzen III. Birkhäuser, Basel, pp 573–575

White PR (1943) Nutrient deficiency studies and an improved inorganic nutrient for cultivation of excised tomato roots. Growth 7:53–65

13 In Vitro Culture and the Production of Steroidal Saponin in *Solanum aculeatissimum* Jacq.

T. IKENAGA[1], T. OYAMA[2], S. KIKUTA[2], and R. HANDAYANI[2]

1 Introduction

Solanum aculeatissimum Jacq. (family Solanaceae) is a perennial shrub that originated in tropical America. It is now distributed throughout the tropical and subtropical zones of the world. It has been used in China to treat bronchitis and rheumatism and as a decorative plant in Japan. *Solanum aculeatissimum* grows 30–90 cm in height, and the leaves and stems are covered with spines. The flowers are white, their globular (1.3–1.8 cm in diameter) fruits change from green to white, yellow and orange during growth (Fig 1). The fruit has about 200 seeds that easily germinate in the wild. This plant accumulates steroidal alkaloids in the leaves, stems and fruits (Kadkade et al. 1979), and steroidal saponins in the roots (Saijo et al. 1983).

High levels of steroidal saponins, aculeatiside A and aculeatiside B (Fig. 2), have been isolated from the root of this plant (Saijo et al. 1983). Natural steroidal sapogenin, diosgenin, is extracted from tubers of the *Dioscorea* species and is used in the commercial production of steroidal hormones, female steroidal hormones and steroidal pharmaceuticals (Ikegami 1981; Imada et al. 1981; Tal et al. 1984; Rokem et al. 1985). Since nuatigenin and isonuatigenin, obtained by the hydrolysis of aculeatiside A and aculeatiside B (Fig. 2), can be easily transformed into pregnenolone (Tschesche et al. 1964; Imada et al. 1981; Saijo et al. 1983), it is conceivable that *S. aculeatissimum* could be a more useful source of pregnane derivatives than *Dioscorea* species (Imada et al. 1981), as it contains such steroidal saponins at a high level. However, this plant has never been cultivated, and the properties and seasonal variations in growth and steroidal saponins production are not yet known.

Our earlier experimental results (Ikenaga et al. 1988a) showed that more than 10% of the dry weight of roots cultivated in the field was steroidal saponins, aculeatiside A and B. Since this content is similar to or greater than the steroidal saponin content in *Dioscorea* species (Imada et al. 1981), *S. aculeatissimum* would be a profitable source for steroidal hormone materials.

[1]Faculty of Environmental Sciences, Nagasaki University, 1–14 Bunkyo-machi, Nagasaki 852-8521, Japan
[2]Faculty of Pharmaceutical Science, Nagasaki University, 1–14 Bunkyo-machi, Nagasaki 852-8521, Japan

Biotechnology in Agriculture and Forestry, Vol. 51
Medicinal and Aromatic Plants XII (ed. by T. Nagata and Y. Ebizuka)
© Springer-Verlag Berlin Heidelberg 2002

Fig. 1. *Solunum aculeatissimum.* (Ikenaga 1995)

2 In Vitro Approaches

2.1 Tissue Culture

The results of a grafting experiment between *S. aculeatissimum* and *Lycopersicon esculentum*, suggest that the steroidal saponins, aculeatiside A and aculeatiside B, are synthesized in the roots (Ikenaga et al. 1990). Hairy roots of these plants under various conditions revealed that they produce steroidal saponins (Ikenaga et al. 1995).

Solasonine, solamargine and solasodine have been detected in callus tissue (Kadkade and Madrid 1977; Kadkade and Rolz 1977a,b). Nabeta et al. (1985) obtained free sterol and steroidal ketones from intact leaves, callus tissue and suspension cells and obtained steroidal saponin in the roots of plants regenerated from callus (Nabeta 1993). Recently, Stattman et al.(1994) have reported the creation of interspecific somatic hybrids between *S. khasianum* and *S. aculeatissimum* (Stattmann et al. 1994) but they did not note compounds produced by these plants.

Therefore, this chapter focuses on steroidal saponin production by the callus of *S. aculeatissimum* (Ikenaga et al.2000)

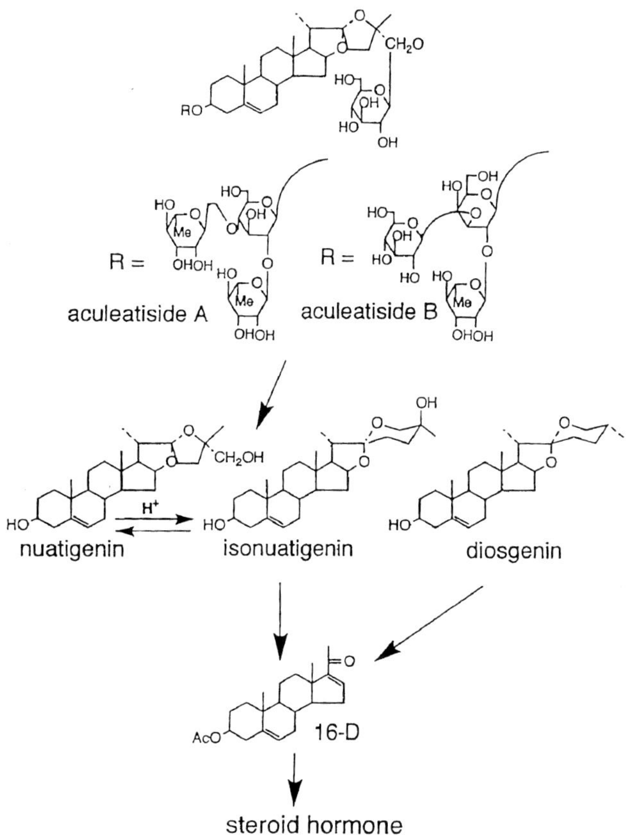

Fig. 2. Schematic representation of steroidal hormone synthesis. (Ikenaga et al. 1995)

2.2 Callus Culture

2.2.1 Time Course of Callus Growth and Production of Steroidal Saponins

To check the relationship between callus growth and steroidal saponin production, a callus was subcultured on MS medium supplemented with 0.1 ppm 2,4-dichlorophenoxyacetic acid (2,4-D) and 1 ppm benzyl adenine (BA) in darkness.

Callus growth was very slow, running into the log phase after 2 weeks and into the stationary phase after 8 weeks. Total steroidal saponin (aculeatiside A and B) content per dry weight was highest at the middle of log phase (after 5 weeks; Fig. 3).

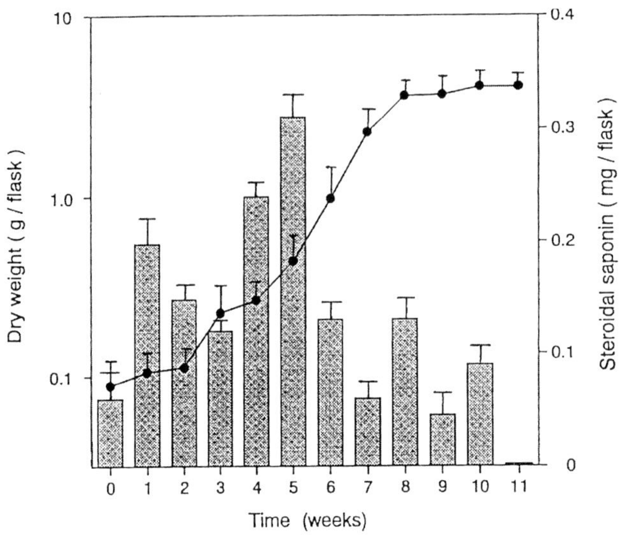

Fig. 3. Time course of callus growth and steroidal saponin (aculeatiside A + B) production in *S. aculeatissimum* cultured with MS medium containing 0.1 ppm 2,4-D, 1 ppm BA and 3% sucrose, in darkness at 25 °C. Values of dry weight and steroidal saponins content are the mean of five observations with standard error. (Ikenaga et al. 2000)

2.2.2 Effects of Cytokinins

After cultivation for 5 weeks, the callus was harvested and the effects of culture conditions on the growth of callus and the production of steroidal saponins were analyzed.

1. *BA*: When a callus was cultured on MS medium containing 0.01, 0.1, 1 and 10 ppm BA without auxin in darkness, growth was maximal in the medium containing 1 ppm BA, and the steroidal saponin content was highest in the medium containing 10 ppm BA (Table 1).
2. *Kinetin and light*: Hairy roots of this plant produced more steroidal saponins in the light than in the dark (Ikenaga et al. 1995). Therefore, the effects of kinetin (Kin) concentrations (0.01, 0.1, 1 and 10 ppm) and light on callus growth and steroidal saponin production were further examined under light and dark conditions. In order to discuss the effects of Kin on callus growth and steroidal saponins production, a control was established by culturing a callus in MS medium containing 10 ppm BA, since these conditions were previously shown to maximize the production of steroidal saponins.

Callus growth was not affected by Kin concentrations under the light, but it was stimulated with higher Kin concentrations in darkness (Fig. 4). Under the light, the callus formed shoots irrespective of Kin concentrations, and the

Table 1. Effects of BA concentration on the production of steroidal saponins in callus of *S. aculeatissimum*. Callus was cultured for 5 weeks with MS medium containing different concentrations of BA and 3% sucrose, in darkness at 25°C

BA (ppm)	Dry weight (g/flask)	Steroidal saponins (mg/g dry weight)	
		Aculeatiside A	Aculeatiside B
10	0.28 ± 0.01	0.10 ± 0.03	0.09 ± 0.02
1.0	0.58 ± 0.03	0.08 ± 0.01	0.07 ± 0.01
0.1	0.37 ± 0.02	0.01 ± 0.004	0.01 ± 0.003
0.01	0.37 ± 0.03	0.005 ± 0.001	0.005 ± 0.001
None	0.37 ± 0.01	0.01 ± 0.002	0.01 ± 0.002

Values of dry weight and steroidal saponins content are the mean of five observations with standard error.

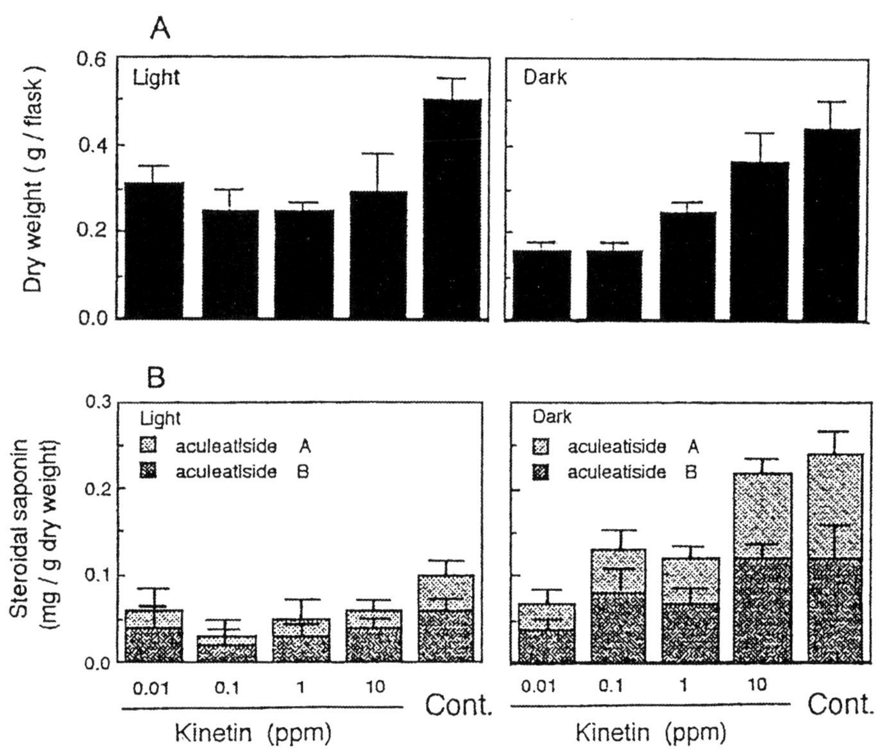

Fig. 4. **A** Effect of various KIN concentrations on growth and **B** steroidal saponins production of a callus of *S. aculeatissimum*. Callus was cultured for 5 weeks with MS solid medium containing three different concentrations of BA and 3% sucrose, in darkness or light at 25 °C. Values of dry weight and steroidal saponins content are the mean of five observations with standard error. Control: the callus was cultured with MS medium containing 10 ppm BA. (Ikenaga et al. 2000)

content of steroidal saponins was lower than in darkness. In darkness, the production of steroidal saponins was stimulated by Kin in a dose-dependent manner. However, the contents of the saponins were always lower at all Kin concentrations than that of the control (10 ppm BA; see Fig. 4). This fact indicated that the addition of 10 ppm BA to MS medium was a favorable condition for steroidal saponin production.

2.2.3 *Effects of Auxins*

The effects of three kinds of auxins [α-naphthaleneacetic acid (NAA), indol butyric acid (IBA) and 2,4-D] at various concentrations on callus growth and steroidal saponin production were examined. A callus cultured on MS medium containing 10 ppm BA was used as the control as above.

Callus growth was improved in proportion to concentration of NAA and IBA, but was suppressed by addition of 2,4-D (Fig. 5). Steroidal saponin contents were only significantly higher than those of the control, when the callus was cultivated in MS medium supplemented with 0.1 ppm NAA (Fig. 5). Steroidal saponins contents did not differ in media with varied concentrations of IBA. Their contents were severely decreased with higher concentrations of 2,4-D and only a trace amount of them was detected in the callus grown on 1 ppm 2,4-D medium. These results showed that addition of 2,4-D to medium inhibited the growth of callus and steroidal saponin production.

2.2.4 *Effects of Combination of NAA and BA*

When the callus was cultivated on media containing various combinations of NAA (0.05, 0.1, 0.5 ppm) and BA (5, 10, 15 ppm) in darkness, callus growth was most enhanced on media containing 0.1 or 0.5 ppm NAA and 10 ppm BA. Callus growth was suppressed when BA concentration was higher than 10 ppm (Fig. 6). Steroidal saponin contents were highest at a BA concentration of 10 ppm, and decreased at higher or lower concentrations of BA in the same manner as callus growth (Fig. 6). These results indicated that the optimal hormone condition for production of steroidal saponins was the combination of 10 ppm BA and either 0.05 or 0.1 ppm NAA.

2.2.5 *Effects of Carbon Source*

The effect of nine sugars (fructose, galactose, glucose, lactose, maltose, mannose, rhamnose, sucrose and xylose) of 3% concentration were examined as carbon sources.

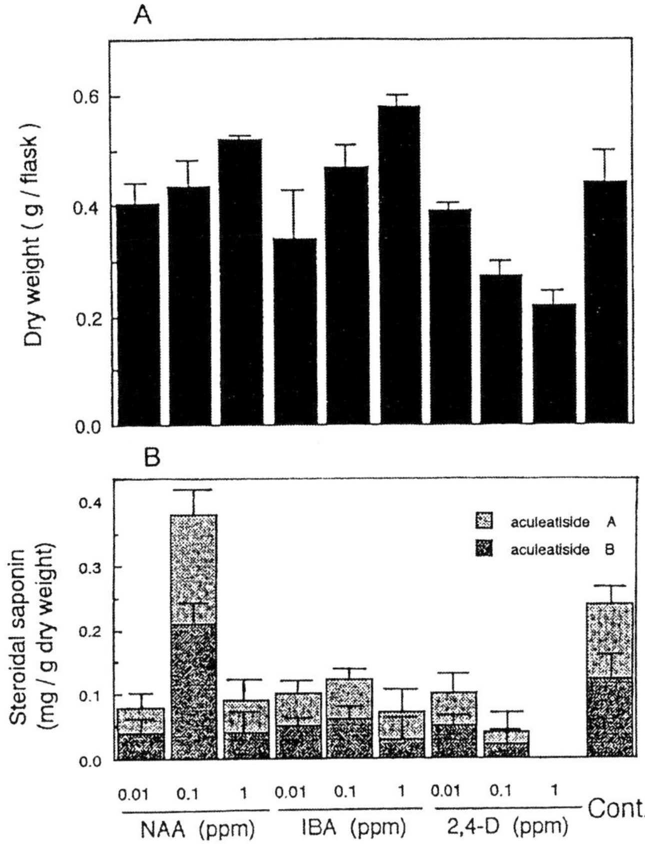

Fig. 5. Effects of several kinds and concentrations of auxins on **A** growth and **B** steroidal saponin production of a callus of *S. aculeatissimum*. Callus was cultured for 5 weeks with MS solid medium containing three kinds and three different concentrations of auxin and 3% sucrose, in darkness or light at 25 °C. Values of dry weight and steroidal saponin content are the mean of five observations with standard error. Control: the callus was cultured with MS medium containing 10 ppm BA. (Ikenaga et al. 2000)

The callus was cultured in darkness at 25 °C and when glucose was added to the medium containing 0.1 ppm NAA and 10 ppm BA, callus growth improved. However, steroidal saponin content was the highest in the culture medium containing fructose (Fig. 7). This result indicates that fructose was the most suitable carbon source for steroidal saponin production.

Therefore, the optimal culture conditions for production of steroidal saponins were MS basal medium containing the combination of 0.05 or 0.1 ppm NAA and 10 ppm BA, with fructose as a carbon source, in darkness at 25 °C. Under these optimal conditions, the callus produced 0.8% (dry weight) steroidal saponins, or 0.32% aculeatiside A and 0.48% aculeatiside B.

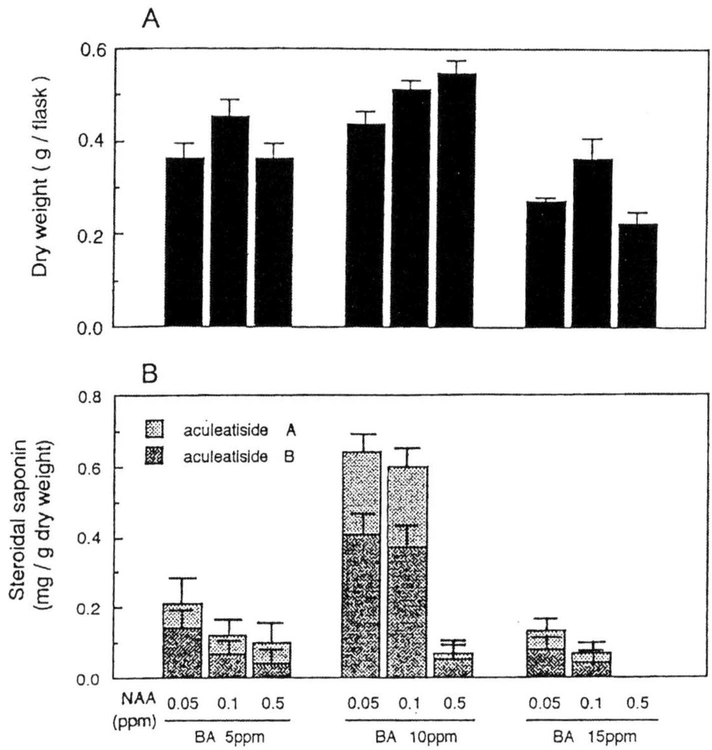

Fig. 6. Effects of combinations of BA and NAA on **A** growth and **B** steroidal saponin production of a callus of *S. aculeatissimum*. Callus was cultured for 5 weeks with MS solid medium containing three different concentrations of NAA and BA, and 3% sucrose, in darkness at 25 °C. Values of dry weight and steroidal saponin content are the mean of five observations with standard error. Control: the callus was cultured with MS medium containing 10 ppm BA. (Ikenaga et al. 2000)

3 Summary and Conclusions

Callus of *S. aculeatissimum* was induced from the epicotyl.

The relation between culture conditions and steroidal saponin production by callus was studied. The results indicate that the callus produces the steroidal saponins, aculeatiside A and B. The concentration of the products was highest in the middle of the log phase of growth.

The optimal conditions for maximum aculeatiside production in this experiment were basal medium of MS including NAA (0.05 or 0.1 ppm) and BA (10 ppm), fructose as a carbon source, and growth (at 25 °C) in the dark. Under these optimal conditions, callus produced 0.8% (per dry weight) total steroidal saponin, 0.32% aculeatiside A and 0.48% aculeatiside B.

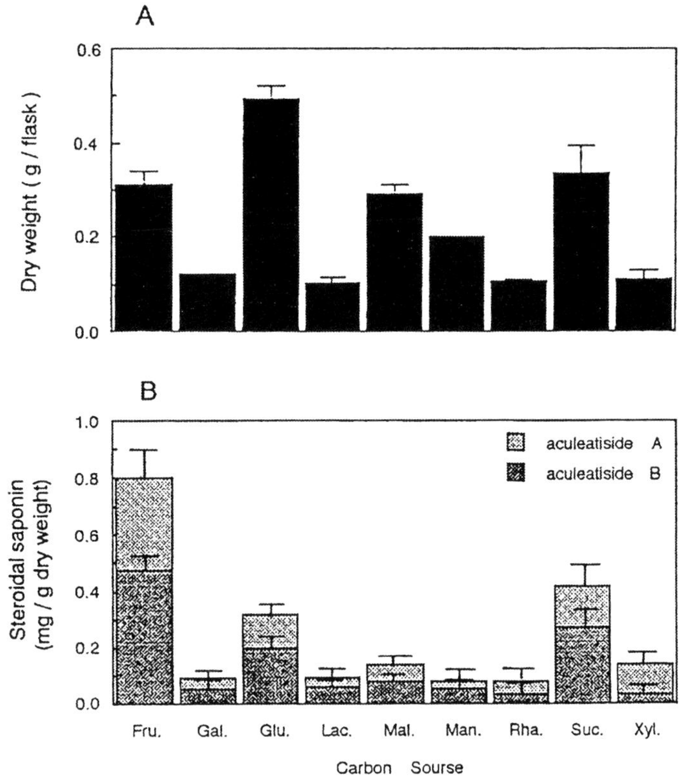

Fig. 7. Effects of various carbon sources on **A** growth and **B** steroidal saponin production of callus of *S. aculeatissimum*. Carbon sources used were fructose (*Fru*), galactose (*Gal*), glucose (*Glu*), lactose (*Lac*), maltose (*Mal*), mannose (*Man*), rhamnose (*Rha*), sucrose (*Suc*) and xylose (*Xyl*). Callus was cultured for 5 weeks with MS solid medium containing 0.1 ppm NAA and 10 ppm BA, and 3% various carbon sources, in darkness at 25 °C. Values of dry weight and steroidal saponin content are the mean of five observations with standard error. (Ikenaga et al. 2000)

4 Protocol

4.1 Callus Culture

When the epicotyl of a seedling from sterilized seeds of *Solanum aculeatissimum* Jacq. (Solanaceae) was cultivated on MS (Murashige and Skoog 1962) medium containing 1 ppm 2,4-dichlorophenoxyacetic acid (2,4-D), 5 ppm kinetin (Kin), 3% sucrose and 0.8% agar in darkness at 25 °C, a callus was induced. The induced callus was subcultured five times on the MS medium containing 0.2 ppm α-naphthaleneacetic acid (NAA) and 5 ppm benzyl adenine (BA) in darkness prior to use in the following experiments. The pH of the media was adjusted to 5.8 before autoclaving. Growth was determined as dry weight. For dry weight determination, the callus was harvested, then weighed after drying at 40 °C for 48 h.

Callus (ca. 0.5 g) was transferred to a 100- or 300-ml Erlenmeyer flask containing 40 or 100 ml MS basal medium supplemented with two kinds of cytokinins, three kinds of auxins, or the combined NAA and BA, and with 3% sucrose and 0.8% agar.

The callus was cultured in darkness or under the continuous light (fluorescent white light: ca. 60 $\mu E\, m^{-2} s^{-1}$).

4.2 Determination of Steroidal Saponins

S. aculeatissimum callus was dried at 40 °C for 48 h, then powdered mechanically and re-dried at 40 °C for 2 h.

In the experiment on callus growth and steroidal saponin production over time, when the callus dry weight was more than 0.5 g per flask, crude extracts for quantitative analysis were obtained by adding approximately 0.5 g of this callus powder to 100 ml methanol. Following filtration, the residue was extracted again in the same manner. The combined filtrate was evaporated to less than 10 ml, transferred into a volumetric flask and diluted to a final volume of 10 ml. When the callus dry weight was less than 0.5 g per flask, approximately 0.1 g powder was extracted twice with 20 ml ethanol. The filtrates were combined, evaporated and transferred into a 5-ml volumetric flask.

In other experiments, 0.1 g of the callus powder was used regardless of the weight of the dry callus and extracted twice with 20 ml methanol. Then, the filtrates were combined, evaporated and transferred into a 5-ml volumetric flask as described above.

Following filtration on a Millipore filter (0.22 μm), the filtrate was used for HPLC analysis. Aculeatiside A and B in the extract were identified by HPLC using a C-8 reverse phase column with 30% acetonitril as the mobile phase, as described previously (Ikenaga et al. 1988b).

References

Ikegami N (1981) Steroid production in industry. J Syn Org Chem 39:433

Ikenaga T, Kikuta S, Itimura K, Nakashima K, Matubara T (1988a) Growth and production of steroid saponin in *Solanum aculeatissimum* during one vegetation period. Planta Med 1988: 140–142

Ikenaga T, Kikuta S, Nakashima K (1988b) High-performance liquid chromatographic determination of steroidal saponin, aculeatiside A and B, in *Solanum aculeatissimum* Jacq. Chem Pharm Bull 36:416–418

Ikenaga T, Kikuta S, Kistuki M, Yamada M, Nakashima K (1990) Production of steroid saponin in grafts: *Solanum aculeatissimum* and *Lycopersicon esculentum*. Hortic Sci 25:1657–1658

Ikenaga T, Oyama T, Muranaka T (1995) Growth and steroidal saponin production in hairy root cultures of *Solanum aculeatissimum*. Plant Cell Rep 14:418–422

Ikenaga T, Handayani R, Oyama T (2000) Steroidal saponins production in callus cultures of *Solanum aculeatissimum* Jacq. Plant Cell Rep 19:1240–1244

Imada Y, Ishikawa H, Nishikawa D (1981) Development of new steroid fermentation. Nippon Nogeikagaku Kaishi 55:713–721

Kadkade PG, Madrid TR (1977) Glyco-alkaloids in tissue cultures of *Solanum aculeatissimum*. Naturwissenschaften 64:147

Kadkade PG, Rolz C (1977a) Steroidal alkaloids of *Solanum acculeatissimum*. Phytochemistry 16(7):112

Kadkade PG, Rolz C (1977b) The solasodine content of *Solanum acculeatissimum*. Lloydia 40(2):217

Kadkade PG, Recinos JA, Madrid TR (1979) Studies on the distribution of glycoalkaloids in *Solanum acculeatissimum*. Planta Med 37:70–72

Murashige T, Skoog F (1962) A revised medium for rapid growth and bioassays with tobacco tissue culture. Physiol Plant 15:473–497

Nabeta K (1993) XXIII *Solanum aculeatissimum* Jacq.: in vitro culture and the production of secondary metabolites. In: Bajaj YPS (ed) Biotechnology in agriculture and forestry, vol 24. Medicinal and aromatic plants V. Springer, Berlin Heidelberg New York, pp 329–341

Nabeta K, Matsubara T, Sugisawa H (1985) Production of free sterols and steroid ketones in plant tissue cultures of *Solanum aculeatissimum* Jacq. Nippon Nogeikagaku Kaishi 59:909–911

Rokem JS, Tal B, Goldberg I (1985) Methods for increasing diosgenin production by *Dioscorea* cell in suspension cultures. J Nat Prod 48:210–222

Saijo R, Fuke C, Murakami K, Nohara T, Tomimastu T (1983) Two steroidal glycosides, aculeatiside A and B from *Solanum aculeatissimum*. Phytochemistry 22:733–736

Stattmann M, Gerick E, Wenzel G (1994) Interspecific somatic hybrids between *Solanum khasianum* and *S. aculeatissimum* produced by electrofusion. Plant Cell Rep 13:193–196

Tal B, Rokem JS, Gressel J, Goldberg I (1984) The effect of chlorophyll-bleaching herbicides on growth, carotenoid and diosgenin levels in cell suspension cultures of *D. deltoidea*. Phytochemistry 23:1333–1335

Tschesche R, Richert KH (1964) über Saponin der spirostanoleihexi nuarigenin, ein Cholegenin – Analogon des Pflanzenreiches. Tetrahedron 20:387–398

14 *Solanum chrysotrichum* (Schldl): Distribution, Identification of the Antifungal SC-1 Saponin, in Vitro Propagation, Cell Suspension Cultures, Large-Scale Production of SC-1

M.L. VILLARREAL[1], L. ALVÁREZ[2], and R. QUINTERO[1]

1 Introduction

1.1 Distribution and Importance of the Plant

Solanum chrysotrichum (Schldl.) of the Solanaceae family belongs to a group of plants commonly known as "sosas" throughout the highlands of Chiapas, México. These plants are used for the treatment of dermatological infections and other skin ailments. Among this group, and as a result of extensive ethnobotanical investigations, two species, *Solanum chrysotrichum* and *S. lanceolatum,* are particularly noted, as revealed by the highest index of citation (Zurita and Zolla 1986; Lozoya and Aguilar 1987). The two species are described by traditional healers as the most effective herbal remedies for the treatment of skin infections. According to popular nosologies considered as "skin infections", water extracts from the leaves of *S. chrysotrichum* constitute the specific treatment for tiñae (*tiña pedis*), scabies and other mycosis (Zurita and Zolla 1986).

S. chrysotrichum is distributed in the states of Chiapas, Hidalgo and Michoacán, in Mèxico, where names such as *sosa, berenjena and cuxpeal* are given to this plant respectively. Among the highland Mayas of Chiapas, it is known as *"kúx peul"* in tzotzil, *"k´uxbal chíx"* in tzeltal, and *"pajutiek"* in chol. It is an erect perennial herb which may grow up to 2 m in height, with spiny stems. The leaves are rough to the touch, 20–30 cm long and 10 cm wide, covered with reddish, rigid, large hairs, and deeply divided margins. The flowers are white with a star appearance (Fig. 1). This plant normally grows at roadsides in temperate weather.

Data given from the healers of Chiapas on the herbal medication of *S. chrysotrichum,* indicated that the fresh leaves from the plant are boiled in water, and administered topically as plasters or poultices. However, sometimes the infusions are given orally, or as douches (Zurita and Zolla 1986).

[1]Centro de Investigación en Biotecnología, Universidad Autónoma del Estado de Morelos, México

[2]Centro de Investigaciones Quìmicas. Universidad Autónoma del Estado de Morelos, México

Biotechnology in Agriculture and Forestry, Vol. 51
Medicinal and Aromatic Plants XII (ed. by T. Nagata and Y. Ebizuka)
© Springer-Verlag Berlin Heidelberg 2002

Fig. 1. Adult specimen *of Solanum chrysotrichum* (Schldl.)

1.2 Pharmacological Investigations

Initial pharmacological investigations with *S. chrysotrichum* leaves were performed using organic solvent extracts from the plants collected in Chiapas. The extracts were evaluated for their ability to inhibit the in vitro growth of several Gram positive and Gram negative bacteria, as well as the yeast *Candida albicans*, and the dermatophyte *Microsporum gypseum*. The MIC (minimal inhibitory concentration) values from the plant extracts against the last two microorganisms were promising, since these pathogens show some resistance to actual prescribed drugs. In another series of studies, the activity of the extracts was specifically evaluated against the main causative agents of athlete's foot: *Tricophyton mentagrophytes, Tricophyton rubrum* and *Microsporum gypseum*, exhibiting important antifungal actions against the three dermatophytes in culture (Lozoya et al. 1991).

A pilot clinical study using organic extracts from the leaves of *S. chrysotrichum* was conducted at the Mexican Social Security Institute (Regional Hospital of Cuernavaca Morelos, México). A group of 18 ambulatory patients with diagnosis of *tiña pedis* (mainly produced by *Tricophyton mentagrophytes)* was selected and treated with a cream containing 5% of the methanolic extract of the leaves, which was applied topically twice a day during 4 weeks of treatment. A similar number of infected patients were considered as a control group, and treated with miconazole that was applied in the same way. The results showed that after 1 week of treatment, 42% of the patients receiving the plant extract recovered, while no cure was observed with miconazole during the same period. Remission of the symptoms was observed in both groups at the end of each complete treatment (García-Cruz 1988).

2 Isolation and Identification of the Antifungal Compound

The antifungal constituent (SC-1) of *S. chrysotrichum* leaves was obtained by bioautographic assay-guided fractionation and purification of the methanolic extract, using *Trichophyton mentagrophytes* as the biological monitor, since this dermatophyte is the main causative agent of athlete's foot disease.

The procedures outlined for the isolation and purification of SC-1 are shown in Fig. 2. Thus, the methanolic extract of this plant was resuspended in water and extracted successively with n-hexane, chloroform, and ethyl acetate to yield an active ethyl acetate-soluble fraction against the growth of *T. mentagrophytes*, as well as the inactive n-hexane, chloroform and aqueous fractions. The ethyl acetate-soluble fraction was chromatographed on a VLC (vacuum liquid chromatography) column to yield five fractions, which were evaluated for antifungal activity.

Chromatographic separation of the most active fraction (Fr. 4), using $CHCl_3$-MeOH (85:15) as eluent afforded a solid containing a mixture of saponins as determined by ^1H and ^{13}C NMR analyses. This mixture was separated by HPLC, using a microporasil silica gel column and eluted with $CHCl_3$–MeOH–H_2O (70:29:1) for 20 min at a flow rate of 0.8 ml/min and the eluent monitored with an RI detector. The first major peak (SC-1, R_t = 5.04 min) was characterized as 3-*O*-[β-D-quinovopyranosyl-(1→6)-β-D-glucopyranosyl-(1→6)-β-D-glucopyranosyl] chlorogenin (1) which showed a MIC = 40 μg/ml against *T. mentagrophytes.*

Compound 1 was obtained as a colorless amorphous powder, m.p. 157 160 °C, [α] −9.8 °C (ca. 0.24, pyridine). The molecular formula of 1 was determined as $C_{45}H_{74}O_{18}$ from the FAB mass spectrum and the ^{13}C-NMR spectrum.

The structure of SC-1 (1) was established on the basis of spectral analyses, mainly proton and ^{13}C-NMR, including two-dimensional techniques, i.e. ^1H-^1H COSY, HMQC and HMBC.

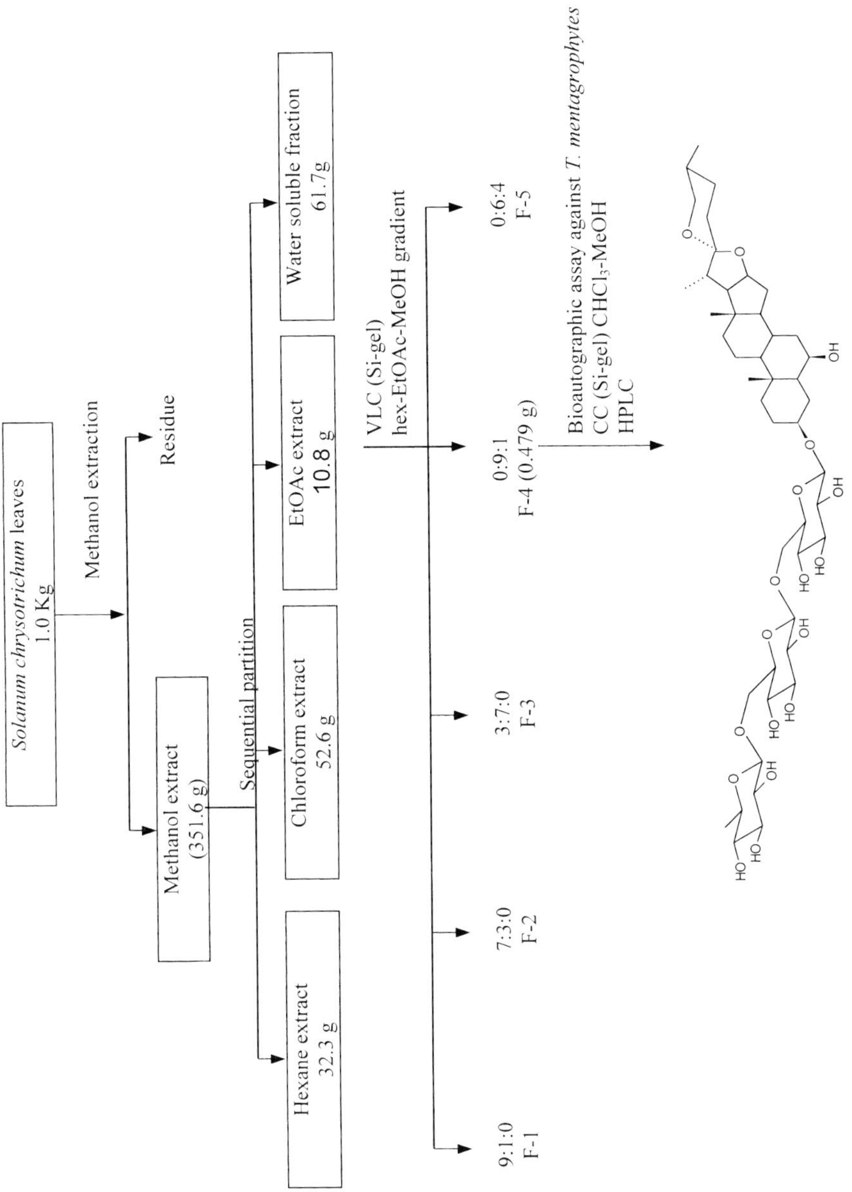

SC-1 (MIC *T.m.* 0.04 mg/ml)

Fig. 2. General procedure for the isolation and identification of SC-1 from *Solanum chrysotrichum*

Acid hydrolysis afforded chlorogenin, quinovose and glucose identified by means of TLC.

^{1}H-NMR and ^{13}C-NMR of 1 (Table 1) showed the presence of a terminal β-D-quinovopyranosyl unit and two 6-substituted β-D-glucopyranosyl units together with a chlorogenin moiety. The HMQC spectrum correlated all the ^{1}H resonances with those of corresponding carbons. The detection of all possible two- and three-bond inter- and intraresidual correlations of the anomeric protons, in the HMBC spectrum, confirmed unambiguously the glycosidic linkages and the assignments for proton and carbon resonances as shown in Table 1. Finally, FAB mass spectrum, showing the quasi-molecular ion peak [M + H]+ at m/z 903 and the fragment ion peaks at 755 $[M - \text{quinovose} + H]^+$, 593 $[M - \text{quinovose} - \text{glucose} + H]^+$, 431 $[M - \text{quinovose} - 2 \text{ X glucose} + H]^+$, 310 $[\text{quinovose} + \text{glucose} + H]^+$, and 147 $[\text{quinovose} + H]^+$, confirmed the sequence of sugar moieties. Thus, the chemical structure of SC-1 was established as

Table 1. ^{1}H- and ^{13}C-NMR spectral data of SC-1 (500 and 125 MHz, pyridine-d_5)[a]

	δ_H	δ_C		δ_H	δ_C
1		37.76	3-Glucose		
2		32.09	1′	4.85, d (7.5)	106.46
3		79.24	2′	4.00, t (7.5)	72.66
4		32.26	3′	4.15, t (8.5)	78.44
5	1.42, m	51.36	4′	3.72–3.81, m	70.63
6	4.32, m	69.40	5′	4.22, m	78.29
7a	1.27, m	41.47	6a′	3.72–3.81, m	66.97
7b	2.51, dd (3.8, 12.0)		6b′	4.32, m	
8		34.24			
9		53.93	(6→1)-Glucose		
10		36.71	1″	4.88, d (7.5)	105.75
11		21.27	2″	4.05, t (7.5)	72.94
12		40.06	3″	4.17, t (9.5)	78.69
13		40.76	4″	3.72–3.81, m	71.22
14		62.79	5″	4.22, m	78.29
15		33.19	6a″	3.72–3.81, m	66.54
			6b″	4.32, m	
16	4.47, q (7.5)	81.09			
17		41.41	(6→1)-Quinovose		
18	0.83, s	16.60	1‴	4.79, d (7.0)	106.34
19	0.83, s	13.57	2‴	4.48, t (8.5)	76.91
20		42.46	3‴	4.16, t (9.0)	76.01
21	1.14, d (7.0)	14.82	4‴	3.72–3.81, m	75.58
22		109.60	5‴	3.72–3.81, m	74.51
23		26.23	6‴	1.64, d (6.5)	18.78
24		26.12			
25		27.50			
26α	4.05, dd (2.0, 11.0)	65.02			
26β	3.35, d (11.0)				
27	0.87 d (6.5)	16.24			

[a] Values in parentheses are in Hz.

3-O-{β-D- quinovopyranosyl (1→6)-β-D-glucopyranosyl (1→6)-β-D-glucopyranosyl} chlorogenin (1).

3 In Vitro Approaches

3.1 Plant Induction and Callus Formation from Axillary Buds

The micropropagation of *Solanum chrysotrichum* as well as the development of undifferentiated calluses were performed using axillary buds as explants, growing in Murashige and Skoog (MS) medium, supplemented with various growth regulators (Villarreal and Muñóz 1991). The cultures were maintained in an incubator room at 28±2 °C with a daily photoperiod of 16 h at a light intensity of approximately 25 $\mu mol\,m^{-2}\,s^{-1}$. The explants were obtained from aseptic plantlets that were germinated from seeds of plant specimens collected in Chiapas, México. Voucher specimens were authenticated at the New York Botanical Garden. Induction of rooted plants was initiated only when indole-3 acetic acid (IAA) was present as an auxin, in combination with either of two cytokinins: kinetin (Kin) or benzyladenine (BA); however, the combination of IAA (0.1 mg/l + 0.2 mg/l BA) was found to be best suited for morphogenesis purposes (Table 2). Recent investigations with other *Solanum* species, *S. khasianum*, *S. nigrum*, *S. peruvianum*, and *S. pennelli*, also documented rooted plant induction using this hormonal combination (Kartha et al. 1974; Tal et al. 1977). Adaptation of in vitro-derived rooted plants was high (94%). Twelve months after they became adapted, the plants flowered. Field cultivation of *S. chrysotrichum* micropropagated plants for research purposes is actually being conducted. The plants are growing in fertilized soils at 28 °C with a mean biomass yield from leaves of 1 ton/10,000 m^2

Alternatively, callus tissue formation was influenced in the presence of naphthalene acetic acid (NAA) at low concentrations, independently of

Table 2. Hormonal effects on rooted plants production of *S. chrysotrichum* in MS medium

Hormone combination (mg/l)	Percentage of rooted shoots Days in culture			
	15	30	45	x
MS1 IAA:Kin (0.2:0.1)	0	21.4	39.3	20.2
MS2 IAA:Kin (0.1:0.2)	3.5	6.9	27.5	12.6
MS3 NAA:Kin (0.1:0.2)	0	10.3	6.9	5.7
MS4 IAA:BA (0.1:0.2)	3.5	51.7	65.5	40.2
MS5 NAA:BA(0.1:0.2)	3.4	3.4	3.4	3.4
F.cal	–	–	–	3.5+
MSE	–	–	–	187.6

Table 3. Hormonal effects on callus induction of *S. chrysotrichum* in MS medium

Hormone combination (mg/l)	Percentage of calluses Days in culture			
	15	30	45	x
MS1 IAA:Kin (0.2:0.1)	21.4	21.4	14.0	19.0
MS2 IAA:Kin (0.1:0.2)	13.8	17.2	20.6	17.2
MS3 NAA:Kin (0.1:0.2)	72.4	72.4	89.6	78.1
MS4 IAA:BA (0.1:0.2)	10.4	10.4	6.8	9.0
MS5 NAA:BA(0.1:0.2)	69.0	75.8	79.3	74.4
F.cal	–	–	–	104.31+
MSE	–	–	–	32.9

the added cytokinin, a situation reported for some other members of Solanaceae (Norton and Boll 1954; Table 3). Addition of Kin increased the callus weight.

3.2 Cell Suspension Cultures.

Cell suspension cultures of *S. chrysotrichum* were established from friable calluses using an inoculum of 3 g/l dry weight. The suspensions were initiated in 250-ml Erlenmeyer flasks containing 100 ml MS media supplemented with a constant Kin concentration (2 mg/l), and augmented with four different auxins MS1 [2 mg/l 2,4 dichlorophenoxyacetic acid (2,4-D)], MS2 (0.5 mg/l NAA), MS3 (1.3 mg/l IAA), and MS4(1 mg/l 2,4,5-trichlorophenoxyacetic acid [2,4,5 T]). The flasks were incubated in a New Brunswick rotary shaker at 130 rpm in a batch mode over 25 days, and with culture conditions already described for callus growth. The kinetic parameters on growth and metabolite production were established with the following variables that were measured every 5 days (three replicates): dry weight, fresh weight, pH, cell viability, medium carbohydrates, and SC-1 concentration. Fresh weight was determined as filtered biomass, dry weight after lyophilization, and cell viability using the fluorescein diacetate method (Widholm 1972). Identification and quantification of SC-1 were carried out by HPLC analysis using an R1–71 Merck refractive index detector, a Lichrosfer Si60 258◊4 mm, and a 5-mm Merck column; with a mobile phase of methanol-chloroform-water (29:70:1), and a flow rate of 1.2 ml/min. Retention times of SC-1 peaks (2.20 min) from cell cultures and wild plant material were compared using co-chromatography. Authentication of SC-1 in the cultures was corroborated by infrared comparison of the in vitro spectra, with the ones obtained from wild plants. Cell growth was registered in the four utilized media; however, MS1 was selected since a finer and more homogeneous suspension was obtained. Under these conditions, the maximum cell biomass was 12.9 g/l dry weight, which represents a 5.6-fold increase from the

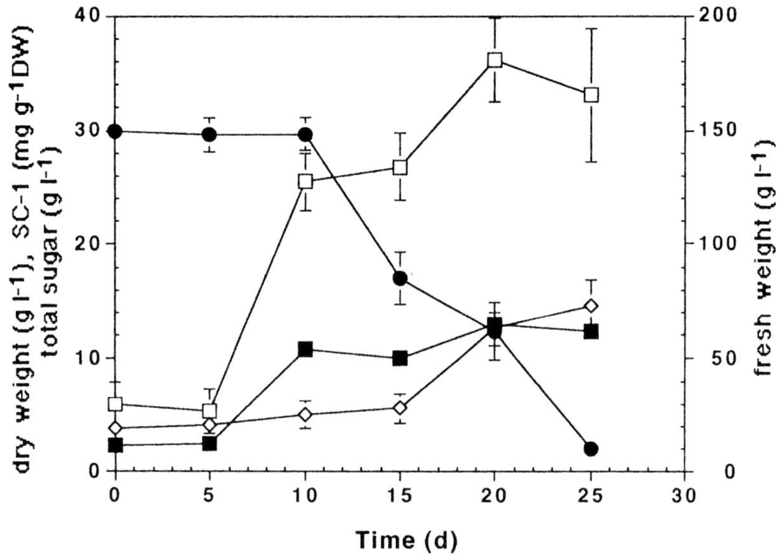

Fig. 3. Typical growth, SC-1 accumulation and carbohydrate uptake by *Solanum chrysotrichum* cell suspension cultures: *open squares* fresh weight; *filled squares* dry weight; *filled circles* total sugars; and *open diamonds* SC-l. (Villarreal et al. 1997a)

inoculum. The specific growth rate was 0.15 day^{-1}. The maximum concentration of SC-1 was 14.6 mg/g dry weight. The culture reached stationary phase after 10 days, even though a high level of sugar (ca. 22 g/l) still remained in the medium (Fig. 3). Doubling times based on fresh and dry weights were 4.5 and 5.0 days respectively (Villarreal et al. 1997a). These values are within average values reported for other species of the genus (Nigra et al. 1990; Alvarez et al. 1993), as well as other plant cell suspension cultures (Kutsal and Caglar 1992). Total carbohydrate uptake was closely associated with the increase in dry biomass. Sucrose was completely converted to glucose and fructose by day 10. All the glucose was consumed after 25 days in culture, whereas fructose levels were above 7 g/l (Fig. 4). Production of SC-1 was non-growth associated as observed from a relatively low product accumulation before cell concentration had reached 10 g/l dry weight (Fig. 5). This pattern has been reported for the accumulation of a number of terpenoid compounds by other plant species (Charlwood and Charlwood 1991). Influence of inoculum size (2.3 and 5%) over cell growth as well as SC-1 accumulation were investigated. As shown in Fig. 6, the highest concentration of biomass was reached with the 5% inoculum, while the highest SC-1 accumulation was obtained when using a 2% inoculum (Villarreal et al. 1997a). These results are in accordance with some earlier reports (Fujita et al. 1981; Alvarez et al. 1993), which demonstrated that the maximum yields for the secondary metabolites were obtained when the smallest inoculum size was used.

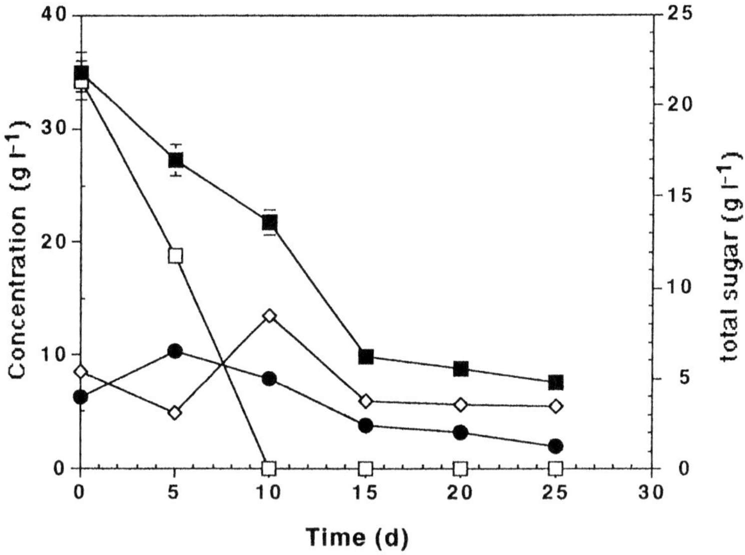

Fig. 4. Carbohydrate consumption in S. *chrysotrichum* suspension cell culture grown in MS medium: *filled squares* total sugars; *opened squares* sucrose; *filled circles* glucose; *open diamonds* fructose. (Villarreal et al. 1997a)

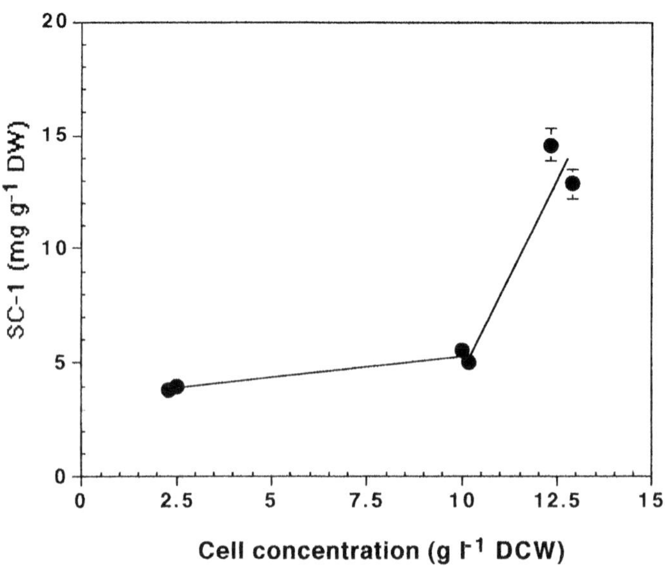

Fig. 5. Non-growth associated accumulation of SC-1 in cell suspension cultures of S. *chrysotrichum*

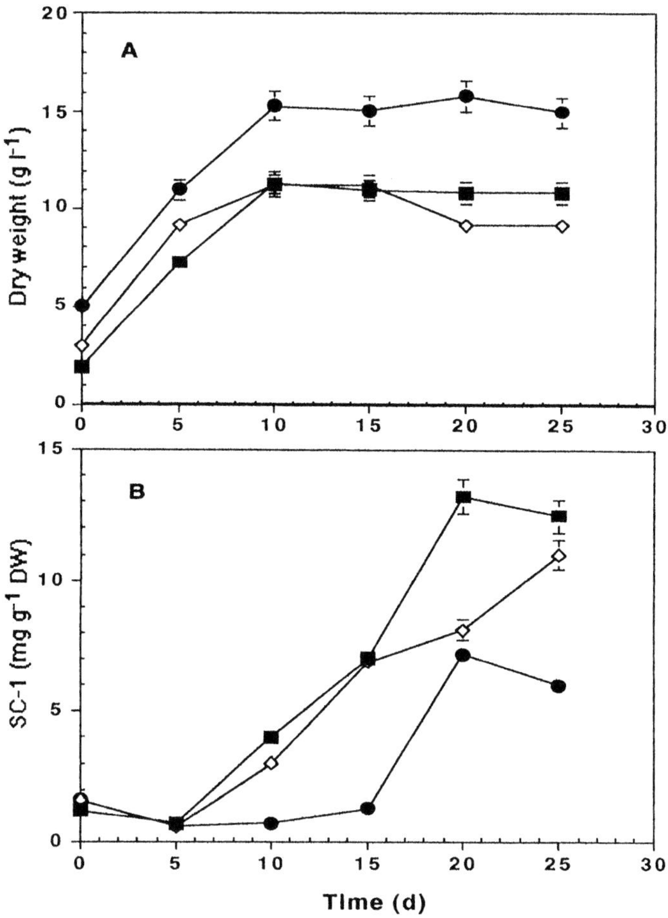

Fig. 6. **A** Effects of inoculum size on growth and **B** SC-1 accumulation of *S. chrysotrichum* cell suspension cultures: *filled squares* 2%; *open diamonds* 3%; and *filled circles* 5%. (Villarraeal et al. 1997a)

The effect of sucrose concentration (30, 45, 60 and 90 g/l) upon growth and metabolite production was also studied. Results after 20 days in culture indicated that maximum accumulation was obtained with sucrose concentrations between 30 and 45 g/l, while maximum cell growth was registered when a concentration of 60 g/l was used. The highest level of sucrose (90 g/l) provoked a decrease in biomass, presumably by changes in the osmolarity of the cultivated cells.

Different concentrations of nitrates, ammonium, phosphates and 2,4-D on cultivated suspensions were investigated using an orthogonal design with four factors and three levels. The results showed that neither variations in nitrate, phosphate or 2,4-D produced any positive results over cell growth and SC-1 accumulation, and that only the highest ammonium concentration (three times the usual) showed important increments in order to reach 14 mg/l biomass, and 15 mg/g dry weight of SC-1 after 20 days in culture.

Experiments in shake flasks show that cell suspension cultures of *S. chrysotrichum* are able to accumulate the SC-1 saponin at levels above 14 mg/g dry weight, which are 50 times higher than the maximum values obtained for the leaves of field-grown plants (0.282 mg/g dry weight) that were measured using similar analytical procedures.

3.2 Large-Scale Production of the Bioactive Saponin

Two different cell lines of *Solanum chrysotrichum*: *ccvx* (cotyledon derived), and *ccvz* (hypocotyl derived) were grown in 10-l airlift bioreactors in order to obtain higher biomasses, and to increase the productivity levels. The cells were cultivated in a batch mode for 3 to 4 weeks using two inocula of 2 and 3 g/l dry weight. A draw-fill batch culture mode was also investigated by harvesting 50% of the cell culture, and replacing it with fresh medium. The cell cultures grew in bioreactors as an homogeneous white to yellow suspension (Fig. 7).

Microscope examination of the cultures showed a variety of cell types with phenotypic differences in size, shape and vacuolar development. Batch growth and accumulation of SC-1 over 21 days in culture of both cell lines, using 2 g/l dry weight as inoculum, showed a maximum biomass concentration of 14.1 and 5.9 g/l dry weight for *ccvx* and *ccvz* respectively (Fig. 8). Cell suspensions displayed doubling times of 3.15 and 6.9 days respectively. Accumulation of SC-1 in bioreactor experiments was non-growth associated, and reached maximum values of 21 and 19 mg/g for *ccvx* and *ccvz* respectively, versus 14 mg/g obtained for *ccvx* in shake-flask experiments (Villarreal et al. 1997b). With 3 g/l inoculum and the same culture conditions described above, maximum biomass concentrations were 14.6 and 7.7 g/l dry weight for *ccvx* and *ccvz*, showing doubling times of 5.3 and 5.8 days respectively. Maximum SC-1 concentration for *ccvx* and *ccvz* were 23 and 20 mg/g dry weight after 17 and 24 days in culture (Villarreal et al. 1997b). Even though SC-1 accumulation reached approximately the same value for the two clones, *ccvx* exhibited a higher biomass concentration. Inoculum size at the two concentrations did not influence biomass accumulation or SC-1 concentration in the bioreactor cultures. Sucrose consumption in the bioreactor was relatively fast, after 12 days practically all the sucrose was hydrolyzed. The concentration of glucose and fructose increased initially, glucose consumption initiated after day 6, while fructose started to be used only after 10 days in culture. Figure 9 presents the growth profile and SC-1 production of *S. chrysotrichum* cells when a draw-fill batch culture mode was introduced. Interestingly, a steady state in specific SC-1 concentration of about 25 mg/g dry weight was obtained during the second

Fig. 7. Cell suspension cultures of S. *chrysotrichum* at 15 days grown in bioreactors

stage of the culture. The productivity reached in the bioreactors was between 2.33 and 2.0 times higher than in shake-flask cultures (Villarrreal et al. 1997a). The increments in productivities are due to higher biomasses, probably obtained since a better oxygen supply is reached in the bioreactors. It was also remarkable that the use of a draw-fill mode in the bioreactor cultures allowed a 2.25 higher productivity in the second stage, compared to the first stage. Such a high productivity was reached as early as 4 days after the addition of fresh medium. These results demonstrate that the use of the draw-fill batch culture mode with *S. chrysotrichum* cell suspensions can importantly increase productivity, eliminating dead periods such as the time required to sterilize the bioreactor as well as initial lag phases in the cultures.

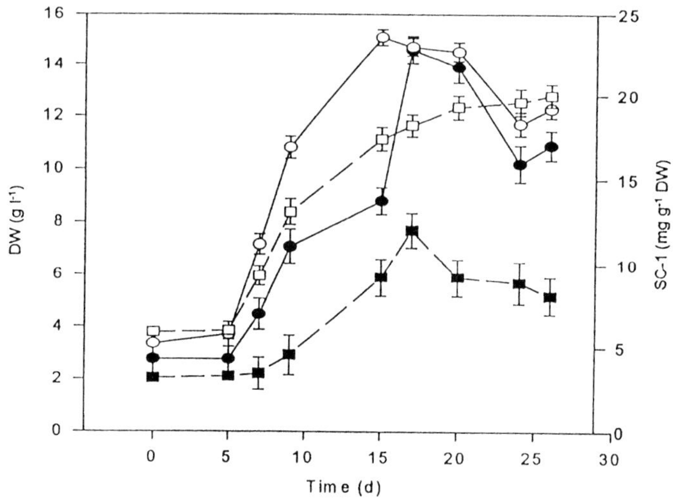

Fig. 8. Typical biomass (*filled circles, ccvx, filled squares, ccvz*) and SC-1 accumulation (*open circles, ccvx, open squares, ccvz*) in cell suspension cultures of *Solanum chrysotrichum* grown in 10-l airlift bioreactors

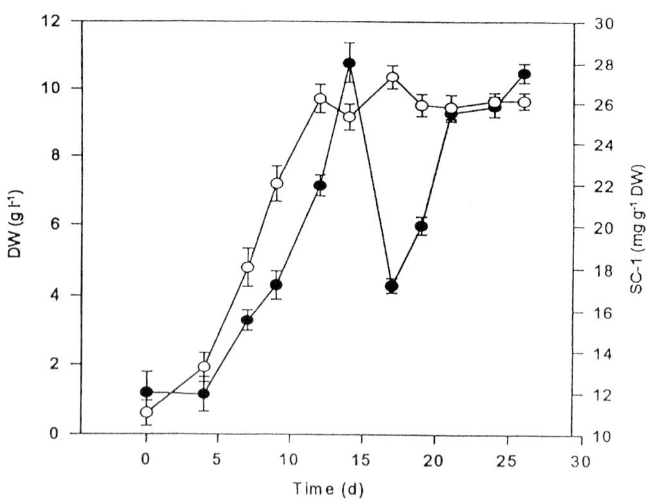

Fig. 9. Cell concentration (*filled circles*) and SC-1 accumulation (*open circles*) in *S. chrysotrichum* (*ccvx*) in a draw-fill cell culture mode

4 Conclusion and Prospect

From traditional medicine knowledge, it was possible to select a plant species used to treat skin mycosis. The pharmacological value of this plant was demonstrated, and we were able to purify and isolate an antifungal spirostanol saponin named SC-1, which represents a new molecule. The in vitro-presented experiments demonstrate that the establishment of cell suspension cultures from *Solanum chrysotrichum* is able to produce SC-1 with higher yields than those obtained for wild plants. Even though *S. chrysotrichum* is widely used as a popular medicine and grows in a very limited region in Mexico, the plant has never been cultivated, and the biotechnological approaches described herein, could allow the permanent and controlled production of the antifungal medicine. The cell suspension cultures of this plant were scaled to 10 l, demonstrating that the airlift bioreactors, that were used for this purpose, were adequate. The productivities were higher at a 10-l volume, and increased by 60% when using a draw-fill mode in the bioreactor. The obtained productivity values in bioreactors are within those that had been considered promising for other plant species. The results with the in vitro culture systems of *S. chrysotrichum* are feasible alternatives for the bioproduction of SC-1. These procedures demonstrate, for the first time, the production of an antimycotic chemical for human use in bioreactors, from an important popular plant remedy.

5 Protocol

Through a bioassay-guided fractionation, the new antimycotic steroidal saponin named SC-1 was isolated and identified from the leaves of *Solanum chrysotrichum*.

On the basis of spectral analysis (proton and ^{13}C-NMR including two-dimensional techniques), the structure of SC-1 was established as 3-*O*-{β-D-quinovopyranosyl (1→6)-β-D-glucopyranosyl (1→6)-β-D-glucopyranosyl} chlorogenin (1).

In vitro cultivation of *S. crysotrichum* was initiated in order to obtain a controlled production of plant specimens as well as bioproductive biomasses from cell cultures. The micropropagation of *Solanum chrysotrichum* was performed using axillary buds as explants in a defined MS medium augmented with a combination of indole-3-acetic acid (0.1 mg/l) + benzyladenine (0.2 mg/l). The bioproduction of SC-1 by individual cells of this species was initiated through liquid cell cultures. Batch suspension cultures obtained from hypocotyl-derived calluses were grown for 25 days in shake flasks containing Murashige and Skoog (MS) medium. The effects of inoculum size, as well as different concentrations of sucrose, nitrate, ammonium, phosphate and 2,4 dichlorophenoxyacetic acid (2,4-D), were evaluated. The maximum SC-1 production (14 mg/g) was 50 times that of field-grown plants reached after 20 days in culture using a 2% inoculum and complete medium supplemented with 2 mg/l 2,4-D + 2 mg/l kinetin, and sucrose between 30 and 45 g/l.

Scale-up of batch suspensions in bioreactors was investigated. Two cell lines *ccvx* (cotyledon derived) and *ccvz* (hypocotil derived) of *Solanum chrysotrichum* were cultivated in a 10-l airlift bioreactor for 3 to 4 weeks using two inocula of 2 and 3 g/l dry weight. A draw-fill batch culture mode was also investigated. Higher biomass growth (14.6 g/l) and productivity levels of SC-1 (0.041 g/l/day) in the draw-fill mode in bioreactors were obtained 4 days after harvesting 50% of the culture broth, and replacing it with fresh medium.

References

Alvarez L, Pérez MC; González JL; Navarro V, Villarreal ML; Olson J (2001) SC-1, an antimycotic spirostan saponin from *Solanum chrysotrichum*. Planta Med 67:372–374

Alvarez MA, Nigra HM, Giuletti AM (1993) Solasodine production by *Solanum eleagnifolium* Cav. in vitro cultures: influence of plant growth regulators, age and inoculum range scale production. Nat Prod Lett 2(1):9–19

Charlwood BV, Charlwood KA (1991) Terpenoid production in plant cell cultures. In: Harborne JB, Thomas Barberan FA (eds) Ecological chemistry and biochemistry of plant terpenoids .Clarendon, Oxford, pp 95–132

Fujita Y, Suga C, Marimato T (1981) Production of shikonin derivatives by cell suspension cultures of *Lithospernum reythrorhizon*. A new medium for the production of shikonin derivatives. Plant Cell Rep 1:61–63

García Cruz M (1988) Tratamiento de la *Tiña pedis* con *Solanum chrysotrichum*. Thesis, Facultad de Medicina IMSS-IPN, Cuernavaca, Mor

Kartha AK, Michaykuk MR, Kao KN, Gamborg OL, Constabel F (1974) Callus formation and plant regeneration from mesophyll protoplasts of rape plants. Plant Sci Lett 3:256–271

Kutsal T, Caglar A (1992) Biotechnological advances in developing countries. In: Vardar-Sukan F, Sukan SS (eds) Recent advances in biotechnology. Kluwer, Dordrecht, pp343–457

Lozoya X, Aguilar A (1987) Encuesta sobre el uso actual de plantas en la Medicina Tradicional Mexicana. Rev Méd IMSS (México) 25:283–291

Lozoya X, Navarro V, García M, Zurita M (1991) *Solanum chrysotrichum* (Schdl). A plant used in México for treatment of skin mycosis. J Ethnol 36:127–132

Nigra HM, Alvárez MA, Giuletti AM (1990) Effect of carbon and nitrogen sources on growth and solasodine production in batch suspension cultures of *Solanum eleagnifolium*. Plant Cell Tissue Organ Cult 21:55–60

Norton JP, Boll WG (1954) Callus and shoot formation from tomato roots in vitro. Science 119: 220–221

Tal M, Dettan K, Heikin II (1977) Morphogenetic potential of cultured leaf sections of cultivated and wild species of tomato. Ann Bot 41:937–941

Villarreal ML, Muñóz J (1991) Studies on the medicinal properties of *Solanum chrysotrichum* tissue culture. Callus formation and plant induction from axillary buds. Arch Med Res 22: 128–133

Villarreal ML, Arias C, Feria-Velasco A, Ramírez OT, Quintero R (1997a) Cell suspension culture of *Solanum chrysotrichum* (Schldl.). A plant producing an antifungal spirostanol saponin. Plant Cell Tissue Organ Cult 50:39–44

Villarreal ML, Arias C, Vega J, Feria-Velasco A, Ramírez OT, Nicasio P, Rojas G, Quintero R (1997b) Large-scale cultivation of *Solanum chrysotrichum* cells: production of the antifungal saponin SC-1 in 10 l airlift bioreactors. Plant Cell Rep 16:653–656

Widholm J (1972) The use of fluorescent diacetate and phenosatramine for determining viability of cultured plant cells. Stain Technol 47:189–194.

Zurita M, Zolla C (1986) Enfermedades dermatológicas en la Medicina Tradicional de México. Bol Oficina Sanitaria Panamericana 101:339–344

15 In Vitro Culture and the Production of Secondary Metabolites in *Solanum khasianum*

U. MÜHLENBECK, A. KORTENBUSCH, and W. BARZ[1]

1 Introduction

1.1 Distribution and Importance of the Plant

Solanum khasianum (Fig. 1) which originates from India, is a widespread plant in Asia. The stem and leaves of this bushy annual or short-lived perennial plant are packed with spines (Mann 1978).

The plant is of commercial and medicinal importance because of its high content of steroidal alkaloids such as solasodine and its triglycosides solasonine and solamargine. For the pharmaceutical industry, solasodine has become increasingly interesting as a valuable starting compound for the synthesis of steroid hormones, such as cortisone and oral contraceptives (Mann 1978). It can be readily converted to 16-dehydropregnenolone acetate, a key intermediate in the synthesis of steroid drugs, and therefore might serve as an alternative to the hitherto extensively used diosgenin (Mann 1978). In addition to the use of solasodine as raw material for the synthesis of steroid hormones, the solasodine glycosides, solamargine and solasonine, show biological activities, which might be of medicinal value. The antitumoral effects of a *Solanum sodomaeum* glycoalkaloid extract, mainly consisting of the glycoalkaloids solamargine and solasonine, have been described and this preparation has been recently used for treatment of certain skin cancers (Cham et al. 1987, 1991). The antitumoral activity of this preparation is largely due to solamargine, which is thought to be bound by lectins in the cancer cell membranes (Cham and Daunter 1990). Furthermore, the solasodine-based glycoalkaloids can be toxic or inhibitory to a wide range of organisms, e.g., several fungi (Fewell et al. 1994), the flagellate parasite *Trypanosoma cruzi* (Chataing et al. 1998), larvae of insects (Weissenberg et al. 1998), molluscs (Alzerreca and Hart 1982) and the nematode *Setaria cervi* (Ghosh et al. 1994). Solamargine has also been reported to exhibit antiviral activities (Thorne et al. 1985).

The toxicity of glycoalkaloids is mainly due to their lytic effects on biological membranes. The disruption of the membrane appears to result from the

[1]Institut für Biochemie und Biotechnologie der Pflanzen, Westfälische Wilhelms-Universität, Hindenburgplatz 55, 48143 Münster, Germany

Biotechnology in Agriculture and Forestry, Vol. 51
Medicinal and Aromatic Plants XII (ed. by T. Nagata and Y. Ebizuka)
© Springer-Verlag Berlin Heidelberg 2002

Fig. 1. *Solanum khasianum*. The stem and leaves have numerous spines

ability of the steroidal alkaloids to complex with free sterols of the membranes (Roddick et al. 1992). With regard to the solasodine-based glycoalkaloids, solamargine is the more disruptive compound compared to solasonine, but together the two compounds showed a marked synergistic effect (Roddick et al. 1990). In addition, steroidal glycoalkaloids can inhibit the acetylcholinesterase, a key postsynaptic enzyme in the transmission of the nerve impulse (Roddick 1989). However, solasonine and solamargine were much less inhibitory than α-solanine and α-chaconine (Roddick 1989).

1.2 Production of Steroidal Glycoalkaloids

The steroidal alkaloid solasodine and related compounds have been reported to occur in more than 200 species within the Solanaceae (Weiler et al 1980). Besides *S. khasianum*, *S. aviculare* and *S. laciniatum* are the most promising species for the commercial production of solasodine (Mann 1978). Solasodine-producing species are cultivated in many parts of the world, especially in the former USSR (*S. laciniatum*), India (*S. khasianum*), New Zealand (*S. aviculare*), eastern Europe and China (Mann 1978).

With regard to *S. khasianum*, the solasodine content ranges from 2.5 to 5% based on dry weight of the fruits, the most promising variety being Chatterjeeanum which is probably not *S. khasianum* but *S. viarum* Dunal (Mann 1978).

Only a few data are available on the commercial yield of solasodine. In India the productivity of this plant was reported as 72.4–106 kg/ha (Sudiarto 1989). In Indonesia, based on figures from cultivation experiments, the yield of solasodine was reported to amount to 37.5–75 kg ha^{-1} $year^{-1}$ (Sudiarto 1989).

Solasodine glycosides are mainly found in the fruit of the plant. The vegetative parts of the plant contain only traces of this steroidal alkaloid (Mann 1978). In the fruit the solasodine glycosides are present in the extracellular mucilage surrounding the seeds. In contrast to most other species, in *S. khasianum* the solasodine content does not decrease during fruit ripening (Khanna and Murty 1972). The presence of solasodine in ripe fruits is advantageous for agriculture because the timing of the harvest is not decisive for an optimal glycoalkaloid yield (Mann 1978). However, the aforementioned presence of spines at the stem and leaves is an enormous problem for the harvesting of the plant. Therefore, besides improving the solasodine content, reducing spininess is a main aim of plant breeders. The attempts to reduce the spines include techniques like gamma-ray irradiation of seeds and, in recent years, somatic hybridization (Bhatt 1975; Stattmann et al. 1994).

1.3 Chemical Structure of Steroidal Glycoalkaloids

The chemical structures of *Solanum* steroidal glycoalkaloids are shown in Fig. 2. These compounds are classified into two groups differing in their steroidal bases, namely the solanidanes and the spirosolanes, of which solanidine and solasodine are representatives, respectively (Goodwin and Mercer 1983). A series of sugar residues are attached to the oxygen at C-3 via β-glycosidic bond. In planta, the steroidal alkaloids occur mostly as glycosides. The triglycosides, like solamargine and solasonine, are the most common form (Mann 1978). In the case of solamargine, the aglycone is linked with glucose and two molecules of rhamnose (chacotriose), while solasonine possesses glucose, galactose and rhamnose (solatriose) as sugar moieties. The 3-*O*-glycosylation probably occurs after the synthesis of the steroid aglycone (Maga 1994). Moreover, a UDP-glucose:solasodine glucosyltransferase from eggplant has recently been partially purified and characterized (Paczkowski et al. 1997)

The biosynthesis of the aglycone starts from acetyl CoA, which is converted to mevalonic acid. Via a mevalonic acid pathway cholesterol, a key intermediate for various biosynthetic pathways in plants, is formed. Cholesterol is then converted through a series of reactions including the insertion of a nitrogen into the steroid skeleton to solasodine. A detailed review on the biosynthesis is presented by Mann (1978).

2 In Vitro Approaches

2.1 Review of Cell Culture Studies on *Solanum khasianum*

Reports about tissue or cell culture studies on *S. khasianum* are very rare. Callus cultures are normally cultured on MS (Murashige and Skoog 1962) or

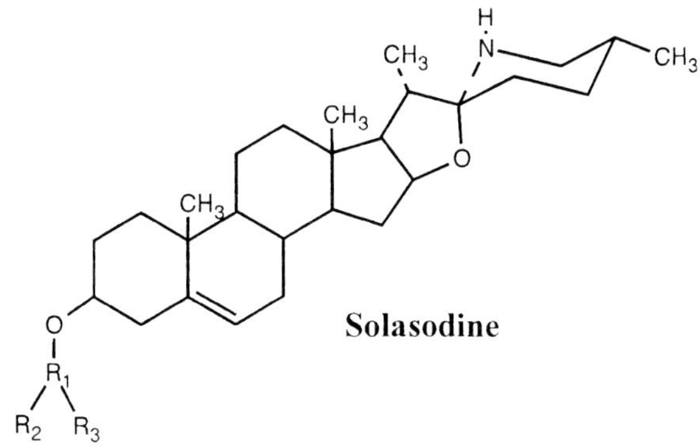

	R_1	R_2	R_3
Solasodine	H	-	-
α-Solasonine	D-Galactose	D-Glucose	L-Rhamnose
α-Solamargine	D-Glucose	L-Rhamnose	L-Rhamnose

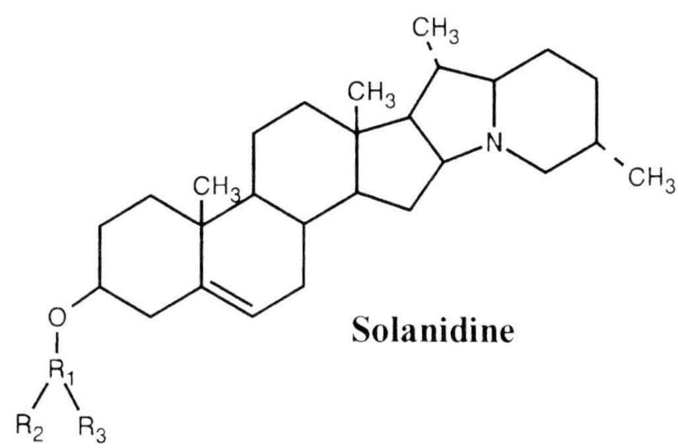

	R_1	R_2	R_3
Solanidine	H	-	-
α-Solanine	D-Galactose	D-Glucose	L-Rhamnose
α-Chaconine	D-Glucose	L-Rhamnose	L-Rhamnose

Fig. 2. Structures of spirosolanes (solasodine) and solanidanes (solanidine)

B5 (Gamborg et al. 1968) culture medium. In callus cultures obtained from *S. khasianum* seedlings, solasodine quantities in the range of 0.048–0.067% (reference is not specified in the paper) have been found (Aminuddin 1978). Investigations of Kokate and Radwan (1979) showed that undifferentiated callus cultures contain only traces of glycoalkaloids, whereas cultures starting to generate rootlets produce a large quantity of these compounds (up to 5.2% of dry weight). However, the major fraction of the isolated steroidal alkaloids could not be identified and solasonine and solanidine occurred only in trace amounts. The regeneration of *S. khasianum* from leaf-derived callus has been reported by Maheshwari and Bhalsing (1997). Shoot regeneration was induced by addition of 6-benzylaminopurine (BAP) to the MS-medium, while generation of roots was induced with α-naphthaleneacetic acid (NAA) supplementation. In addition, the solasodine content of these callus cultures amounted to 2% of dry weight. Although the solasodine contents obtained by the different research groups differ considerably, cell cultures do not seem to be a suitable system to produce solasodine on a large scale, because the solasodine content is much smaller than in the fruits of the plant. On the other hand, hairy root cultures of *Solanum aviculare* produce steroidal alkaloids at amounts as high as in the fruits of the plant (Subroto and Doran 1994). Kittipongpatana et al. (1998) demonstrated that in *S. aviculare* the solasodine content was much higher in hairy root cultures than in callus and cell suspension cultures, the latter showing the smallest solasodine content. To our knowledge, hairy root cultures of *S. khasianum* have not yet been investigated with regard to solasodine content. In *S. aviculare* cell cultures, the solasodine content could also be increased by addition of the cell-cycle inhibitor hydroxyurea (Mak and Doran 1993).

The use of in vitro cultures for improvement of plant species through new breeding techniques is gaining importance over classical breeding. In the case of *S. khasianum*, a high solasodine content, less spines as well as stress and disease resistance are desirable. One of the modern breeding methods to increase the genetic variability is the interspecific somatic hybridization through protoplast fusion. For this technique, it is necessary to regenerate plants from protoplast cultures. In the case of *S. khasianum*, regeneration of plants from protoplasts could be achieved by standard procedures (Kowalczyk et al. 1983; Serraf et al. 1988). Sihachakr et al. (1988) obtained somatic hybrid plants of *S. khasianum* and *S. melongena* by electrofusion of protoplasts. With the same method, Stattmann et al. (1994) succeeded in fusing protoplasts of *S. khasianum* and *S. aculeatissimum*. The regenerated hybrid plants showed intermediate phenotypes and produced fertile pollen and berries. Nevertheless, the hybrid plants died of a disease caused by an unidentified microorganism after they had been subjected to field trials in Indonesia (G. Wenzel, pers. comm.).

2.2 Establishment of Cell Suspension Cultures

For our investigations, callus cultures of *S. khasianum* C.B. Clarke were established from different parts of sterile plantlets, namely stem, hypocotyl and leaf. They were cultivated on MS-medium supplemented with 3% sucrose and 1 mg/l each of 2,4-dichlorophenoxyacetic acid (2,4-D) and kinetin (Mühlenbeck

1996). The stem-derived calluses served as starting material for photomixotrophic cell suspension cultures since they grew better than the calluses derived from the roots and leaves.

The cell suspension cultures were grown in the aforementioned MS medium on rotary shakers at 25 °C under continuous white light. The sucrose concentration was reduced stepwise from 3 to 1.5%. Every 14 days ca. 5 g (wet weight) of the cells was transferred into fresh culture medium. Moreover, heterotrophic cell suspension cultures were initiated by transferring the photomixotrophic cultures to the dark.

The cells formed aggregates of about 5 mm in diameter. Therefore, upon transfer into fresh medium, these aggregates were reduced to small pieces with a sterile spoon. Attempts to reduce aggregation by changing the concentration of phytohormones or by using sieves during inoculation were unsuccessful.

Growth characteristics of the photomixotrophic cell cultures were investigated over a period of 17 days in cultures growing with 3% sucrose. The fresh weight (fresh wt.) of these cultures increased threefold during this period. However, by the 17th day of the growth cycle, the stationary growth phase was not yet reached. The dry weight showed an increase by a factor of 2.3 over the same period. Furthermore, the ratio between dry and fresh weight decreased between day 13 and day 17, indicating an increase in the cellular water content during this time. The chlorophyll content of the cells reached values between 11 and 17 μg/g fresh wt. and the minimum content was observed between days 3 and 7.

The sucrose and phosphate utilization was also measured. The cells utilized ca. 70% of the sucrose during the first day of the growth cycle and phosphate was also consumed almost completely (95% of the initial concentration) within the first day. After that, the sucrose content remained on a level of ca. 8 g/l until the 14th day and then decreased to 3 g/l until day 17.

2.3 Extraction of Steroidal Alkaloids and Phenolic Compounds

2.3.1 Extraction of Solasodine and Its Glycosides

The solasodine content in cell suspension cultures was measured semiquantitatively with thin layer chromatography using solasodine and solamargine as standards (Kortenbusch and Barz, unpubl. results). The extraction was performed by both methanolic and acidic extraction. For methanolic extraction dried or fresh cells were extracted three times with methanol at 60 °C. The combined methanolic fractions were dried on a rotary evaporator and the residue was resuspended in small volumes of methanol. Acidic extraction was performed according to Ehmke and Eilert (1986). The cells were extracted three times with 0.2 N HCl at 60 °C. The combined aqueous supernatants were acidified to 1 N HCl (final concentration) and hydrolyzed for 2 h at 95 °C. After alkalization to pH 9 with 25% NH_4OH, the hydrolyzed extract was extracted three times with chloroform. The organic phase was then evaporated to dryness and the residue was redissolved in methanol. The methanolic and acidic extracts were analyzed and quantified by TLC on silica gel using different solvent systems and different detection reagents (Stahl 1967). As a rule, the methanolic extracts were developed in chloroform/methanol/water (20:10:1) and the acidic extracts in chloroform/methanol/water (50:10:1). The most sensitive dye reagent was the Car-Price reagent with a detection limit of 0.25 μg for solasodine and 0.5 μg for solasonine. However, solasodine could be found neither constitutively in cell suspension cultures nor after elicitation with a yeast polysaccharide elicitor. On the other hand, stimulation of solasodine production by addition of fungal elicitors was reported for cell suspension cultures of *Solanum elaeagnifolium* (Paniego et al. 1995) and *S. surattense* (Malpathak and David 1992). In our studies, different parts of the plant were also analyzed. In contrast to cell cultures, the fruits were found to contain ca. 4.5% of solasodine glycosides or 2.1% solasodine per gram dry weight, respectively, whereas solasodine could not be detected in leaves and roots. In addition, the solasodine content of fruits of three hybrid plants, which had been produced through protoplast fusion of *S. khasianum* and *S. aculeatissimum* (Stattmann et al. 1994), was also determined. All three hybrids clearly contained less solasodine (ca. 50%) than *S. khasianum*.

2.3.2 *Extraction of Phenolic Compounds*

We especially analyzed elicitor-induced effects on phenolic constituents. Elicitation of the cell culture and extraction of phenolic compounds from the cell culture medium were conducted as described by Mühlenbeck et al. (1996).

The most evident reaction after elicitation was the massive accumulation of nine phenolic compounds in the cell culture medium. Figure 3A,B shows HPLC chromatograms of extracts of the non-elicited control cultures and the elicited cultures, respectively. The structures of the compounds were elucidated by UV, mass and NMR spectroscopy. They were identified as the *cis*- and *trans*-isomers of *N*-p-coumaroyloctopamine, *N*-feruloyltyramine, *N*-p-coumaroyltyramine and *N*-feruloyltyramine as well as α-hydroxyacetovanillone (Fig. 4). The feruloyl conjugates and α-hydroxyacetovanillone were also found to accumulate constitutively in the culture medium, however, in much smaller quantities. Since the appearance of *cis*-isomers of cinnamates is very limited in plants, we investigated the formation of these isomers. Therefore, the cell cultures which normally grew under light were transferred to the dark and then elicited. In these cell cultures elicited in the dark, only the *trans*-isomers were found. Transfer of the cell-free culture medium containing the *trans*-isomers to light again led to the formation of *cis*-isomers. These results demonstrate that the formation of the *cis*-isomers is a light-induced isomerization process (Mühlenbeck et al 1996). Moreover, kinetic studies revealed a transient time course of the accumulation of the cinnamoylamides with

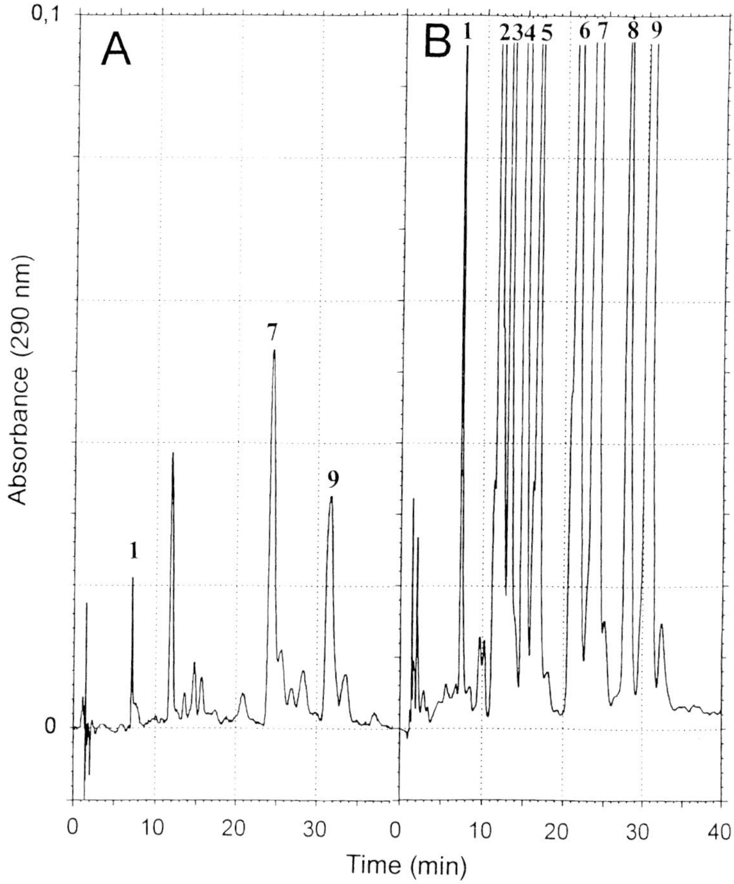

Fig. 3. HPLC chromatograms of phenolic compounds accumulating in the culture medium of photomixotrophic *S. khasianum* cell cultures. **A** A medium extract of a control culture, **B** an extract of an elicited culture. The signals **1**–**9** represent the isolated compounds (cf. Fig. 4)

maximum accumulation between 6 and 24 h depending on the compound. The decrease in the compounds was shown to be a peroxidase-mediated reaction and it is hypothesized that they are oxidatively polymerized.

With regard to their biological function, the hydroxycinnamoylamides seem to play an important role in plant defence against pathogens especially in some species within the Solanaceae, where they have been found to be incorporated in cell walls (Clarke 1982; Beimen et al. 1992; Hohlfeld et al. 1996). It is assumed that the incorporation leads to a reinforcement of the plant cell wall and, therefore, prevents the pathogen from penetrating into the plant cell. In our *S. khasianum* cell suspension cultures, the *trans*-isomers of the hydroxycinnamoyltyramines could also be detected in alkaline cell wall hydrolysates

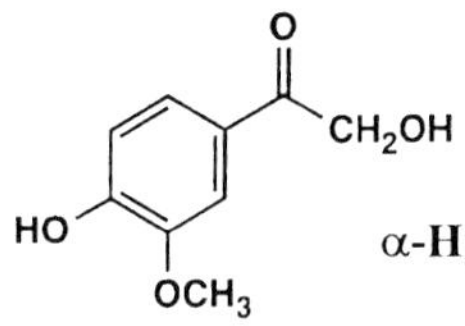

α-Hydroxyacetovanillone (1)

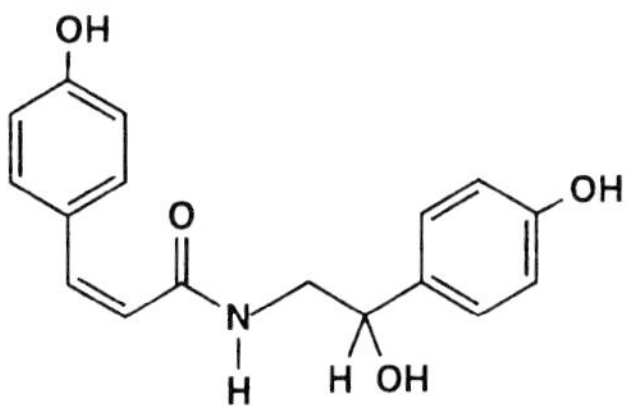

cis-p-Coumaroyloctopamine (2)

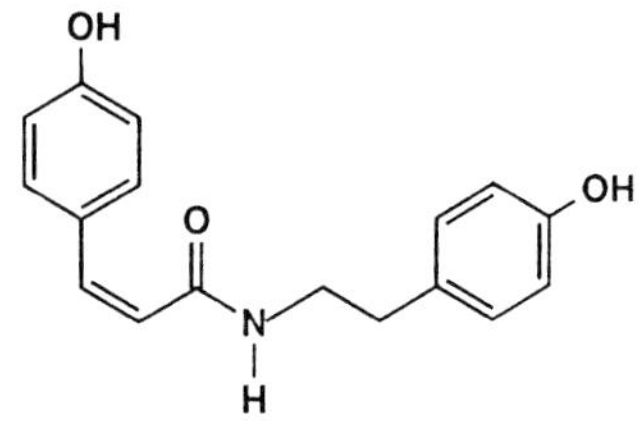

cis-p-Coumaroyltyramine (6)

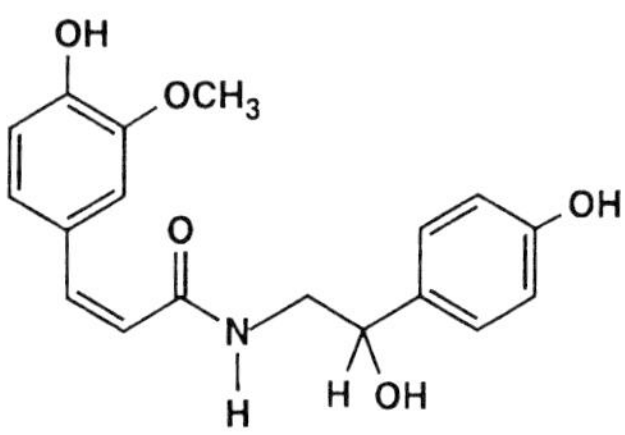

cis-Feruloyloctopamine (3)

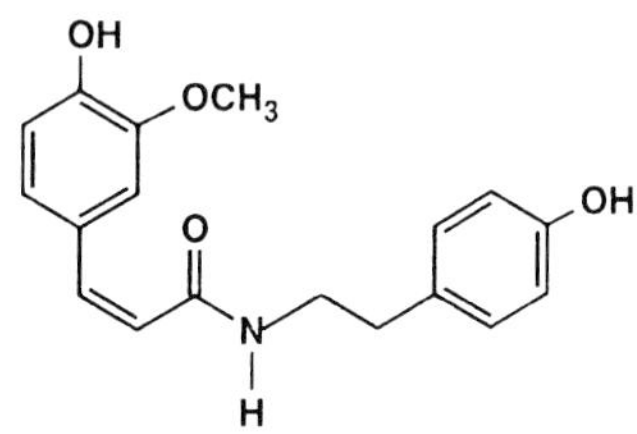

cis-Feruloyltyramine (7)

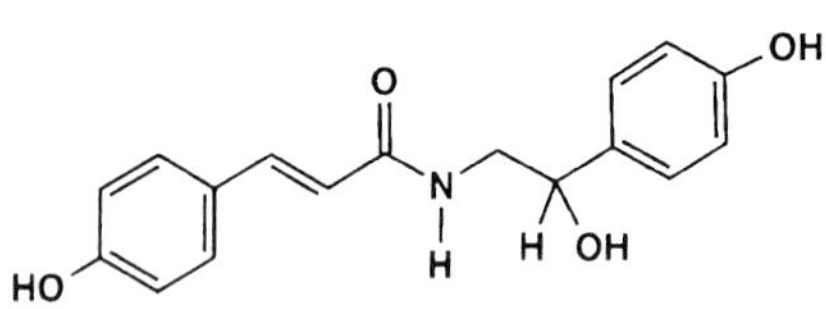

trans-p-Coumaroyloctopamine (4)

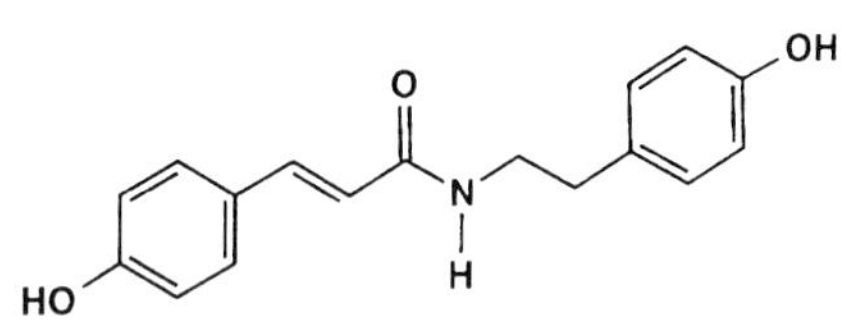

trans-p-Coumaroyltyramine (8)

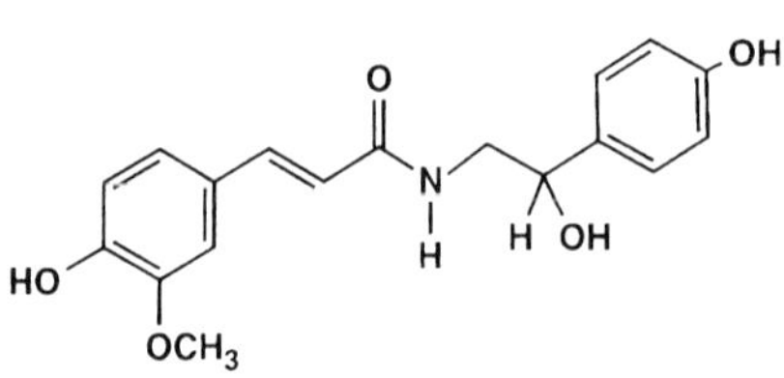

trans-Feruloyloctopamine (5)

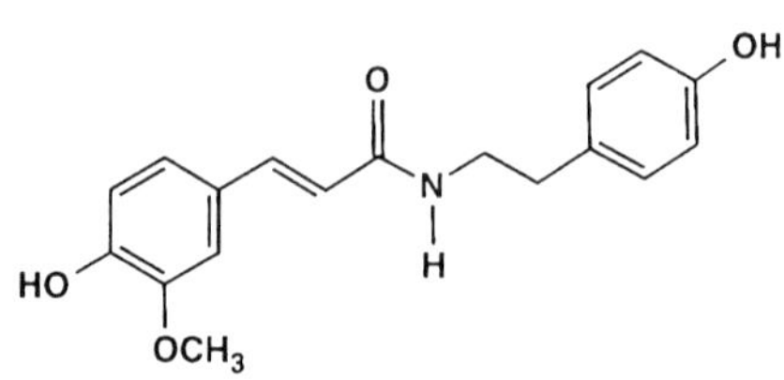

trans-Feruloyltyramine (9)

Fig. 4. Structures of the nine isolated phenolic compounds from the *S. khasianum* cell culture medium

besides p-OH-benzaldehyde, vanillin, *trans*-p-coumaric acid and *trans*-ferulic acid (unpubl. results). Furthermore, the tyramine conjugates were found as intracellular constituents (unpublished results). Besides their function in the plant cell wall modification, antifungal and antiviral activities have been described for cinnamoyltyramines (Martin-Tanguy et al. 1976; Grandmaison et al. 1993).

In addition to their role in plant-pathogen interaction, the cinnamoyltyramines are reported to also have pharmacological properties. Okuyama et al (1986) showed that these compounds, which are present in a Chinese drug made from *Allium bakeri* Reg, have inhibitory effects on human platelet aggregation. Furthermore, pharmacotoxic effects of p-coumaroyl and feruloyltyramine have been reported. Yamamoto et al. (1991) showed that these two compounds, which were isolated from *Canabis sativa*, caused hyperthermia and motor incoordinations in mice, after intracerebroventicular injection. The authors assume that these amides may be responsible for some pharmatoxicity of marihuana. Cavin et al. (1998) isolated the *cis*- and *trans*-isoforms of feruloyltyramine from *Tinospora cripsa*. These compounds have been shown to exhibit pronounced antioxidative and radical scavenging activities.

α-Hydroxyacetovanillone (Fig. 4), which was also found in the *S. khasianum* cell culture medium, is known as a wound-inducible signal molecule, which activates the virulence genes of the crown gall disease-causing bacterium *Agrobacterium tumefaciens* (Spencer and Towers 1991). Acetophenones as well as their α-hydroxyl derivatives have especially been found in wound exudates of species within the Solanaceae. In our studies we could demonstrate that the hydroxylation of the α-C of acetovanillone is catalyzed by a cytochrome P450 monooxygenase (Mühlenbeck and Barz 1997). Recent investigations have further shown that acetophenones also express antifungal activities (Kokubun et al. 1994; Kokubun and Harborne 1995; Curir et al. 1996).

3 Conclusion and Prospect

Based on published data and the experience gathered in the senior author's laboratory (Beimen et al. 1992; Mühlenbeck et al. 1996; Mühlenbeck and Barz 1997) with *S. khasianum* and other Solanaceae plants, the former plant represents a good species for cell culture work. Establishment of callus and cell suspension cultures as well as regeneration processes of protoplasts/cells/tissue explants to plants appear to proceed smoothly. Therefore, attempts to achieve the most desired breeding aims (spineless leaves, increased solasodine content, improved resistance towards bacterial and fungal pathogens) can be strongly recommended.

The antimicrobial defence reactions of Solanaceae obviously comprise cell wall modifications by peroxidative polymerization of cinnamoylamides as a prominent feature. Several species when studied in elicitor-treated cell cultures or infected leaves express this pathway at a high level. The adherent mechanisms of gene activation and the full pattern of other induced defense reactions can be studied well at the cell culture level. The availability of heterotrophic and photosynthetically active cell cultures further allows the investigation of those reactions, which in a plant are only expressed in a particular organ (cf. Beimen et al. 1992).

With regard to steroids and their biosynthesis, more intensive investigations are required in order for these compounds to accumulate in a cell culture.

References

Alzerreca A, Hart G (1982) Molluscicidal steroid glycoalkaloids possessing stereoisomeric spirosolane structures. Toxicol Lett 12:151–155

Aminuddin (1978) Solasodine in *Solanum khasianum* cultures. Indian Drugs 16:31–32

Beimen A, Witte L, Barz W (1992) Growth characteristics and elicitor-induced reactions of photosynthetically active and heterotrophic cell suspension cultures of *Lycopersicon peruvianum* (Mill.). Bot Acta 105:152–160

Bhatt B (1975) Induced tetraploidy in curved spine mutant of *Solanum khasianum* Clarke. Curr Sci 44:677–678

Cavin A, Hostettmann K, Dyatmyko W, Potterat O (1998) Antioxidant and lipophilic constituents of *Trinospora crispa*. Plant Med 64:393–396

Cham Be, Daunter B (1990) Solasodine glycosides. Selective cytotoxicity for cancer cells and inhibition of cytotoxicity by rhamnose in mice with sarcoma 180. Cancer Lett 55:221–225

Cham BE, Gilliver M, Wilson L (1987) Antitumor effects of glycoalkaloids isolated from *Solanum sodomaeum*. Planta Med 53:34–36

Cham BE, Daunter B, Evans RA (1991) Topical treatment of malignant and premalignant skin lesions by very low concentrations of standard mixture (BEC) of solasodine glycosides. Cancer Lett 59:183–192

Chataing B, Concepción JL, Lobatón R, Usubillaga A (1998) Inhibition of *Trypanosoma cruzi* growth in vitro by *Solanum* alkaloids: a comparison with ketoconazole. Planta Med 64:31–36

Clarke DD (1982) The accumulation of cinnamic acid amides in the cell walls of potato tissue as an early response to fungal attack. In: Wood RKS (ed) Active defence mechanisms in plants. Plenum Press, New York, pp 321–322

Curir P, Marchesini A, Danieli B, Mariani F (1996) 3-hydroxyacetophenone in carnations is a phytoanticipin active against *Fusarium oxysporum* f. sp. *dianthi*. Phytochemistry 41: 447–450

Ehmke A, Eilert u (1986) Steroidal alkaloids in tissue cultures and regenerated plants of *Solanum dulcamara*. Plant Cell Rep 5:31–34

Fewell AM, Roddick JG, Weissenberg M (1994) Interactions between the glycoalkaloids solasodine and solamargine in relation to inhibition of fungal growth. Phytochemistry 37:1007–1011

Gamborg OL, Miller RA, Ojima K (1968) Nutrient requirements of suspension cultures of soybean root cells. Exp Cell Res 50:151–158

Ghosh M, Sinhababu SP, Sukul NC, Sahu NP, Mahato SB (1994) Antifilarial effect of solamargine isolated from *Solanum khasianum*. Int J Pharmacogn 32:184–190

Goodwin TW, Mercer EI (1983) Introduction to plant biochemistry. Pergamon Press, Oxford, pp 480–527

Grandmaison J, Gyorgy MO, Van Calsteren MR, Furlan V (1993) Characterization and localization of plant phenolics likely involved in the pathogen resistance expressed by endomycorrhizal roots. Mycorrhiza 3:155–164

Hohlfeld H, Scheel D, Strack D (1996) Purification of hydroxycinnamoyl-CoA: tyramine hydroxycinnamoyltransferase from cell suspension cultures of *Solanum tuberosum* L. cv. *Datura*. Planta 199:166–168

Khanna KR, Murty AS (1972) Effect of fruit stage and maturity on glycoalkaloid content in *S. khasianum*. Planta Med 21:182–187

Kittipongpatana N, Hock RS, Porter J (1998) Production of solasodine by hairy root, callus, and cell suspension cultures of *Solanum aviculare* forst. Plant Cell Tissue Organ Cult 52: 133–143

Kokate CK, Radwan SS (1979) Enrichment of *Solanum khasianum* callus generating rootlets with steroidal glycoalkaloids. Z Naturforsch 34c:634–636

Kokubun T, Harborne JB (1995) Phytoalexin induction in the sapwood of plants of the Maloideae (Rosaceae): biphenyls or dibenzofurans. Phytochemistry 40:1649–1654

Kokubun T, Harborne JB, Eagles J (1994) 2′,6′-Dihydroxy-4′-methoxy-acetophenone, a new phytoalexin from the roots of *Sanguisorba minor*. Phytochemistry 35:331–333

Kowalczyk TP, Mackenzie IA, Cocking EC (1983) Plant regeneration from organ explants and protoplasts of the medicinal plant *Solanum khasianum* C.B. Clarke var. *chatterjeeanum* Sengupta (syn. *Solanum viarum* Dunal). Z Pflanzenphysiol 111:55–68

Maga JA (1994) Glycoalkaloids in solanaceae. Food Rev Int 10:385–418

Maheshwari VL, Bhalsing SR (1997) Regeneration of *Solanum khasianum* from leaf derived callus and extraction of solasodine. FASEB J 11:A1107

Mak Y, Doran M (1993) Effect of cell-cycle inhibition on the synthesis of steroidal alkaloids by *Solanum aviculare* plant cells. Biotechnol Lett 15:1031–1034

Malpathak NP, David SB (1992) Stimulation of solasodine production by combining fungal elicitors and immobilized cell suspension cultures of *Solanum surattense* Burm. Biotechnol Lett 14:965–968

Mann JD (1978) Production of solasodine for the pharmaceutical industry. Adv Agron 30:207–245

Martin-Tanguy J, Martin C, Gallet M, Vernoy R (1976) Natural potent inhibitors of tobacco mosaic virus multiplication. CR Acad Sci 282:2231–2234

Mühlenbeck U (1996) Elicitor-induzierte Stimulation des Phenolstoffwechsels in Zellkulturen von *Solanum khasianum*. Doctoral Thesis, University of Münster

Mühlenbeck U, Barz W (1997) Cytochrome P450-dependent formation of α-hydroxy-acetovanillone from acetovanillone in *Solanum khasianum*. Phytochemistry 44:865–867

Mühlenbeck U, Kortenbusch A, Barz W (1996) Formation of hydroxycinnamoyl-amides and α-hydroxyacetovanillone in cell cultures of *Solanum khasianum*. Phytochemistry 42:1573–1579

Murashige T, Skoog F (1962) A revised medium for rapid growth and bioassays with tobacco tissue culture. Physiol Plant 15:473–497

Okuyama T, Shibata S, Hoson M, Kawada T, Osada H, Noguchi T (1986) Effect of oriental plant drugs on platelet aggregation; III. Effect of Chinese drug "xiebai" on human platelet aggregation. Planta Med 52:171–175

Paczkowski C, Kalinowska M, Wojciechowski ZA (1997) UDP-glucose:solasodine glucosyltransferase from eggplant (*Solanum melongena* L.) leaves: partial purification and characterization. Acta Biochem Polon 44:43–54

Paniego NB, Talou JR, Giulietti AM (1995) Influence of fungal elicitors on solasodine production by cell suspension cultures of *Solanum eleagnifolium*. Fitoterapia 66:162–166

Roddick JG (1989) The acetylcholinesterase-inhibitory activity of steroidal glycoalkaloids and their aglycones. Phytochemistry 28:2631–2634

Roddick JG, Rijnenberg AL, Weissenberg M (1990) Membrane-disrupting properties of the steroidal glycoalkaloids solasonine and solamargine. Phytochemistry 29:1513–1518

Roddick JG, Rijnenberg AL, Weissenberg M (1992) Alterations to the permeability of liposome membranes by the solasodine-based glycoalkaloids solasonine and solamargine. Phytochemistry 31:1951–1954

Serraf I, Sihachakr D, Chi NTL, Herbreteau C, Rossignol L, Ducreux G (1988) High rate of plant regeneration from cultured protoplasts of two medicinal plants; *Solanum laciniatum* Ait. and *Solanum khasianum* C.B. Clarke. J Plant Physiol 133:498–501

Sihachakr D, Haicour R, Serraf I, Barrientos E, Herbreteau C, Ducreux G, Rossignol L, Souvannavong V (1988) Electrofusion for the production of somatic hybrid plants of *Solanum melongena* L. and *Solanum khasianum* C.B. Clark. Plant Sci 57:215–223

Spencer PA, Towers GHN (1991) Restricted occurrence of acetophenone signal compounds. Phytochemistry 30:2933–2937

Stahl E (1967) Dünnschichtchromatographie. Springer, Berlin Heidelberg New York

Stattmann M, Gerick E, Wenzel G (1994) Interspecific somatic hybrids between *Solanum khasianum* and *S. aculeatissimum* produced by electrofusion. Plant Cell Rep 13:193–196

Subroto MA, Doran PM (1994) Production of steroidal alkaloids by hairy roots of *Solanum aviculare* and the effect of gibberellic acid. Plant Cell Tissue Organ Cult 38:93–102

Sudiarto (1989) Morphological and agronomic characters of solasodine yielding plants in relation to crop improvement. In: The Bogor Research Institute for Estate Crops (ed) Proceedings of the Workshop on Biotechnology for the improvement of oil- and steroid-producing plants, Jakarta, Indonesia, pp 121–126

Thorne HV, Clarke GF, Skuce R (1985) The inactivation of herpes simplex virus by some Solanaceae glycoalkaloids. Antiviral Res 5:335–343

Yamamoto I, Matsunaga T, Kobayashi H, Watanabe K, Yoshimura H (1991) Analysis and pharmacotoxicity of feruloyltyramine and p-coumaroyltyramine in *Canabis sativa* L. Pharmacol Biochem Behav40:465–469

Weiler EW, Krüger H, Zenk MH (1980) Radioimmunoassay for the determination of the steroidal alkaloid solasodine and related compounds in living plants and herbarium specimens. Planta Med 3:112–124

Weissenberg M, Levy A, Svoboda JA, Ishaaya I (1998) The effect of some *Solanum* steroidal alkaloids and glycoalkaloids on larvae of the red flour beetle, *Tribolium castaneum*, and the tobacco hornworm, *Manduca sexta*. Phytochemistry 47:203–209

16 In Vitro Culture and the Production of Secondary Metabolites in *Stephania*

Y. Sugimoto[1]

1 Introduction

1.1 Importance and Distribution of the Genus

The genus *Stephania* (Menispermaceae) comprises approximately 50 species distributed from Africa through Asia to Australia. The importance of the genus in traditional medicine in Asia and Africa is well documented (Min et al. 1986; Cao and Zhu 1997). The underground tubers of the vines are generally characterized by powerful pharmacological effects (Blanchfield et al. 1993).

S. abyssinica is a creeper indigenous to southern and eastern Africa. The leaves of this plant are used as a purgative and emetic, whereas the roots are employed in the treatment of roundworm, menorrhagia and boils (Dagne et al. 1993). *S. bancroftii* is used by the aboriginal communities of Australia both as a treatment for diarrhea and as a fish poison (Cao and Zhu 1997).

S. cepharantha (Fig. 1), a perennial plant native to mainland China known by the vernacular name "bei-yan-zi", is commonly used as a folk medicinal herb (Zhen et al. 1994). Decoctions from the tuber of *S. cepharantha* are traditionally used in China to treat a number of diseases including parotiditis, gastric ulcer, leukopenia, alopecia areata and alopecia androgenetica. The major components of this crude drug, known as Cepharanthin preparations, are the bisbenzylisoquinoline (BBI) alkaloids cepharanthine, isotetrandrine and cycleanine (Nakaoji et al. 1997).

S. dinklagei is a climbing shrub of the deciduous forests of both East and West Africa. The roots and stems are used medicinally in Ghana as a vermifuge, an analgesic, an aphrodisiac, a sedative and for the treatment of menorrhagia. The leaves of the plant are used in folkloric medicine as a treatment for infertility in women and impotence in men. The stems are also used as a fish poison (Dwuma-Badu et al. 1980). *S. errecta* has been used in Thai folk medicine as an analgesic and a skeletal muscle relaxant (Likhitwitayawuid et al. 1993).

S. rotunda is a climber indigenous to India and Indochina, where it has been used as a folk medicine for the treatment of pulmonary consumption, dysen-

[1]Arid Land Research Center, Tottori University, 1390 Hamasaka, Tottori 680-0001, Japan

Biotechnology in Agriculture and Forestry, Vol. 51
Medicinal and Aromatic Plants XII (ed. by T. Nagata and Y. Ebizuka)
© Springer-Verlag Berlin Heidelberg 2002

Fig. 1. *Stephania cepharantha*

tery, fever, abdominal ills (tubers), asthma (tubers and stems), ascariasis, dysmenorrhea (stems), indigestion, wounds, headache, sore breasts (leaves) and leprosy (flowers) (Kozuka et al. 1985). *S. glabra* is a glabrous climber with smooth terete stems attaining a length of up to 7 m. It is distributed in the Himalayas, Assam and in the western ghats up to the Nilgiris and Tirunelveli hills, up to an altitude of 2100 m. Its tubers attain a large size, some of them have a diameter of approximately 23 cm and weigh up to 30 kg. They are acrid and it appears that they are employed, like the tubers of *S. rotunda*, in the treatment of pulmonary tuberculosis, asthma and intestinal complaints. They also possess hypoglycemic activity (Soni et al. 1986). Thornber (1970) treated *S. glabra* and *S. rotunda* as the same plant.

S. hernandifolia is a slender twining shrub found in India on the western and eastern coasts, in Cachar, Sikkim, East Bengal and Assam. The roots are used in the treatment of fever, diarrhea, dyspepsia and urinary diseases (Kupchan et al. 1968). An extract of *S. japonica* has been used in China as an antidiarrheal, an antianetus, an antifebrile, a tonic, a diuretic and as a remedy for podagra and cholera (Taga et al. 1984). *S. pierrii*, indigenous to Thailand, is a slender, herbaceous climber with large tubers and round leaves. The tubers are used in folk medicine as a skeletal muscle relaxant and also as an analgesic and tonic under the name "bua bok" (Tantisewie et al. 1989).

The vine *S. suberosa*, native to Thailand, is commonly used for treatment of a variety of ailments under the local name "borapet pungchang" (Patra et al.

1987). *S. sutchuenensis* is indigenous to the southeastern part of China, and its root is used in folk medicine to treat common cold, sore throat and arthritic pain (Wang et al. 1995).

S. tetrandra is indigenous to China. The roots of this plant are the main constituents of the traditional Chinese medicine "fen-fan-ji", which has been in use as an analgesic, diuretic and detumescent for thousands of years in China (Deng et al. 1990). *S. tetrandra* has been reported to have antiphagocytic and antioxidizing effects, and to exhibit effectiveness in clinical and experimental silicosis models. It is known to have the ability to inhibit the production of interleukin-1 and tumor necrosis factor-α which are secreted from human monocytes. The alkaloid tetrandrine has been used as an antiinflammatory and antihypertensive agent. Furthermore, tetrandrine and its derivatives are reported to promote brain function and have been developed as an antimalarial drug and also as a stimulant for hair growth (Kang et al. 1996).

The rhizomes of the Southeast Asia vine *S. venosa*, known in Thailand under the name of "sabu-le-ad" or blood-soap due to its red latex, are sometimes used as a bitter tonic (Charles et al. 1987).

1.2 Alkaloids

The alkaloids of the genus *Stephania* were reviewed by Thornber (1970), as a part of an article on alkaloids of the Menispermaceae. According to the author, 45 alkaloids had been identified from 9 *Stephania* species. The alkaloids in *Stephania* species were further reviewed by Chinese researchers Wang and Zhao (1990) and Yang and Chen (1994). Alkaloids in *S. cepharantha*, *S. glabra*, *S. hernandifolia* and *S. japonica* have been investigated more extensively than others. Based on information drawn from Chemical Abstracts, more than 130 new alkaloids were reported in the period 1969 to 1997. Including known compounds, so far, approximately 300 alkaloids have been identified from about 40 *Stephania* species. Based on structures and functional groups, the alkaloids are assigned to several groups including benzylisoquinoline, bisbenzylisoquinoline (BBI), aporphine, proaporphine, oxoaporphine, morphine, protoberberine and hasubanane.

2 In Vitro Studies

2.1 Review of in Vitro Studies on *Stephania* Plants

Several in vitro studies on *S. glabra* and *S. cepharantha* were reported by Russian and Japanese researchers. Popov (1985) established suspension cultures of *S. glabra* and studied the effect of some plant growth regulators on morphogenesis of the cells. Davydenkov et al. (1988) developed an analytical method to measure stepharine in the cells. Shamina et al. (1994) improved stepharine production in the suspension cells by physiological and genetic methods.

Suzuki et al. (1992) reported on the propagation of *S. cepharantha* by tissue culture. Itokawa et al. (1973) established callus cultures of *S. cepharantha* and identified three phytosterols and six oxidized derivatives of them. From the same material, they also identified five fluorescent components, three cepharanones, three cepharadiones (Akasu et al. 1974, 1975), two 7-oxoaporphine type compounds (Akasu et al. 1975) and two BBIs, aromoline and berbamine (Akasu et al. 1976). Sugimoto et al. (1988) established root cultures of this plant and identified five BBIs and studied their biosynthesis (Sugimoto et al. 1990, 1993, 1997; Sugimoto and Yamada 1993). They also conducted a series of tracer experiments and demonstrated that BBIs, aromoline and berbamine are composed of four molecules of tyrosine and that tyramine was specifically incorporated into the isoquinoline portions of the alkaloids.

2.2 Establishment of Root Cultures of *Stephania cepharantha*

It takes several years to produce tubers of *S. cepharantha* large enough to use as raw materials for alkaloid extraction. Akasu et al. (1976) reported that the callus of this species produced BBI alkaloids, but not the full spectrum as in the whole plant. Only aromoline and berbamine were produced.

Sterilized segments of *S. cepharantha* tuber, harvested in April, were inoculated on Linsmaier-Skoog (LS) agar medium containing 10 μM naphthaleneacetic acid (NAA) and 1 μM benzyladenine (BA). Calluses and adventitious roots were formed 1 month after incubation in the dark at 26 °C. The callus was subcultured on the same medium and on LS agar medium containing 1 μM NAA and 1 μM BA. Adventitious roots differentiated when callus was cultured on medium containing 10 μM NAA and 1 μM BA; whereas shoots differentiated on medium containing 1 μM NAA and 1 μM BA (Fig. 2). Roots differentiated spontaneously from the callus irrespective of its age. No adventitious roots were induced, however, when segments of tuber, harvested in the middle of October, were cultured under the same conditions. The season of harvest of the tuber was significant.

Roots obtained from tuber and from callus were cultured separately on a rotary shaker (85 rpm) in the dark at 26 °C in liquid media containing various concentrations of indole-3-butyric acid (IBA). LS and Gamborg B5 basal media were used. A total of 50 root cultures, cultured under various conditions, were established. The media used are shown in Table 1. In spite of slow growth, the gradual browning of the culture media caused a transfer of the root culture to fresh media at weekly intervals. After 3 months of successive subculturing, the roots were harvested and dried. Alkaloids were extracted from the dried roots with MeOH and analyzed.

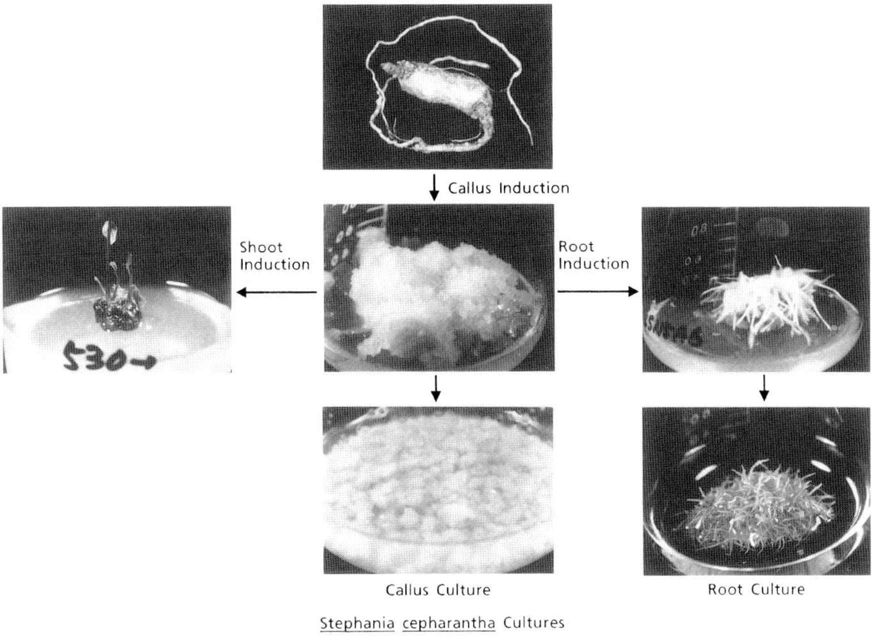

Fig. 2. *Stephania cepharantha* cultures

Table 1. Effect of the basal medium and IBA on alkaloid production in cultured *S. cepharantha* roots

Basal medium	IBA conc. (M)	No. of cell lines	Alkaloid content (% dry wt.)	
			Aromoline (4)	Berbamine (6)
LS	10^{-5}	5	0.197 ± 0.109	0.156 ± 0.054
	10^{-6}	7	0.392 ± 0.068	0.284 ± 0.059
B5	5×10^{-5}	3	0.176 ± 0.078	0.221 ± 0.028
	10^{-5}	10	0.462 ± 0.254	0.556 ± 0.231
	7.5×10^{-6}	7	0.563 ± 0.231	0.520 ± 0.197
	5×10^{-6}	6	0.444 ± 0.197	0.513 ± 0.149
	10^{-6}	12	0.637 ± 0.241	0.546 ± 0.220
Tuber[a]			not detected	0.262

[a] Harvested in October.

Production of at least five BBI alkaloids were confirmed. Homoaromoline, isotetrandrine, berbamine and cycleanine were identified by comparing their retention times on high-performance liquid chromatography (HPLC) and their mass spectra with the respective values of authentic samples.

Thin layer chromatography (TLC) of cultured root extracts indicated that another base, which behaves chromatographically like a BBI alkaloid, was one of the main constituents present. The base was isolated as a powder and identified as aromoline based on its R_f value on TLC, melting point and mass spectrum.

In all root cultures, the main alkaloids were aromoline (0.1–1.0% dry wt.) and berbamine(0.1–1.0% dry wt.), the other alkaloids constituted less than one-tenth for aromoline and berbamine. Alkaloid production in calluses was less than one-tenth of that in cultured roots and therefore further work concentrated on root cultures.

Aromoline and berbamine contents of roots cultured under various conditions are shown in Table 1. In many cultures, contents of both alkaloids were higher than those in tubers. B5 medium was more suitable than LS medium for the root culture of this plant. In both LS and B5 media, a low concentration (1 μM) of IBA stimulated the production of alkaloids, but root growth was extremely slow (data not shown). A high concentration (50 μM) of IBA inhibited alkaloid production. Consequently, B5 medium containing 10 μM IBA proved to be the best for alkaloid production but, in this medium as well as in the other media tested, the growth of roots was poor. Further studies on culture conditions are described in the following section.

The correlation between the aromoline and berbamine contents in each culture is shown in Fig. 3. Interestingly, the value for the correlation coefficient was very high, $r = 0.89$. The high correlation coefficient suggests that the production of aromoline and berbamine in cultured roots depends on the production of the common precursor, *N*-methylcoclaurine. The low production of the other BBIs may be due to low enzyme activity which precludes the conversion of aromoline and berbamine to the further modified, cepharanthine, cepharanoline, homoaromoline, isotetrandrine and so on.

2.3 Effect of Culture Parameters on BBI Alkaloid Production

In the previous section, the production of BBI alkaloids in cultured roots of *S. cepharantha* was described. As illustrated in the case of shikonin production by suspension cultures of *Lithospermum erythrorhizon* (Yamada and Fujita 1983), culture conditions may drastically increase product formation. Cultured *S. cepharantha* roots produced and secreted brown compounds, possibly phenolics, which may be inhibitory to their growth. In this investigation, various cultural parameters were examined to find conditions which would suppress this browning and increase the production of BBI alkaloids.

2.3.1 *Effects of Gibberellin*

B5 medium was shown to be more suitable than LS medium for the root culture of *S. cepharantha*. However, in both media, the cultured roots and culture media gradually turned brown. To circumvent this phenomenon, various plant

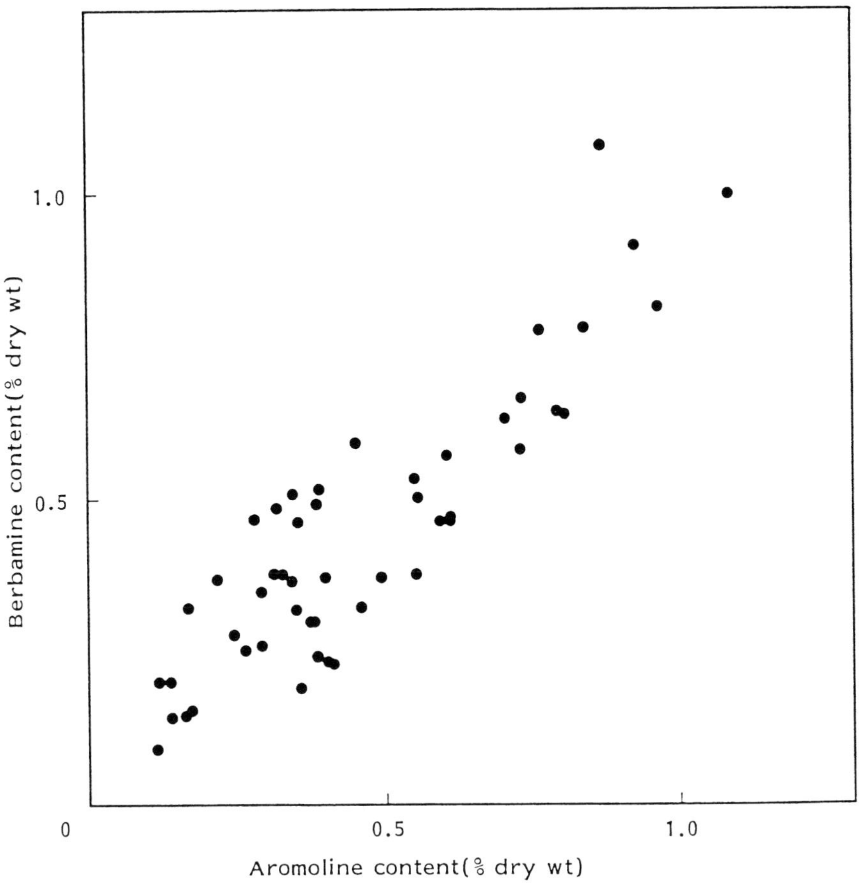

Fig. 3. Correlation between aromoline and berbamine contents (r = 0.89)

hormones were tested and gibberellic acid (GA_3) was found to suppress the browning of the roots and media. GA_3 also stimulated root growth and the production of alkaloids. Various concentrations of GA_3 in B5 medium were tested for effects on root growth and major BBI alkaloid, aromoline and berbamine, contents (Fig. 4). The optimal GA_3 concentration was 1 μM, at which concentration the roots doubled in dry weight and alkaloid contents increased about 1.5 times that observed in B5 medium without GA_3.

2.3.2 Effects of Inorganic Components in the Culture Medium

The form of the nitrogen source sometimes facilitates alkaloid formation. Alkaloid formation in two modified B5 media was studied. In one medium (SB5 medium), the concentration of the major elements, $(NH_4)_2SO_4$, KNO_3 and

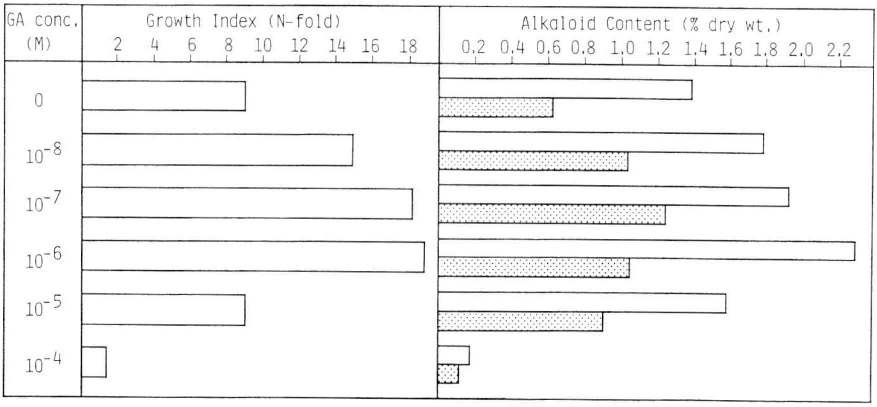

Fig. 4. Effects of GA_3 concentration on root growth and alkaloid formation. Roots (0.1 g fresh wt.) were cultured for 30 days in 100-ml flasks containing 25 ml of modified B5 medium with various concentrations of GA_3. *Open bars* Aromoline; *dotted bars* berbamine

Table 2. Effects of the culture media on root growth and alkaloid formation in *S. Cepharantha* root cultures

Basal medium	No. of replicates	Growth index	Alkaloid Content (% dry wt.)	
			Aromoline	Berbamine
B5	7	2.3	0.493	0.174
SB5	5	4.0	1.031	0.362
NB5	9	2.7	1.464	0.614

Roots (0.1 g fresh wt.) were cultured for 14 days in 50-ml flasks containing 10 ml of the given medium. Each value is the mean of the given replicates.

$MgSO_4$, was reduced to half that of the original B5 medium. In another B5 modification (NB5 medium), $(NH_4)_2SO_4$ was eliminated. B5 medium without modification of the inorganic components was included as a control. All media contained 2% sucrose, 10 μM IBA and 1 μM GA. Root growth and the major alkaloid contents in these media are shown in Table 2. The B5 medium, with no alteration in the inorganic components, resulted in slow root growth and low alkaloid contents. The roots displayed a slow growth in the NB5 medium but a high alkaloid production was displayed. In the SB5 medium, a higher root growth was achieved. However, alkaloid production was less than in the NB5 medium.

2.3.3 *Effects of Carbon Source*

The effects of carbon source on root growth and alkaloid formation were studied. Sucrose, glucose and fructose promoted alkaloid formation but lactose

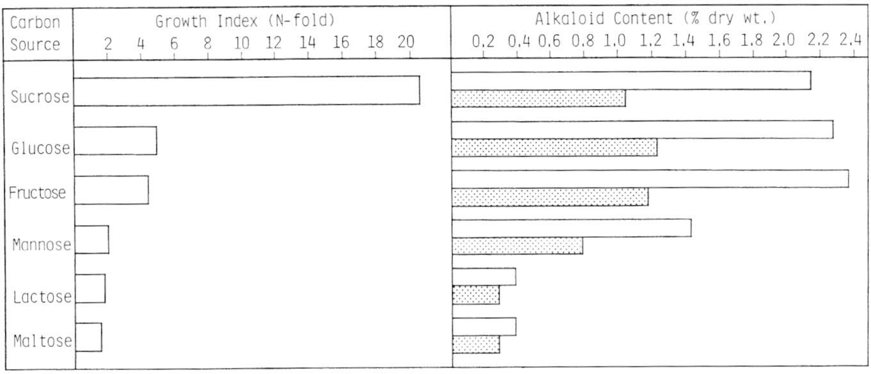

Fig. 5. Effects of carbon source on root growth and alkaloid formation. Roots (0.1 g fresh wt.) were cultured for 31 days in 100-ml flasks containing 25 ml of modified B5 medium with 3% each carbon source. *Open bars* Aromoline; *dotted bars* berbamine

and maltose caused the reduction of alkaloid content. Roots grew well only when sucrose was supplied. The other sugars did not promote root growth (Fig. 5).

The effects of sucrose concentrations on cell growth and alkaloid formation were studied. In SB5 medium which contained 10 μM IBA and 1 μM GA, high concentrations of sucrose reduced root growth and alkaloid production. The optimal sucrose concentration for alkaloid production was 3% (Fig. 6). Increasing sucrose concentrations to more than 3% increased root thickness and decreased their water and alkaloid contents.

2.3.4 *Effects of Cytokinin*

SB5 medium containing 10 μM IBA, 1 μM GA and 3% sucrose was further supplemented with benzyladenine (BA) at 1 nM to 10 μM. High rates of BA increased root thickness and reduced growth and alkaloid production (Fig. 7). Isopentenyladenine gave similar results to BA.

2.3.5 *Optimum Medium for Alkaloid Production*

As described above, SB5 medium supplemented with 3% sucrose, 10 μM IBA and 1 μM GA was found to be the optimum medium for BBI alkaloid production. A typical time course of root growth and alkaloid content by roots cultured in this medium is shown in Fig. 8. Cultured roots increased about 20-fold in 30 days. Alkaloid production increased with time and was maximal at 40 days after subculturing with no further increase on extension of the culture period to 50 days. Aromoline content of the cultured roots reached more than

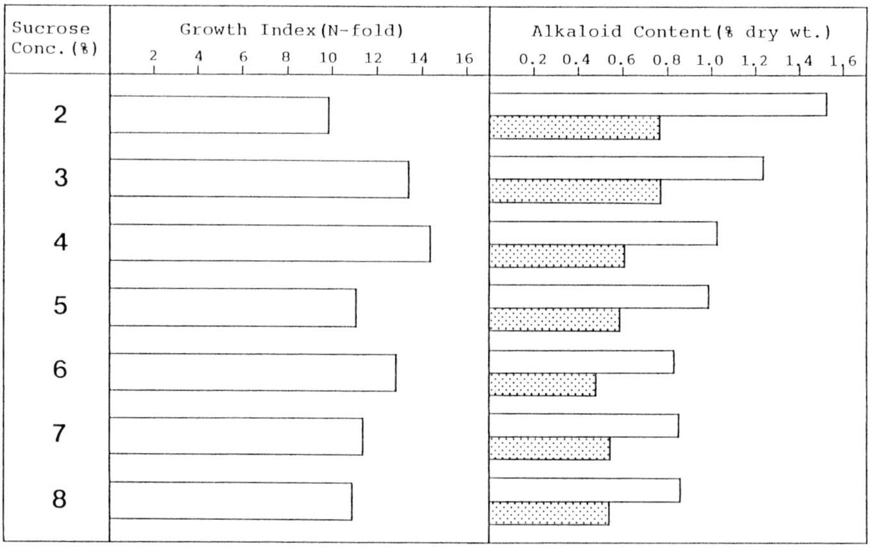

Fig. 6. Effects of sucrose concentration on root growth and alkaloid formation. Roots (0.2 g fresh wt.) were cultured for 25 days in 100-ml flasks containing 25 ml of modified B5 medium with various concentrations of sucrose. *Open bars* Aromoline; *dotted bars* berbamine

1% and berbamine content more than 0.5% of dry wt. In the intact plants, aromoline was not detected and berbamine content was 0.26% of dry wt. (Table 1). Hence, the aromoline and berbamine contents of the cultured roots are much higher than those of the original plants. Moreover, unlike in *Berberis* cell cultures (Cassels et al. 1987), the bisbenzylisoquinoline alkaloids formed were not degraded in cultured roots of *S. cepharantha*.

2.4 Comparison of BBI Alkaloids in Cultured Roots and Intact Plants

In the previous section, production of five BBI alkaloids; aromoline, homoaromoline, berbamine, isotetrandrine and cycleanine, was demonstrated. Aromoline has been identified in many plant sources; *Abuta splendida*, *Pycnarrhena longifolia*, *Triclisia patens* (Menispermaceae), *Daphnandra aromatica*, *D. tenuipes*, *Doryphora aromatica* (Monimiaceae), *Thalictrum lucidum*, *T. rugosum*, *T. thunbergii* (Ranunculaceae) and *Berberis orthbotrys* (Berberidaceae) (Guha and Mukherjee 1979; Schiff 1983). For *S. cepharantha* (Menispermaceae), Sugimoto et al. (1988) identified aromoline in cultured roots and Akasu et al. (1976) in calluses, but there has been no report on the identification of aromoline in the intact plant.

Aromoline is a probable precursor of homoaromoline, cepharanoline and cepharanthine, all of which have been identified in the intact *S. cepharantha*

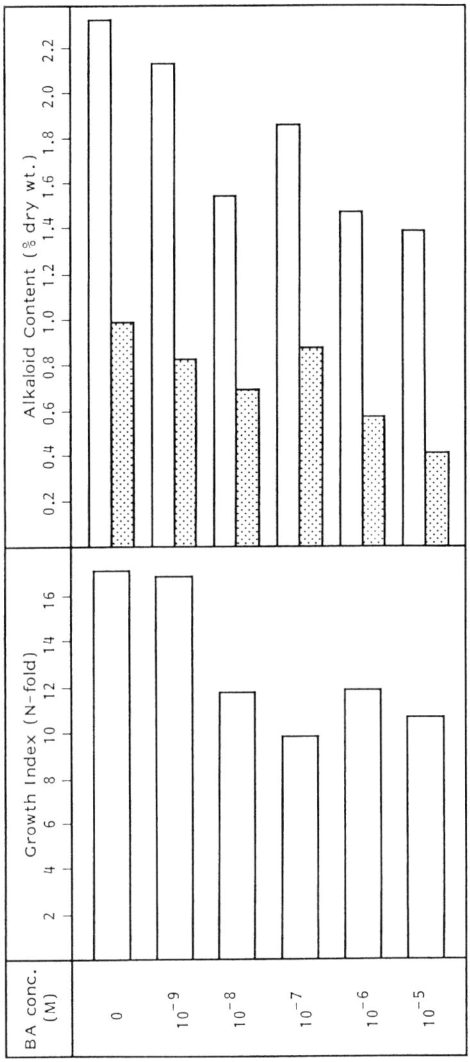

Fig. 7. Effects of BA concentration on root growth and alkaloid formation. Roots (0.1 g fresh wt.) were cultured for 30 days in 100-ml flasks containing 25 ml of modified B5 medium with various concentrations of benzyladenine. *Open bars* Aromoline; *dotted bars* berbamine

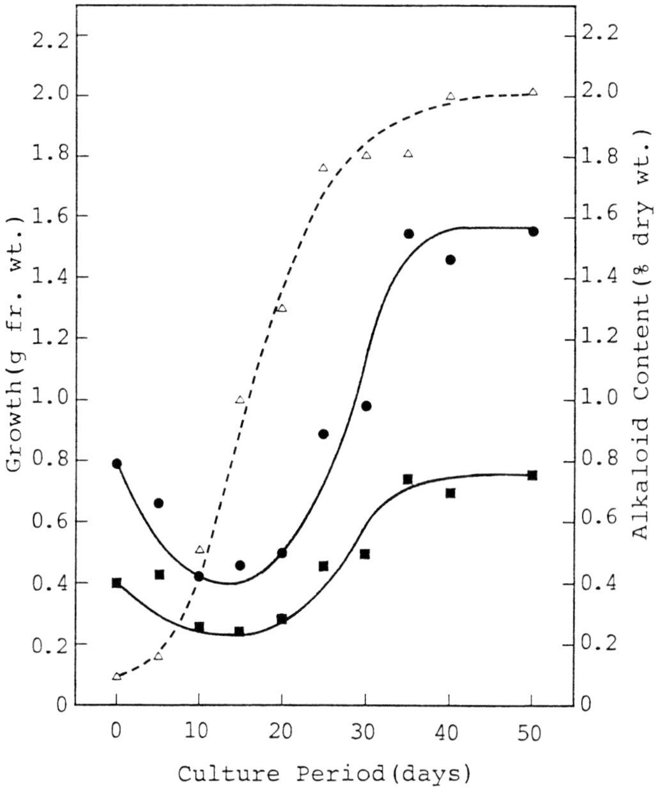

Fig. 8. Time course of root growth and alkaloid content in cultured roots of *S. cepharantha*. *Filled circles* Aromoline; *filled squares* berbamine; *open triangles* root growth

plant. Aromoline is also the main constituent in cultured *Stephania* roots; therefore, it was expected to be present in the intact plants, specifically in the roots. In this section, the BBI alkaloids present in cultured *S. cepharantha* roots are compared with those in the roots and tuber of the intact plant. Stability of alkaloid production in cultured roots and the biogenesis of BBI alkaloids in *S. cepharantha* are also discussed.

Aromoline constitutes a significant proportion of the alkaloids in the basic fraction of cultured roots of *S. cepharantha*. Slightly yellowish crystals were formed when MeOH was added to the dried root residue. These crystals were identified as aromoline by their chromatographic behaviour (HPLC and TLC), melting point and NMR spectroscopy. The absolute configuration was determined by measuring the optical rotation. The yield was 1.2% on a dry weight basis.

Changes in aromoline and berbamine contents in root cultures that were maintained for more than 10 years are shown in Fig. 9. The high, positive correlation between contents of the two alkaloids was maintained. However, the

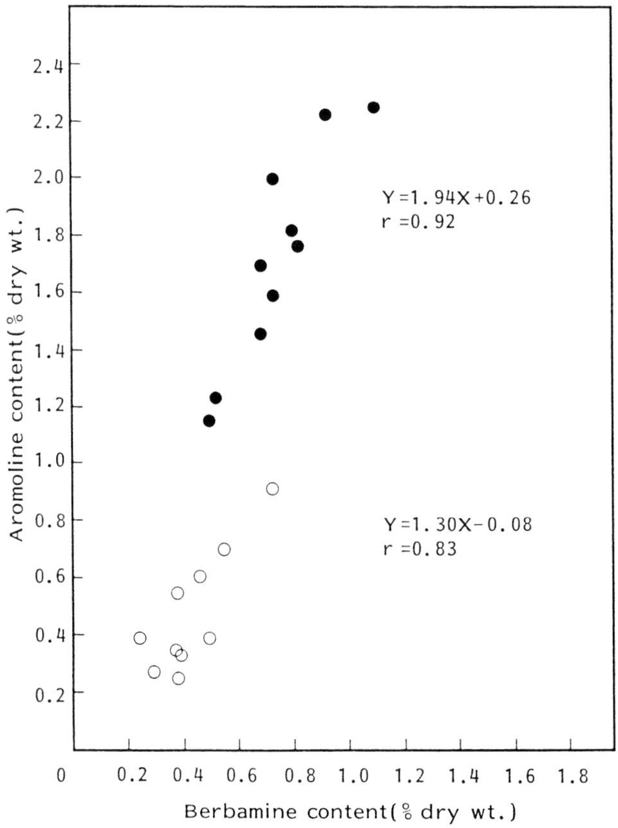

Fig. 9. Correlation between aromoline and berbamine contents in *S. cepharantha* root cultures. *Open circles* 3 months after induction; *filled circles* 15 months after induction

contents of the alkaloids and their ratio changed with time. The ratio of aromoline to berbamine was about 1.3:1, 3 months after induction. After 15 months or more, the ratio increased to about 2:1. HPLC analysis showed that aromoline accounted for about 60% of the basic fraction prepared from cultured roots. Aromoline and berbamine might be derived from a common progenitor berbamunine. Intramolecular oxidative coupling of the progenitor between C-7-OH and C-8′ gives aromoline. Alternatively, intramolecular oxidative coupling of the progenitor between C-8 and C-7′-OH followed by 7-*O*-methylation affords berbamine.

Aromoline is the probable precursor of homoaromoline, cepharanoline and cepharanthine. These three alkaloids, derived from aromoline, were identified in intact *S. cepharantha* plants. However, aromoline was not reported. A reinvestigation of BBI alkaloids in the aerial parts, tuber and roots of *S. cepharantha* (Table 3) showed that the aerial parts did not contain alkaloids; whereas several alkaloids were identified in the underground parts. The presence of

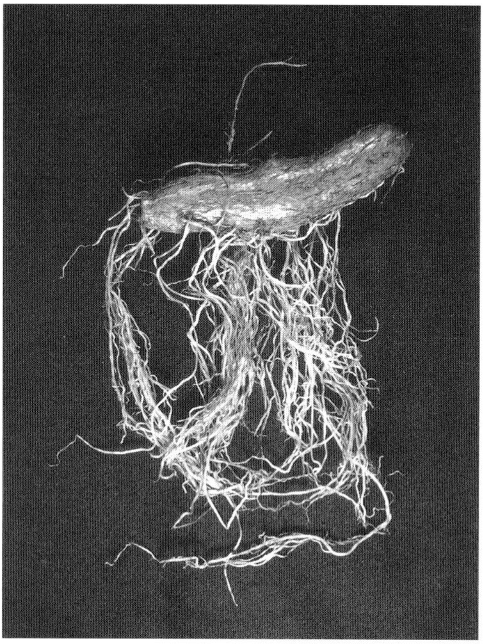

Fig. 10. *S. cepharantha* tuber harvested during the growing season

Table 3. Comparison of bisbenzylisoquinoline alkaloids in *S. cepharantha*

Source	Alkaloid content (% dry wt.)				
	Aromoline	Berbamine	Homoaromoline + cepharanoline	Isotetrandrine + cycleanine	Cepharanthine
Tuber[a]	n.d.	0.262	0.115	0.532	0.300
Roots[b]	0.191	0.466	0.118	0.186	0.029
Cultured roots[c]	2.326	0.980	0.256	0.071	n.d.

[a] Harvested in October.
[b] Harvested in May.
[c] Harvested after 30 days of culture.

aromoline in *S. cepharantha* roots was confirmed by HPLC, TLC and MS. Previous investigations of BBI alkaloids in *S. cepharantha* mainly dealt with its tubers. In the growing season, tuber bears many roots (Fig. 10). However, very few roots are present on *Stephania* tubers during harvest (from October to November in Japan). For that reason, aromoline, a main alkaloid in the roots, might not have been found in the intact plant.

The alkaloid composition of cultured *S. cepharantha* roots was similar to that of the intact roots, but differed markedly from that of the tuber. A comparison of the alkaloid composition of the roots with that of the tuber of *S. cepharantha* showed aromoline and berbamine were present mainly in the roots; whereas, the further modified cepharanthine and isotetrandrine were present in the tuber. Such a skewed distribution of alkaloids is evidence that in *S. cepharantha* the early steps of BBI alkaloid biosynthesis take place in the roots, after which the alkaloids formed are transported to the tuber where they are further modified and stored. The tuber showed very little seasonal changes in the quality and quantity of the BBI alkaloids.

2.5 Biosynthesis of BBI Alkaloid

The BBI alkaloids are structurally constructed of two monomeric benzylisoquinoline units joined by one or more ether linkages. Tracer experiments using whole plants, excised branches or cultured cells showed that coclaurine and its *N*-methyl derivatives act as BBI precursors (Bhakuni et al. 1980, 1987; Stadler et al. 1987). Stadler et al. (1988), using *Berberis stolonifera* cell cultures, demonstrated that tyrosine and tyramine were incorporated into the BBI skeleton, and that tyrosine was incorporated to an equal extent into both the benzyl and the isoquinoline portions. On the other hand, tyramine was almost exclusively incorporated into the isoquinoline portion. The same group (Stadler and Zenk 1993) isolated a new cytochrome P-450 enzyme which catalyzes the oxidative coupling of (*R*)- and (*S*)-*N*-methylcoclaurine to afford the BBI alkaloid berbamunine. In our laboratory, cultured *S. cepharantha* roots were fed with ^{14}C-labeled tyrosine, tyramine and dopamine, and ^{13}C-labeled tyrosine and tyramine. Their incorporation into BBIs was studied in an attempt to elucidate the role of these precursors in BBI biogenesis in cultured roots.

2.5.1 Tracer Experiments with ^{14}C-Labeled Precursors

Roots were harvested 10 days after applying ^{14}C-labeled tyrosine, tyramine and dopamine to 15-day-old cultured roots of *S. cepharantha*. Incorporation of each substrate was 87–93% as determined by counting the remaining radioactivity in the medium. Sufficient radioactivity was recovered in methanolic extracts of roots treated with tyrosine and tyramine. However, poor recovery was obtained from roots treated with dopamine. In the crude alkaloids, high radioactivity was associated with tyrosine treatment, whereas low radioactivity was recovered from tyramine and dopamine treatment (Table 4).

HPLC analysis of crude alkaloids extracted from cultured roots incubated with ^{14}C-labeled tyrosine showed that the BBI alkaloids were sufficiently labeled in proportion to the UV absorption (Fig. 11). Three main peaks attributable to highly ^{14}C-labeled compounds are present. Two of the peaks correspond to the BBI alkaloids aromoline and berbamine. However, the third was attributable to an unknown compound. The unidentified compound was thought to be a key intermediate in BBI biosynthesis. The unidentified

Table 4. Incorporation of potential precursors into cultured roots of *S. cepharantha*

Precursor	Incorporation (%)	Radioactivity recovered ($\times 10^3$ dpm)	
		MeOH extract	Crude alkaloids
[U-^{14}C] tyrosine	93.0	578	100
[7-^{14}C] tyramine	91.3	654	6.6
[7-^{14}C] dopamine	87.0	97	1.1

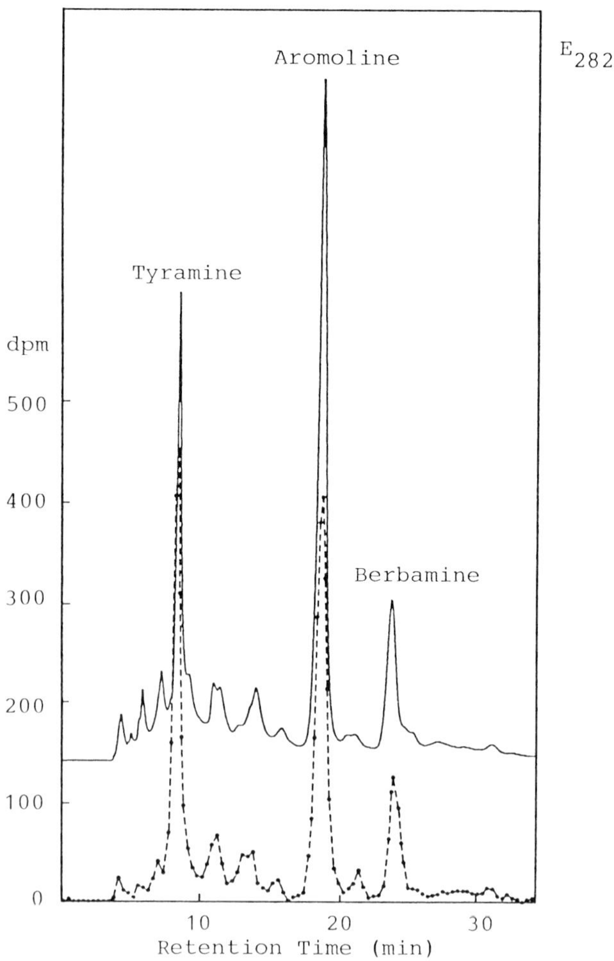

Fig. 11. HPLC separation of the basic fraction extracted from cultured roots of *S. cepharantha* administered L-[U-^{14}C] tyrosine. *Dotted lines* Radioactivity of the fractions; *solid lines* absorbance at 282 nm

compound was isolated in a separate experiment and identified as tyramine from mass spectrum and chromatographic behavior (HPLC and TLC). To obtain a more precise assessment of the distribution of label derived from tyrosine, feeding experiments with ^{13}C-labeled tyrosine and tyramine were undertaken as described below.

2.5.2 Tracer Experiments with ^{13}C-Labeled Precursors

Preliminary experiments showed that less than 200 ppm of tyrosine or tyramine did not significantly reduce cell growth or alkaloid formation (Table 5). Roots were cultured for 25 days in SB5 medium containing 200 ppm [3-^{13}C] tyrosine or [2-^{13}C] tyramine, prepared by decarboxylation of [3-^{13}C] tyrosine (Sugimoto and Yamada 1993). Roots were harvested and freeze-dried, then alkaloids were extracted and purified. ^{13}C-labeled aromoline and isotetrandrine (berbamine methyl ether) were obtained as crystals. ^{13}C-NMR spectra of ^{13}C-enriched aromoline and isotetrandrine are shown in Figs. 12 and 13, respectively. In the spectra for each of the ^{13}C-enriched alkaloids, four signals were specifically enhanced when [3-^{13}C] tyrosine was fed, while only two signals were enhanced when [2-^{13}C] tyramine was fed as compared with the spectra of the unlabeled compounds. The ^{13}C-NMR spectrum of aromoline was previously assigned by Koike et al. (1982). According to their assignments, the enhanced signals correspond to the carbons at C-4 (δ 28.5), C-α (δ 38.3), C-4′ (δ 24.6) and C- α' (δ 39.4) of aromoline. The ^{13}C-NMR spectrum of isotetrandrine was assigned by our group (Sugimoto et al. 1993) and the enhanced signals correspond to the carbons at C-4 (δ 23.7), C-α (δ 38.6), C-4′ (δ 25.5) and C- α' (δ 37.8). The ^{13}C-enrichments of the signals, calculated on the basis of peak heights, are summarized in Figs. 14 and 15 for tyrosine and tyramine, respectively. These results unequivocally show that tyrosine was incorporated in approximately equal amounts into the isoquinoline and benzyl portions of

Table 5. Effects of tyrosine and tyramine on root growth and alkaloid formation in *Stephania cepharantha* root cultures

Concentration (ppm)	Growth index (*n*-fold)	Alkaloid content (% dry wt.)	
		Aromoline	Berbamine
Control	22.4	2.83	1.29
Tyrosine			
40	17.2	2.83	1.29
200	18.4	2.49	1.10
400	14.3	2.37	1.07
Tyramine			
40	18.0	2.88	1.26
200	21.6	2.64	1.21
400	15.8	2.21	1.08

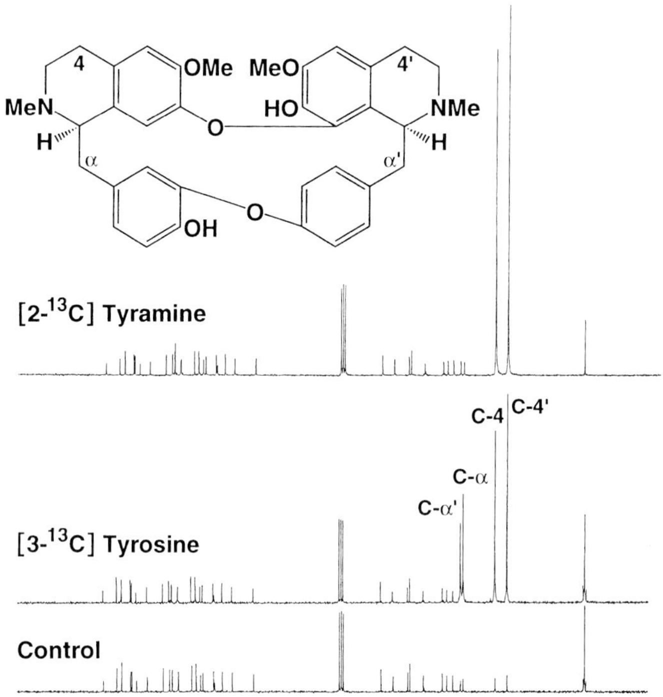

Fig. 12. ^{13}C-NMR spectra of aromoline. The control sample has normalized peak heights for comparison with labeled products

aromoline and berbamine. On the other hand, tyramine was specifically incorporated into the isoquinoline portion.

When 200 ppm of tyrosine or tyramine was fed to *S. cepharantha* root cultures and allowed to be metabolized for 1 month, millimolar quantities of tyramine were detected in the roots, approximately 4 mM for tyrosine and 6 mM for tyramine. However, none of the [2-^{13}C] tyramine administered exogenously was incorporated into the benzyl moiety of aromoline and berbamine, as shown in Figs. 12 and 13, respectively. These results suggest that, in *Stephania* root cultures, the hydroxylation of tyramine proceeds more rapidly than its oxidation or that tyrosine conversion to the benzyl moiety proceeds via its corresponding α-keto acid (Fig. 16).

Because the ratios of ^{13}C enrichments of C-4, C-α, C-4′ and C-α′ in aromoline and berbamine were the same within the experimental limits, both alkaloids are believed to be derived from a common intermediate berbamunine. Moreover, in aromoline and berbamine, the ratios of ^{13}C-enrichments of C-4 and C-α in (*S*) and (*R*) halves was the same within experimental limits. This suggests that the two coclaurine units have one and the same biogenetic origin. According to Rueffer et al. (1981), coclaurines were synthesized as (*S*)-forms by the stereospecific condensation of dopamine and phenylacetaldehydes in *Eschscholtzia* cell cultures, which are rich sources of benzylisoquinolines. From

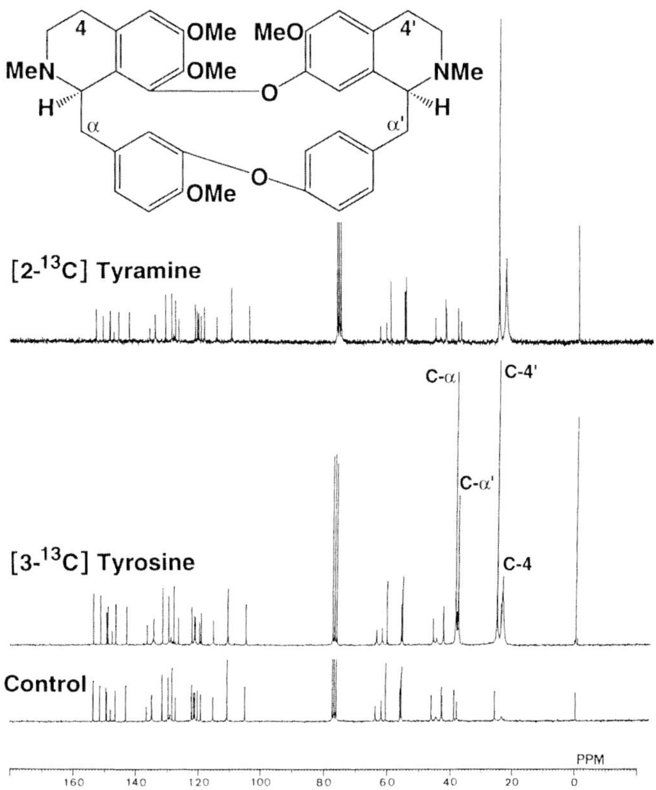

Fig. 13. ^{13}C-NMR spectra of berbamine methyl ether. The control sample has normalized peak heights for comparison with labeled products

their results and ours, (*R*) coclaurines might be derived from the corresponding (*S*) compounds (Fig. 16), probably via dehydro-derivatives as was shown in morphine biosynthesis. The fact that (*S*)-tetrahydroprotoberberine oxidase catalyzed the conversion of (*S*) tetrahydrobenzylisoquinolines to the corresponding dehydroforms (Amann and Zenk 1987) supports this hypothesis, though no racemization of (*S*)- to (*R*)-coclaurine or vice versa took place in *Berberis stolonifera* cells (Stadler et al. 1988).

3 Protocols

3.1 Extraction of Alkaloids

Harvested roots were freeze-dried, then powdered and soaked overnight in MeOH. This macerated material was centrifuged for 5 min at 3000 rpm. Treatment with MeOH was repeated once more, and the combined MeOH extracts

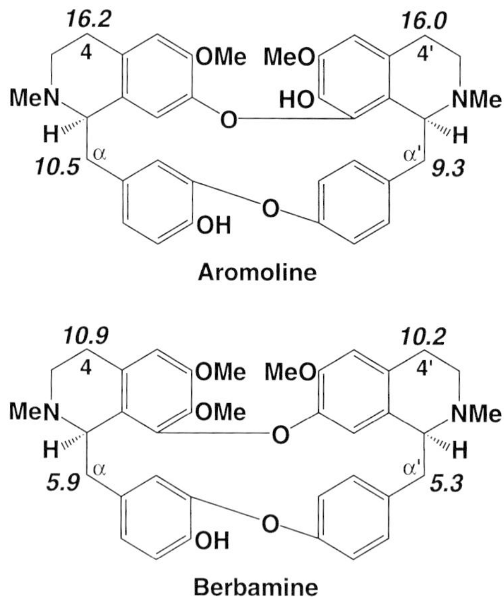

Fig. 14. Incorporation of tyrosine into aromoline and berbamine. Values represent isotopic excess of ^{13}C at each carbon

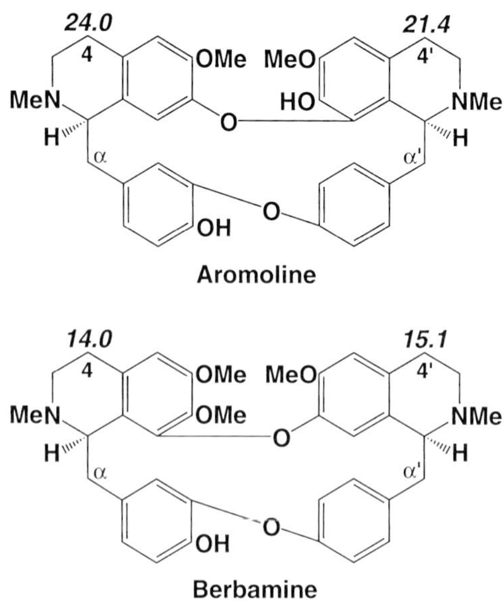

Fig. 15. Incorporation of tyramine into aromoline and berbamine. Values represent isotopic excess of ^{13}C at each carbon

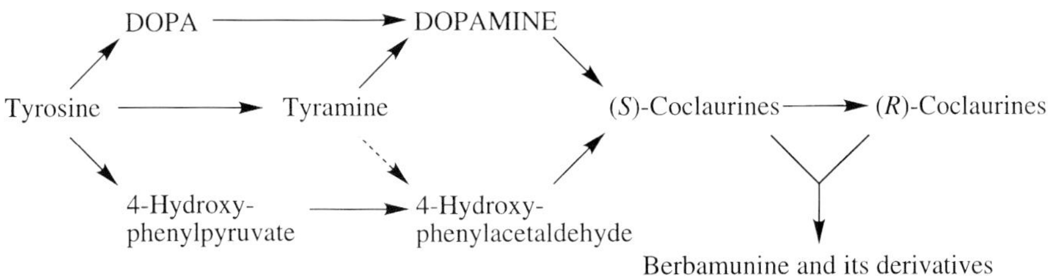

Fig. 16. Biosynthetic sequence leading from tyrosine to the bisbenzylisoquinolines

were evaporated to dryness at 40 °C. The dry residue was dissolved in 2 ml of 3% citric acid, and the acidic aqueous solution was filtered through filter paper into a glass tube and made alkaline (pH 10) with NH_4OH. A 1-ml portion of this alkaline solution was loaded onto an Extrelut column (Merck Art. 15731). After 10 min, 3 ml of $CHCl_3$ was passed through the column twice. The $CHCl_3$ extracts were combined, then evaporated to dryness at 30 °C. The dry residue was dissolved in 0.5 ml MeOH and put through HPLC and TLC.

3.2 HPLC and TLC

Alkaloid contents were measured with an HPLC (Shimadzu LC-4 A) at 35 °C. The stationary phase was Develosil ODS-3 (150 × 4.6 mm) and the solvent 80% MeOH containing 0.2% NH_4OH. The flow rate was 0.4 ml/min. A short pre-column (30 × 4.6 mm) was placed between the injector and the separation column. All the alkaloids were detected by UV at 282 nm. The respective R_ts of aromoline, berbamine, homoaromoline, cepharanoline, isotetrandrine, cycleanine and cepharanthine were 10.0, 11.5, 14.7, 15.0, 18.0, 18.8 and 23.8 min. Peak areas were calculated with a Chromatopac (Shimadzu CR-2 A). TLC was conducted on a 0.25-mm silica gel plate (Merck Art. 5715) with a solvent system of CHCl3-MeOH-NH_4OH (200:50:1). Alkaloids were located by UV illumination and by spraying the plate with modified Dragendorff's reagent. The *Rf* value of aromoline was 0.33 and of berbamine 0.50.

3.3 Tracer Experiments with ^{14}C-Labeled Precursors

L-[U-^{14}C] tyrosine (17.98 GBq/mmol), [7-^{14}C] tyramine (2.07 GBq/mmol) and [7-^{14}C] dopamine (2.07 GBq/mmol) were purchased from Amersham International. ^{14}C-labeled L-tyrosine, tyramine or dopamine (10^7 dpm) was applied to vigorously growing 15-day-old cultured roots of *S. cepharantha*. The roots were allowed to metabolize for 10 days in the dark. Thereafter, the roots were harvested and alkaloids extracted and analyzed as mentioned above. Radioac-

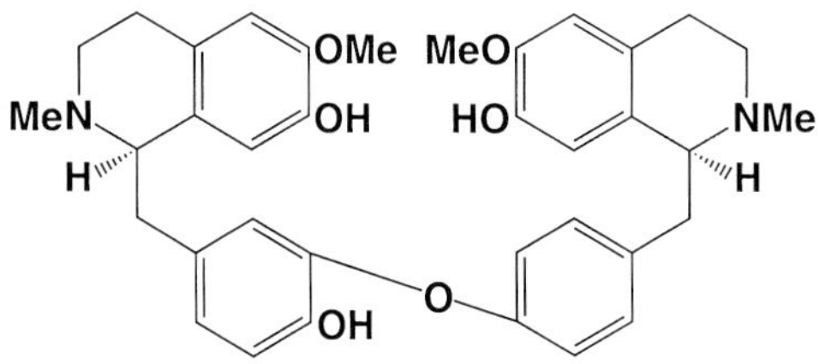

Berbamunine

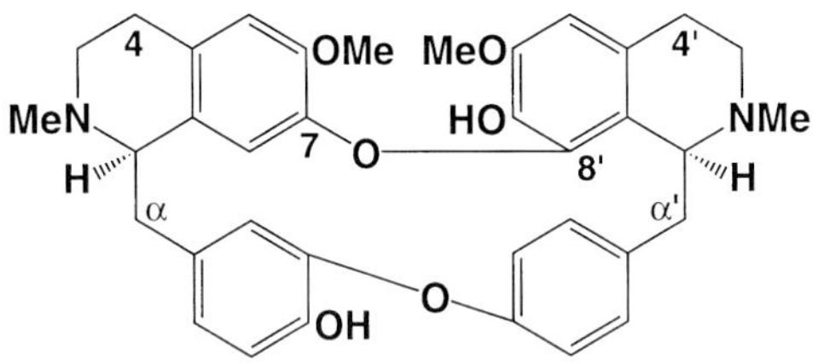

Aromoline

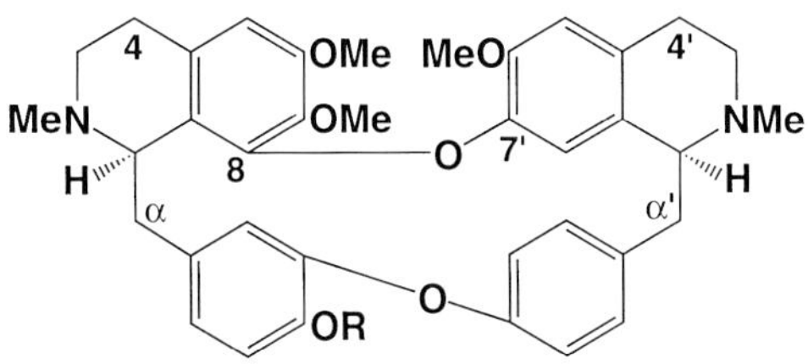

R = H Berbamine
R = Me Isotetrandrine

Structural Formulae

tivity of the fractions separated by HPLC was measured in a toluene-based scintillator.

3.4 Tracer Experiments with ^{13}C-Labeled Precursors

L-[3-^{13}C] tyrosine (99.8 atom%) was purchased from Isotec, Inc., and L-tyrosine decarboxylase (EC 4.1.1.25) from Sigma. [2-^{13}C] tyramine was prepared by decarboxylation of L-[3-^{13}C] tyrosine with L-tyrosine decarboxylase. Roots were cultured for 25 days in modified B5 medium containing 200 ppm L-[3-^{13}C] tyrosine or [2-^{13}C] tyramine, and harvested. Freeze-dried roots were treated as mentioned above, and aromoline was isolated as pale yellow crystals. After removal of aromoline, CH_2N_2 was introduced to the residual MeOH solution to convert berbamine to its methyl ether, isotetrandrine, which was isolated as pale yellow crystals. ^{13}C-Enrichment was calculated from the peak-height analysis of ^{13}C NMR spectra of aromoline and berbamine methyl ether.

4 Summary and Conclusion

The genus *Stephania* comprises approximately 50 species extending in distribution from Africa through Asia to Australia. The underground tubers of the vines are important because of their powerful pharmaceutical effects.

Root cultures of *Stephania cepharantha* were established, using adventitious roots from callus and roots from tuber as starting materials. The cultured roots produced at least five bisbenzylisoquinoline alkaloids; homoaromoline, aromoline, isotetrandrine, berbamine and cycleanine. The main constituents were aromoline and berbamine, and their respective contents reached more than 2 and 1% under optimum culture conditions for alkaloid production. These values are much higher than those in the intact plant.

Feeding experiments with ^{14}C-labeled tyrosine, tyramine and dopamine and ^{13}C-labeled tyrosine and tyramine demonstrated that aromoline and berbamine were composed of four molecules of tyrosine, and two coclaurine units have one and the same biogenetic origin. Moreover, these results suggest that aromoline and berbamine are derived from a common progenitor berbamunine, and that the hydroxylation of tyramine proceeds more rapidly than its oxidation or that tyrosine conversion to the benzyl moiety proceeds via its corresponding α-keto acid in *Stephania cepharantha*.

References

Akasu M, Itokawa H, Fujita M (1974) Four new fluorescent components isolated from the callus tissue of *Stephania cepharantha*. Tetrahedron Lett41:3609–3612

Akasu M, Itokawa H, Fujita M (1975) Oxoaporphine alkaloids from the callus tissue of *Stephania cepharantha*. Phytochemistry 14:1673–1674

Akasu M, Itokawa H, Fujita M (1976) Biscoclaurine alkaloids in callus tissue of *Stephania cepharantha*. Phytochemistry 15:471–473

Amann M, Zenk MH (1987) Preparation of dehydrobenzylisoquinolines by immobilized (*S*)-tetrahydroprotoberberine oxidase from plant cell cultures. Phytochemistry 26:3235–3240

Bhakuni DS, Singh AN, Jain S (1980) Biosynthesis of isotetrandrine. Tetrahedron 36:2149–2151

Bhakuni DS, Jain S, Chaturvedi R (1987) The biosynthesis of the alkaloids of *Cissampelos pareira* Linn. Tetrahedron 43:3975–3982

Blanchfield JT, Kitching W, Sands DPA, Thong YH, Kennard CHL, Byriel KA (1993) Alkaloids from some Australian *Stephania* (Menispermaceae) species. Nat Prod Lett 3:305–312

Cao Z-F, Zhu X-Q (1997) Antioxidant action of tetrandrine: an alkaloid from the roots of *Radix stephania tetrandra*, S Moore. J Sci Food Agric 73:106–110

Cassels BK, Breitmaier E, Zenk MH (1987) Bisbenzylisoquinoline alkaloids in *Berberis* cell cultures. Phytochemistry 26:1005–1008

Charles B, Bruneton J, Pharadai K, Tantisewie B, Guinaudeau H, Shamma M (1987) Some unusual proaporphine and aporphine alkaloids from *Stephania venosa*. J Nat Prod 50:1113–1117
Dagne E, Gunatilaka AAL, Kingston DGI (1993) 4′-*O*-Methylstephavanine from *Stephania abyssinica*. J Nat Prod 56:2022–2025
Davydenkov VN, Tareeva NV, Kir'yanov AA, Bondarenko LT (1988) Qualitative determination of stepharine in the plant cell cultures of *Stephania glabra*. Khim-Farm Zh 22:326–328
Deng J, Zhao S, Lou F (1990) A new monoquaternary bisbenzylisoquinoline alkaloid from *Stephania tetrandra*. J Nat Prod 53:993–994
Dwuma-Badu D, Ayim JSK, Withers SF, Agyemang NO, Ateya AM, El-Azizi MM, Knapp JE, Slatkin DJ, Schiff PL Jr (1980) Constituents of west African medicinal plants. XXVII. Alkaloids of *Rhigiocarya racemifera* and *Stephania dinklagei*. J Nat Prod 43:123–129
Guha KP, Mukherjee B (1979) Bisbenzylisoquinoline alkaloids – a review. J Nat Prod 42:1–84
Itokawa H, Akasu M, Fujita M (1973) Several oxidized sterols isolated from callus tissue of *Stephania cepharantha*. Chem Pharm Bull (Tokyo) 21:1386–1387
Kang H-S, Kim Y-H, Lee C-S, Lee J-J, Choi I, Pyun K-H (1996) Anti-inflammatory effects of *Stephania tetrandra* S. Moore on interleukin-6 production and experimental inflammatory disease models. Mediators Inflammation 5:280–291
Koike L, Marsaioli A, Reis F de AM (1982) Carbon-13 nuclear magnetic resonance spectroscopy and conformational analysis of the daphnoline-repandine class of bis(benzylisoquinoline) alkaloids. J Org Chem 47:4351–4353
Kozuka M, Miyaji K, Sawada T, Tomita M (1985) A major alkaloid of the leaves and stems of *Stephania rotunda*. J Nat Prod 48:341–342
Kupchan SM, Suffness MI, White DNJ, McPhail AT, Sim GA (1968) The isolation and structural elucidation of 4-demethylhasubanonine, a new alkaloid from *Stephania hernandifolia*. J Org Chem 33:4529–4532
Likhitwitayawuid K, Angerhofer CK, Cordell GA, Pezzuto JM (1993) Cytotoxic and antimalarial bisbenzylisoquinoline alkaloids from *Stephania erecta*. J Nat Prod 56:30–38
Min Z-D, Yang H-M, Mizuno M, Tanaka T (1986) Studies on chemical constituents and chemotaxonomy of genus *Stephania* in China. Gifu Yakka Daigaku Kiyo 35:28–35
Nakaoji K, Nayeshiro H, Tanahashi T, Su Y, Nagakura N (1997) Bisbenzylisoquinoline alkaloids from *Stephania cepharantha* and their effects on proliferation of cultured cells from the murine hair apparatus. Planta Med 63:425–428
Patra A, Montgomery CT, Freyer AJ, Guinaudeau H, Shamma M, Tantisewie B, Pharadai K (1987) The protoberberine alkaloids of *Stephania suberosa*. Phytochemistry 26:547–549
Popov YG (1985) Somatic embryogenesis and plant development in *Stephania glabra* tissue culture. Biol Nauki (Moscow) 6:86–89
Rueffer M, El-Shagi H, Nagakura N, Zenk MH (1981) (*S*)-Norlaudanosoline synthase: the first enzyme in the benzylisoquinoline biosynthetic pathway. FEBS Lett 129:5–9
Schiff PL Jr (1983) Bisbenzylisoquinoline alkaloids. J Nat Prod 46:1–43
Shamina ZB, Savina TA, Osipova EA, Popov YG (1994) Improvement of *Stephania glabra* (Roxb) Miers. cell culture productivity. Russian J Plant Physiol 41:778–783
Soni PL, Sharma HW, Bisen SS, Srivastava HC, Gharia MM (1986) Physicochemical studies of starch isolated from forest tubers *Stephania glabra* and *Pueraria thomsonii*. Starch 38:355–358
Stadler R, Zenk MH (1993) The purification and characterization of a unique cytochrome P-450 enzyme from *Berberis stolonifera* plant cell cultures. J Biol Chem 268:823–831
Stadler R, Kutchan TM, Loeffler S, Nagakura N, Cassels B, Zenk MH (1987) Revision of the early steps of reticuline biosynthesis. Tetrahedron Lett 28:1251–1254
Stadler R, Loeffler S, Cassels BK, Zenk MH (1988) Bisbenzylisoquinoline biosynthesis in *Berberis stolonifera* cell cultures. Phytochemistry 27:2557–2565
Sugimoto Y, Yamada Y (1993) Involvement of tyramine in bisbenzylisoquinoline biosynthesis in cultured roots of *Stephania cepharantha*. J Nat Prod 56:1229–1233
Sugimoto Y, Sugimura Y, Yamada Y (1988) Production of bisbenzylisoquinoline alkaloids in cultured roots of *Stephania cepharantha*. Phytochemistry 27:1379–1381

Sugimoto Y, Sugimura Y, Yamada Y (1990) Biosynthesis of bisbenzylisoquinoline alkaloids in cultured roots of *Stephania cepharantha*. FEBS Lett 273:82–86

Sugimoto Y, Kawaminami S, Yamada Y (1993) Biosynthetic relationship of aromoline and berbamine in cultured roots of *Stephania cepharantha*. Phytochem Anal 4:100–102

Sugimoto Y, Uchida S, Inanaga S, Isogai A (1997) Effect of P-450 inhibitors on benzylisoquinoline alkaloid biosynthesis in cultured roots of *Stephania cepharantha* and *Menispermum dauricum*. J Plant Physiol 150:376–380

Suzuki S, Fujino H, Tatsuo Y, Yamazaki N, Yoshizaki M (1992) Rapid propagation of *Stephania cephalantha* Hayata by tissue culture. Jpn J Breed 42:769–777

Taga T, Akimoto N, Ibuka T (1984) Stephadiamine, a new skeletal alkaloid from *Stephania japonica*: the first example of a *C*-norhasubanan alkaloid. Chem Pharm Bull (Tokyo) 32:4223–4225

Tantisewie B, Amurrio S, Guinaudeau H, Shamma M (1989) New bisbenzylisoquinolines from *Stephania pierrii*. J Nat Prod 52:846–851

Thornber CW (1970) Alkaloids of the Menispermaceae. Phytochemistry 9:157–187

Wang X-K, Zhao T-F (1990) The distribution and biological activity of the alkaloids from *Stephania* plants. Chin Pharm J 25:3–6

Wang X-K, Zhao Y-R, Zhao T-F, Che C-T (1995) Further constituents of *Stephania sutchuenensis*. Planta Med 61:99

Yamada Y, Fujita Y (1983) Production of useful compounds in culture. In: Evans DA, Sharp WR, Ammirato PV, Yamada Y (ed) Handbook of plant cell culture, vol 1. Macmillan, New York, pp 717–728

Yang H-M, Chen Y (1994) Chemosystematic notes of genus *Stephania* (Menispermaceae) in China. Acta Bot Sin 36:14–22

Zhen P, Chen P, Wang H (1994) Analysis of tetrandrine and demethyltetrandrine in Chinese medicinal herbs *Stephania tetrandra* S. Moore by high performance liquid chromatography. Fenxi Kexue Xuebao 10:20–23

17 *Trigonella* Species: In Vitro Culture and Production of Secondary Metabolites

P. CHRISTEN[1]

1 Introduction

1.1 The Plant

The Leguminosae (syn. Fabaceae) family is one of the three largest families of flowering plants. There is still no general agreement regarding the number of genera and species. Estimates vary between 590–690 genera and 12,000–17,000 species (Heywood 1971). The family is divided into three subfamilies: Caesalpinioïdeae, Mimosoïdeae and Papilionoïdeae (Brummitt 1992).

The genus *Trigonella* sensu stricto belongs to the latter subfamily and is composed of 75 species (Willis 1973). The name of the genus derives from the Latin *Trigonus*, "three-angled" in reference to the small, triangular appearance of the flower. *Trigonella foenum-graecum* L. (fenugreek) is an erect, annual, herbaceous plant widely distributed in many parts of Asia, Africa and Europe (see Fig. 1). It is 10–50 cm high, sparsely pubescent with leaves pinnately three-foliolate. Leaflets (20–50 × 10–15 mm) are obovate to oblong-oblanceolate and denticulate. Flowers are solitary or in pairs in the axils of the leaves. The calyx is short (6–8 mm) and the corolla (12–18 mm) is yellowish-white tinged with violet at the base. The fruit (legume) (60–110 × 4–6 mm) is linear, somewhat curved, glabrous or glabrescent with longitudinal veins. The seeds (2–6 × 2–4 mm) are quadrangular, somewhat compressed, yellow or pale-brown, finely tuberculate with punctiform hilum (Tutin et al. 1968).

1.2 Chemical Constituents and Their Medicinal Importance

Fenugreek foliage was one of the components of the celebrated Egyptian kuphi incense, a holy smoke used in fumigation and embalming rites. Crushed fenugreek seeds are used in the preparation of poultices for burns, in cosmetics, hair oils, and in cough, diuretic and laxative preparations. They are mucilaginous when mixed with water and may be consumed raw or boiled. Egyptian

[1]University of Geneva, Laboratory of Pharmaceutical Analytical Chemistry, 20, Bd D'Yvoy, 1211 Geneva 4, Switzerland

Biotechnology in Agriculture and Forestry, Vol. 51
Medicinal and Aromatic Plants XII (ed. by T. Nagata and Y. Ebizuka)
© Springer-Verlag Berlin Heidelberg 2002

Fig. 1. *Trigonella foenum-graecum* L. (Botanical Garden, Geneva, Switzerland, B. Renaud)

and Indian women eat the seeds to promote lactation. They are also used locally as a source of yellow dye and as an ingredient of curry powder (Allen and Allen 1981). Commercial samples vary according to their geographic origin. Furthermore, the seeds present a pharmaceutical interest as they contain 1–2% steroidal sapogenins, particularly diosgenin and yamogenin in the oily embryo. These sapogenins are of considerable economic importance as they are used as a starting material for the partial synthesis of oral contraceptives, sex hormones and other steroids. Although the diosgenin yield is lower than in the tubers of certain species of *Dioscorea* (yams), the easy cultivation of fenugreek and its rapid growth, affording seeds 3–4 months after sowing, make the plant a potentially valuable crop for sapogenin production. Diosgenin [(25*R*)-spirost-5-en-3β-ol] and yamogenin [(25*S*)-spirost-5-en-3β-ol] are two epimers. Other steroidal sapogenins such as yuccagenin [(25*R*)-

spirost-5-en-2α,3β-diol], its epimer lilagenin [(25*S*)-spirost-5-en-2α,3β-diol], tigogenin [(25*R*)-5α-spirostan-3β-ol] and its epimer neotigogenin [(25*S*)-5α-spirostan-3β-ol], gitogenin ([(25*R*)-5α-spirostan-2α,3β-diol] and its epimer neogitogenin [(25*S*)-5α-spirostan-2α,3β-diol], smilagenin [(25*R*)-5β-spirostan-3β-ol] and its epimer sarsasapogenin [(25*S*)-5β-spirostan-3β-ol] are also present in smaller quantities (see Fig. 2).

Steroidal sapogenins occur in all parts of the plant. Ortuño et al. (1998) reported maximum levels of diosgenin in young leaves of *Trigonella foenum-graecum*. However, most studies focus on the seeds. Sapogenins occur as their glycosides (saponins) in plant tissues. Besides saponins, the seeds also contain fixed oil, polysaccharides, proteins, flavonoids and various gums which may approximate 23% of the seed content. The alkaloid trigonelline (Fig. 3), a nicotinic acid derivative which has been isolated from all plant parts, is converted into nicotinic acid by roasting.

The characteristic odour of the seed, especially if powdered, is spicy and is mainly due to 3-hydroxy-4,5-dimethyl-2(5H)-furanone, a compound isolated from the volatile constituents of the seeds (Girardon et al. 1986).

2 In Vitro Culture Studies

2.1 Callus and Suspension Cultures

During the last decades, the development of plant cell and tissue culture technology has extended the spectrum of phytochemicals which might be produced by this mean. Interest in saponins as semi-synthetic precursors of steroid hormones and the demonstration that many of these compounds can be produced in vitro, has increased research activity. In particular, numerous in vitro studies have been reported on *Dioscorea* (Culafic et al. 1999) and *Costus* sp. (Indrayanto et al. 1999).

The first report on the production of spirostane derivatives by *Trigonella* tissue cultures was published by Khanna and Jain (1973) and concerned the establishment of static cultures grown on solid Murashige and Skoog (MS) medium (Murashige and Skoog 1962) supplemented with 1 mg/l 2,4-dichlorophenoxyacetic acid (2,4-D). They reported the production of diosgenin, gitogenin and tigogenin along with other sterols. Six-week-old cultures showed a high growth index (GI = final wet weight - initial wet weight/initial wet weight) and the total steroidal content was higher than in the seeds. Khanna et al. (1975a) demonstrated that suspension cultures of *T. foenum-graecum* grown on media supplemented with various concentrations of cholesterol produced higher sapogenin contents than those grown on medium without cholesterol. Eight-week-old callus cultures raised from seeds and grown on solid Revised Tobacco (RT) medium, supplemented with ascorbic acid, showed a high GI and steady increase in intracellular ascorbic acid (Jain

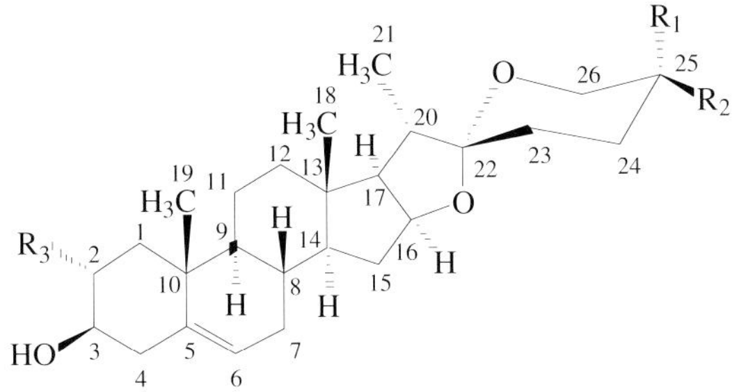

	R_1	R_2	R_3
Diosgenin	H	CH_3	H
Yamogenin	CH_3	H	H
Yuccagenin	H	CH_3	OH
Lilagenin	CH_3	H	OH

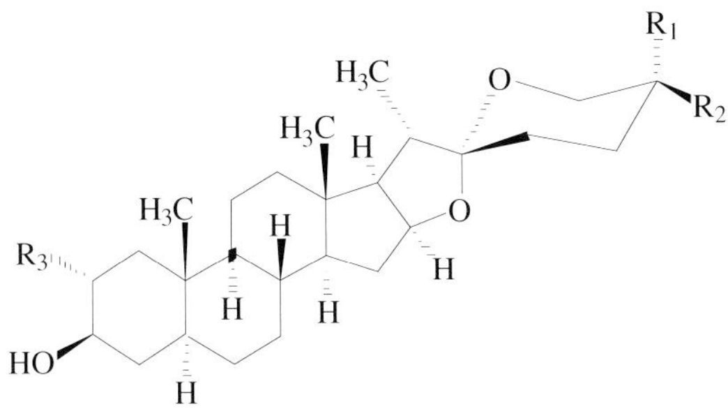

	R_1	R_2	R_3
Tigogenin	H	CH_3	H
Neotigogenin	CH_3	H	H
Gitogenin	H	CH_3	OH
Neogitogenin	CH_3	H	OH

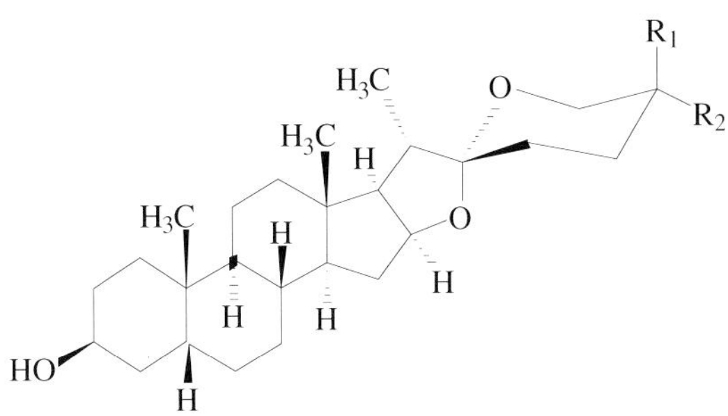

	R_1	R_2
Smilagenin (25R,5β)	H	CH_3
Sarsapogenin (25S,5β)	CH_3	H

Fig. 2. Main sapogenins frequently encountered in *Trigonella foenum-graecum* L.

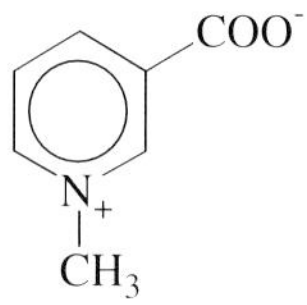

Fig. 3. Trigonelline

et al. 1975). Gupta (1974) investigated the effects of indole-3-acetic acid (IAA), α-naphthaleneacetic acid (NAA) and 2,4-D on growth and cytohistology of fenugreek calli. While IAA (1–5 mg/l) induced better growth and produced bigger cells with large intercellular spaces, NAA (1 mg/l) induced comparatively smaller cells without intercellular spaces. 2,4-D (5 mg/l) retarded the callus growth and induced giant cell formation and the development of multinucleate cells. The addition of 1–5 mg/l IAA, NAA, 2,4-D, kinetin and gibberellic acid (GA_3) to fenugreek cell suspension cultures on growth and production of spirostane derivatives and sterols was investigated by Khanna et al. (1975b). A maximum GI of 14 was obtained in cultures fed with 3 mg/l IAA, while the minimum GI of 3 was obtained with cultures fed with 5 mg/l 2,4-D. Cultures grown on media supplemented with NAA and GA singly showed a decrease in diosgenin and other steroids, whereas tissues grown on media supplemented with IAA or kinetin showed a significant increase in these compounds. Cultures fed with 5 mg/l kinetin presented a remarkable increase in sapogenin content, particularly of diosgenin (2.58%). Other results on the medium optimization for the production of diosgenin by callus cultures of fenugreek were published by Singh et al. (1981). Free and bound amino acid contents from calli were determined by Khanna and Nag (1973). Twelve amino acids were detected, among which alanine, glutamine, leucine, phenylalanine and proline were in free and bound forms. Brain and Lockwood (1976) investigated the effects of medium variation on steroid levels in cultures from fenugreek. Phytosterols and sapogenin contents of static and suspension cultures were examined. The growth of calli on solid MS medium, supplemented with 0.25 mg/l 2,4-D, doubled by adding 0.5 mg/l kinetin and even tripled in the presence of 10% v/v coconut water. However, this increase was at the expense of phytosterols. In suspension cultures, it was demonstrated that the age of culture, nature and dose of auxin and cytokinin can have marked effects on growth and steroid levels. In particular, certain combinations of 2,4-D and kinetin had a synergistic effect, whilst others were antagonistic. However, the sapogenin contents in the cultures were much lower than those in the seeds, whilst the phytosterol contents tended to be higher. Time-course studies were carried out to examine the effects of callus induction from seeds on steroid levels for 12 days (Lockwood and Brain 1976). According to the hormonal supplementation of the medium, important variations were observed in the levels of phytosterols as well as of sapogenins. However, in all cases, phytosterol contents reached higher levels than in the seeds, but the sapogenin content was always lower. To understand this phenomenon, labeled cholesterol was added to cell suspension cultures of *T. foenum-graecum*. It has been suggested that there are alternative pathways from sterol to diosgenin (Brain and Williams 1983). A first pathway is the incorporation of cholesterol and this pathway is predominant when the precursor is added at subculture. The second pathway involves side-chain cleavage before incorporation and takes place when the sterol is added 10 days after subculture. In a similar way, Trisonthi et al. (1980) demonstrated that mevalonic acid promotes the synthesis of steroidal sapogenins in fenugreek tissue, particularly in suspension cultures grown on MS or Miller medium.

More recently, the effects of diniconazole, a triazole-type fungicide with a broad antifungal spectrum, on cell suspension cultures have been studied (Cerdon et al. 1995). This fungicide was added (125 μM) to the fenugreek cell cultures, and, after 21 days, the cell growth decreased by about 20% compared to the control cell growth. Furthermore, diniconazole treatment led to a decrease of about 50% of total sterol content, causing an inhibition of sterol biosynthesis at the 14α-demethylation step leading to accumulation of 14α-methyl Δ^8-sterols. Oncina et al. (2000) reported on the production of diosgenin by callus cultures of *T. foenum-graecum*. Leaf, stem and root calli were established and cultured on different solid growth media (MS, White's basal medium, Gamborg's B5) supplemented with coconut milk, malt extract and NAA. In all cases, MS medium supplemented with 15% (v/v) coconut milk and 3×10^{-6} M NAA was the most suitable medium for callus growth. Diosgenin levels were higher in leaf calli than in stem and root calli and corresponded to about three to five times the levels observed in the calli from other plant organs. Maximum diosgenin levels were attained after 45 days and reached 2.2 mg/g dry weight in leaf calli, 0.74 mg/g in stem calli and 0.60 mg/g in root calli, which represents 22, 10 and 27% of the levels detected in the corresponding organs of the mother plant at 45 days (Ortuño et al. 1998).

Three flavonoids, quercetin, luteolin and vitexin-7-glucoside, were identified from callus cultures established from seedlings grown on solid RT medium supplemented with 1 mg/l 2,4-D (Sharma and Khanna 1977). Growth index and total flavonoid content were maximum (6.5 and 2.57%) in 6- and 4-week-old tissues, respectively.

Undifferentiated tissue cultures from germinating *T. occulta* Delile seeds have been established on solid MS medium, supplemented with 1 mg/l 2,4-D, and their steroidal constituents examined (Jain et al. 1977). Calli were subcultured every 6–8 weeks and were finally harvested after 19 months. The sapogenin and phytosterol fraction reached approx. 2.6% dry wt. Calli contained 0.37% diosgenin, 0.14% gitogenin, 0.05% tigogenin and 0.23% sitosterol. These values are higher than those obtained when analysing the initial seeds.

The pyridine alkaloid, trigonelline, is widely distributed in the plant kingdom and is known for its hypoglycaemic and hypocholesterolaemic activity (Iribbarren and Pomilio 1983). It has been reported to occur in the seeds (0.3–0.4%) of *T. foenum-graecum* (Kuhn and Gerhard 1943). Joshi and Handler (1960) demonstrated the biosynthesis of trigonelline from nicotinic acid and

S-adenosylmethionine. Cell-free extracts of root callus cultures have been reported to catalyze the conversion of nicotinic acid and S-adenosylmethionine to trigonelline in the presence of ATP and $MgCl_2$ (Antony et al. 1975). Maximum trigonelline biosynthesis was observed when the cells were at the exponential growth phase.

Eight-week-old callus tissue cultures of fenugreek, established from seeds on solid RT medium and supplemented with 1 mg/l 2,4-D, produced 4.5% trigonelline. In the presence of 0.5 and 1.0 mg/l nicotinic acid, trigonelline increased to 5.25 and 5.01%, respectively (Khanna and Jain 1972). Comparison of trigonelline contents of seeds, roots, shoots and in vitro cultures was carried out by Radwan and Kokate (1980). The explants were first subjected to shocks with high concentrations of 2,4-D, IAA, indolepropionic acid (IPA), NAA, GA and kinetin to determine which auxin(s) was suitable for the development of calli. Hormonal shocks for 1 h stimulated significantly the growth of the calli. Whereas 2,4-D, IAA, IPA and NAA (10 mg/l) led to some increases in the trigonelline content, GA (1 mg/l) and kinetin (2 mg/l) did not have pronounced effects. Gamborg's B5 (Gamborg 1970) medium, modified by supplementing it with 3 g/l of casein hydrolysate, 2 g/l of yeast extract and 40 g/l of saccharose, was used as growth medium. Four-week-old callus cultures of *T. foenum-graecum* produced 15.6 mg/g dry wt. of trigonelline, which represents 3 to 4 times more trigonelline than the seeds and 12 to 18 times more than the roots and shoots of the parent plants. Four- and 6-week-old suspension cultures produced 38.2 and 44.2 mg/g dry wt., respectively, of trigonelline which is more than twice the amount found in the calli. Proportions of 9–12% of the trigonelline were released into the solid medium. In suspension cultures, one third or more of the trigonelline was dissolved in the liquid medium. Therefore, it appears that the biosynthesis of this compound is favored by its enhanced removal from the site of its biosynthesis in the cells, into the medium. Cultures grown in the presence of 50 mg/l nicotinic acid contained more trigonelline than those grown in the presence of 1 mg/l only of this substrate. At the same time, the proportion of the alkaloid released into the liquid medium increased from 31%, with a low level of nicotinic acid, to 37%, with a high level of the same compound.

Polysaccharides extracted from callus and suspension cultures of *T. foenum-graecum* established from seeds, showed significant differences in their sugar composition when compared to the mucilage polysaccharides of the seeds (Bräutigam and Franz 1985).

Another *Trigonella* species, *T. polycerata* L., is used as green fodder. However, only little phytochemical investigation has been done on this plant. Callus cultures were raised from seeds and maintained on RT medium by frequent subculturings. They were harvested at 2-, 4-, 6-, 8- and 10-week intervals and analyzed for their steroidal content (Kamal and Yadav 1992). Diosgenin and tigogenin were identified. Their amount in the callus cultures showed parallelism with the growth index. The highest diosgenin and tigogenin content was found in 8-week-old calli (0.40 and 0.15%, respectively). Mehra et al. (1996)

investigated the influence of nicotinic acid in the production of trigonelline by callus cultures. Calli were established from seeds and cultured on solid RT medium supplemented with 1 mg/l 2,4-D. Growth index and trigonelline content increase with the age of the callus. Calli fed with various doses of nicotinic acid (250, 500 and 750 mg/l) and harvested at different time intervals (2, 4, 6 and 8 weeks after subcultures) showed a significant reduction of growth and increase in trigonelline content in all the treatment doses. It is reported that the decrease in callus growth may be due to enhanced trigonelline content which controls cell proliferation.

Differentiation in callus cultures established from leaf segments of *Trigonella foenum-graecum* and *T. corniculata* L. were investigated (Sen and Gupta 1979). Explants of the two species formed actively growing calli within 5 weeks of culture on MS medium and White's basal medium (White 1943), both media being supplemented with NAA (0.5 mg/l) and coconut milk (15% v/v). Calli of the two species started differentiation 3–4 weeks following initiation on MS medium containing NAA and coconut milk. However, the replacement of coconut milk by casein hydrolysate increased the number of differentiated organs per callus.

2.2 Protoplast Culture and Organogenesis

Plant protoplasts provide a useful system for transformation studies involving the uptake and expression of exogenous DNA. Root protoplasts of *T. foenum-graecum* were isolated using a simple enzyme mixture composed of 2% w/v Rhozyme, 4% w/v Meicelase and 0.3% w/v Macerozyme (Xu et al. 1982). Protoplasts entered division after 3–4 days of culture and formed cell colonies with a plating efficiency of 6–10%. Tissues derived from protoplasts formed roots only and failed to produce embryoids or shoots on any of the hormone combinations tested. In another study, mesophyll protoplasts were obtained after the treatment of leaves with purified cellulase (Shekhawat and Galston 1983). Up to 70% plating efficiency could be obtained. The protoplast-derived colonies developed into rapidly growing green calli which in turn produced leafy shoots by the addition of 6-benzylaminopurine and zeatin (0.1 mg/l each) but without auxin.

2.3 Hairy Root Cultures

2.3.1 Establishment of Hairy Root Cultures

Plant Material. Seeds of *Trigonella foenum-graecum*, obtained from Hänseler (Herisau, Switzerland), were surface sterilized for 6 min with sodium hypochlorite solution (1.5% w/v available chlorine) supplemented with 2 drops Triton X-100 (Fluka, Switzerland) and rinsed three times with sterile distilled water. They were left for germination on agar plates containing 0.5% sucrose at 25 °C in the light, with a daily 16-h photoperiod (Osram-L-Fluora 77R; 700 lux). Plantlets were transferred onto solid half-strength McCown's woody plant (WP) medium (Lloyd and McCown 1980) supplemented with 3% sucrose (1/2 WP 3) and maintained at 25 °C in the light (Merkli et al. 1997).

Bacterial Strain. *Agrobacterium rhizogenes* strain A_4 ($pRiA_4$) was used for the induction of hairy roots. The bacteria were grown on YMB solid medium (Hooykaas et al. 1977) and subcultured monthly.

Transformation with *Agrobacterium rhizogenes*. Stems of 2-week-old plantlets were inoculated with *A. rhizogenes* A_4. Approximately 4 weeks after infection, hairy roots appeared at the infection sites. They were excised and transferred to solid 1/2 WP 3 medium containing 0.25 g/l Cefotaxine and 1 g/l Ampicillin (Sigma, USA) to eliminate the excess of bacteria. The axenic hairy roots thus obtained were maintained in hormone-free 1/2 WP 3 (50 ml in 250 ml conical flasks) and cultured in the dark at 25 °C on a gyratory shaker at 80 rpm. They were subcultured at 4-week intervals.

The anatomy of nontransformed and transformed roots was compared microscopically. Whereas the transverse section of both root systems was similar, surface characters presented some differences. In particular, root hairs of transformed roots are fairly abundant (Fig. 4). They are unicellular with a spherical head. Occasionally, some hairs are branched forming two arms, each ended by an ovoid head. Hairs of nontransformed roots are less abundant, significantly shorter and without branching.

2.3.2 Opine Assay

Agrobacterium rhizogenes A_4 is an agropine-type strain ($pRiA_4$) inducing the production of agropine and mannopine. These opines are specifically found in *Agrobacterium*-mediated transformed tissues. To prove transformation, the opines were extracted, separated by paper electrophoresis according to the method described by Petit et al. (1983) and visualized with an alkaline silver nitrate reagent (Trevelyan et al. 1950). Silver nitrate-positive compounds were identified by comparison with authentic agropine and mannopine.

2.3.3 Extraction, Identification and Quantification of Diosgenin

The fresh hairy roots were lyophilized, powdered and extracted essentially as described by Sauvaire and Baccou (1978). Approximately 200 mg of powdered dry root material was extracted and hydrolyzed by refluxing for 5 h with 1 M sulfuric acid in 70% isopropanol. The solution was filtered, adsorbed on an Extrelut (Merck, Germany) column for 15 min and diosgenin was eluted with

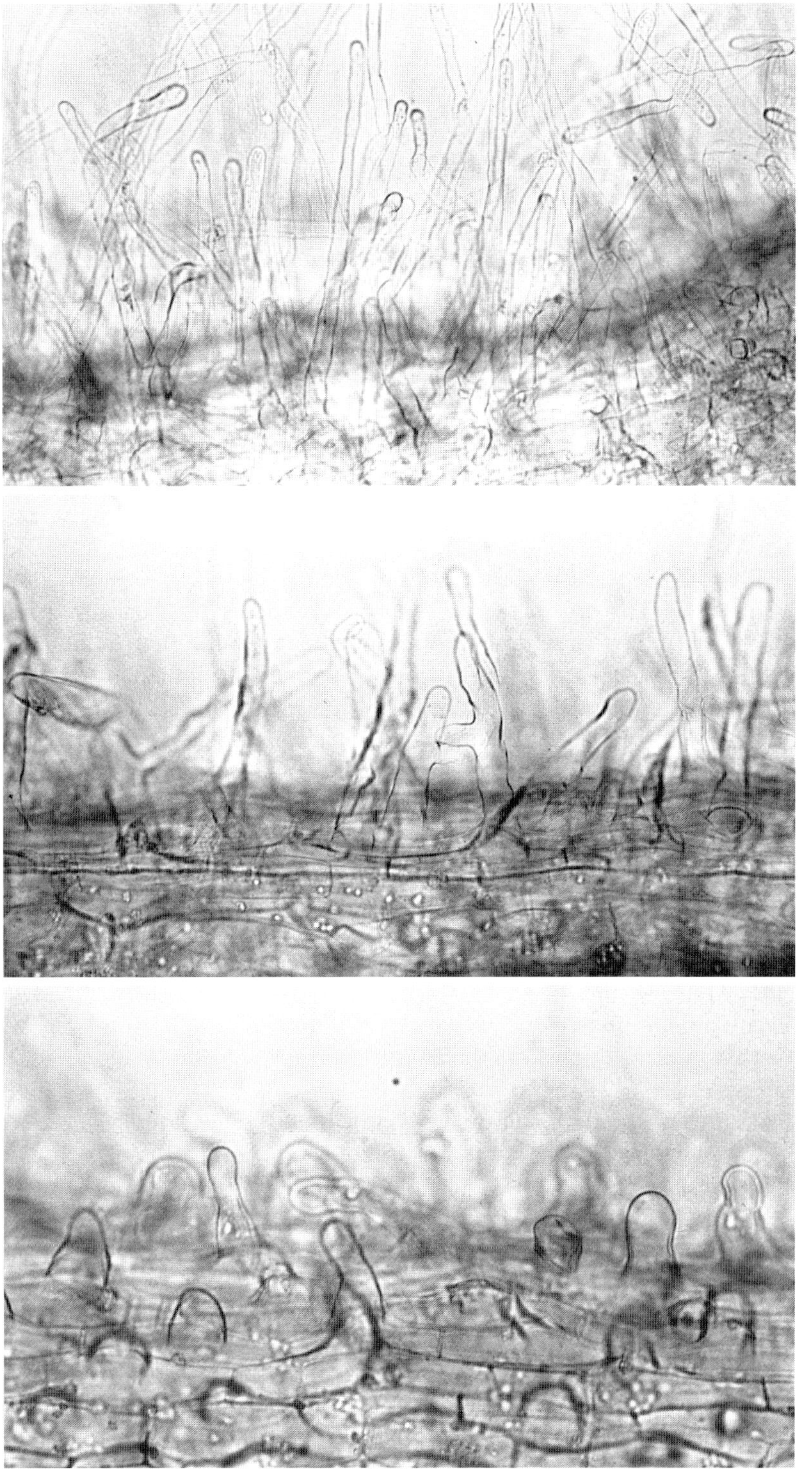

a

b

c

Fig. 4. Numerous root hairs attached to fragments of epidermis. **a** and **b** In transformed roots, **c** in normal roots. (1 cm = 25 μm)

n-hexane. The organic solvent was evaporated to dryness. This procedure allowed us to hydrolyze diosgenin without formation of its degradation product, spirosta-3,5-diene. Root tissues of non transformed 8-month-old plants grown in the field were extracted as described above. After filtering off cell debris, the liquid media were lyophilized and extracted using the same procedure.

The authenticity of diosgenin in root extracts was confirmed by thermospray liquid chromatography-positive ion mass spectrometry (TSP LC-MS) and comparison with reference material (Merkli 1996).

The quantification was carried out by reversed-phase high-performance liquid chromatography after derivatization with benzoyl chloride (Higgins 1976). Typically, 2 ml of anhydrous pyridine (Merck, Germany) and 0.1 ml of benzoyl chloride (Fluka, Switzerland) were added to the dried extract. The mixture was heated at 80 °C for 30 min; then, after the addition of 2 ml of MeOH to transform the excess of benzoyl chloride into methyl benzoate, it was heated again at 100 °C for 30 min. After cooling to room temperature, 10 ml of dichloromethane, 20 ml of water and 2 ml of concentrated hydrochloric acid were added. The organic phase was washed successively with water to eliminate pyridine, with a saturated sodium carbonate solution to eliminate benzoyl chloride, and finally twice with water. The dichloromethane layer was evaporated to dryness and the residue was dissolved in chloroform. Vitamin K_1 (2-methyl-3-phytyl-1,4-naphthoquinone, Fluka, Switzerland) was used as internal standard.

HPLC analyses were performed using a Waters 600 E Multisolvent Delivery System (Milford, USA) instrument equipped with a Model 484 Variable Wavelength UV Detector coupled to a Perkin-Elmer LCI-100 Integrator (Norwalk, USA). Samples (20 µl) were introduced via a Waters 700 Satellite Wisp Autosampler. A stainless steel Nucleosil 100–5 C-18 column (250 × 4 mm; Macherey-Nagel, Germany) fitted with a Nucleosil 120–5 C-18 guard column (11 × 4 mm) was used. An isocratic acetonitrile-water mixture (92:8) was used as the mobile phase for 60 min with UV detection at 230 nm. The flow rate was set at 2 ml/min and the column temperature was maintained at room temperature.

A stock of standard solution of diosgenin benzoate (5 mg/ml) was prepared in chloroform. Working standard solutions were obtained by diluting the stock standard solution with chloroform. A calibration curve reporting peak area ratio of diosgenin benzoate/vitamin K_1 as a function of diosgenin benzoate concentration was established in the range of 25–250 µg/ml.

2.3.4 *Optimization of the Culture Medium*

The effects of 14 different culture media supplemented with various concentrations of sucrose were investigated for hairy root growth and diosgenin content during 35 days. The transformed roots were cultured in WP, MS and

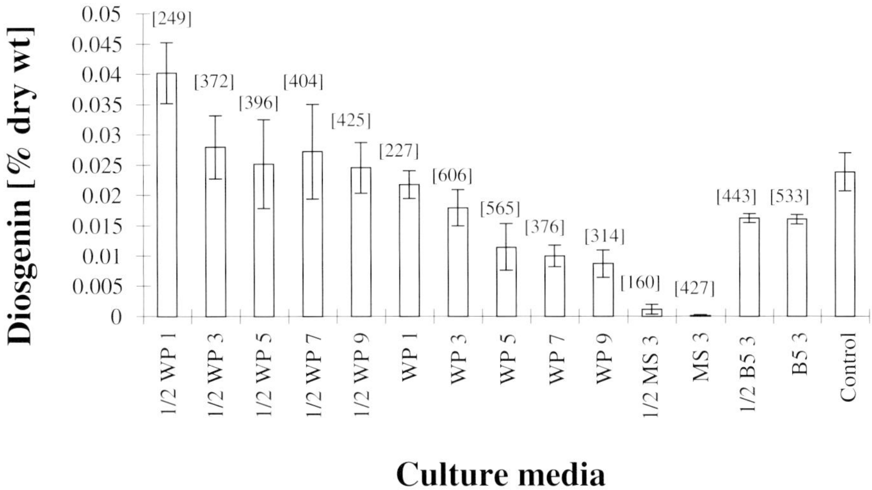

Fig. 5. Diosgenin content of *Trigonella foenum-graecum* hairy roots cultured in different liquid media for 35 days. Untransformed plants (control) were cultured in the field for 8 months. Numbers in brackets indicate the dry weight (mg)

Gamborg's B5 media and compared with 8-month-old nontransformed plants grown in the field. All media were hormone-free and adjusted to pH 5.9 before autoclaving (Fig. 5).

For all experiments, about 100 mg fresh wt. of hairy roots was inoculated into 50 ml liquid medium in 250-ml conical flasks and cultured in the dark at 25 °C on a gyratory shaker at 80 rpm. The contents of three to five flasks of each culture were harvested and the fresh and dry wt. after lyophilization was determined individually, and analyzed by HPLC as described above.

The fastest growth (606 mg dry wt.) was observed in WP medium supplemented with 3% sucrose and corresponded to a GI of 80. In half-strength WP (1/2 WP) media, the diosgenin content was higher than in the nontransformed roots (control). The highest concentration of diosgenin was observed in 1/2 WP1 and reached 0.040% dry wt. This is about twice the amount detected in the control roots (0.024% dry wt.). Increasing the sucrose concentration of 1/2 WP medium induced rapid growth but increasing the strength of the WP medium reduced the diosgenin content significantly. Half- and full-strength MS media were not suitable for the production of diosgenin. In MS 3 medium, diosgenin was not detectable. Gamborg's B5 media were neither suitable for hairy root growth nor for diosgenin biosynthesis.

2.3.5 Time-Course Study

In a further experiment, the kinetics of root growth and diosgenin production were investigated in WP medium supplemented with 3% sucrose for 50 days

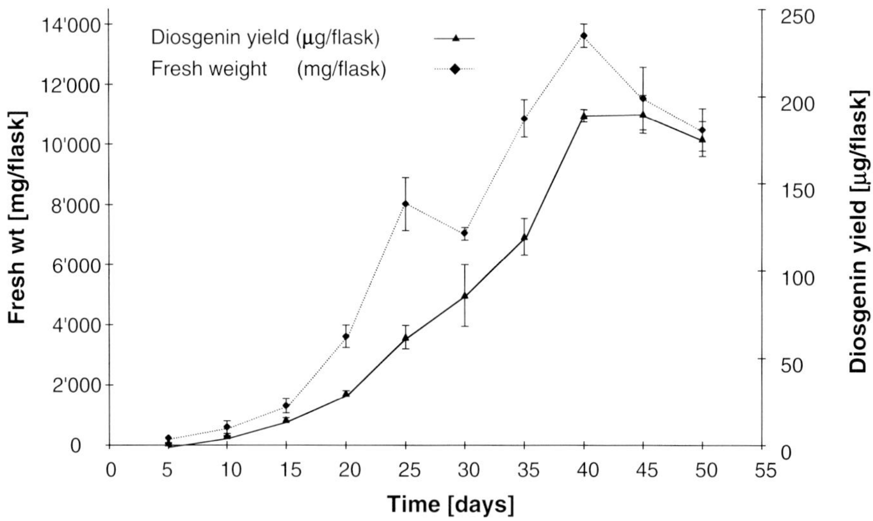

Fig. 6. Time-course study of growth and diosgenin production in *Trigonella foenum-graecum* hairy roots cultured in WP3 liquid medium

(Fig. 6). At 5-day intervals, three flasks were harvested, and growth and diosgenin production were measured. The fresh weightt of the hairy roots increased from an original inoculum of ca. 100 mg to 13.6 g/flask (GI: 136). The maximum biomass was reached after 40 days of culture. The diosgenin content appeared to be closely related to the root growth and reached 0.028% dry wt. after 50 days, representing a mean accumulation rate of 5.6 $\mu g\, g^{-1}\, day^{-1}$. Diosgenin production increased relatively slowly during the first 20 days, followed by a rapid rise over the next 25 days to reach a value of about 190 μg/flask after 45 days. Afterwards, both the fresh weight and the diosgenin yield decreased. The culture medium was also examined for diosgenin, but none was detected during the 50 days of culture, suggesting that it was all retained within the tissues.

2.3.6 Addition of Chitosan

Chitosan, a polymer of β-1,4-D-glucosamine, is known to be an effective inducer of secondary metabolites in plants (Funk and Brodelius 1990). It is obtained by alkaline hydrolysis of shellfish chitin. Chitosan (Sigma, USA) purification was carried out by dissolution in 6% w/v acetic acid, centrifugation (10,550 *g*, 15 min) and precipitation by addition of 5 N NaOH. This procedure was repeated twice. The chitosan residue was washed three times with distilled water and freeze-dried. In the first experiment, chitosan in concentrations ranging from 0.5–30 mg/l was added to WP liquid medium

Table 1. Growth and diosgenin content of *Trigonella foenum-graecum* hairy roots, cultured for 35 days in WP 3 liquid medium supplemented with different concentrations of chitosan. Each value is the mean of three determinations. (Merkli et al. 1997)

Chitosan concentration (mg/l)	Hairy root growth (mg dry wt.)	Diosgenin content (% dry wt.)
Control (no chitosan added)	606	0.018
0.5	713	0.015
1.0	740	0.009
2.5	753	0.017
5.0	684	0.022
10.0	579	0.026
20.0	685	0.054
30.0	702	0.040

Table 2. Growth and diosgenin content of *Trigonella foenum-graecum* hairy roots, cultured for 35 days in half-strength WP 1 liquid medium supplemented with different concentrations of chitosan. Each value is the mean of three determinations. (Merkli et al. 1997)

Chitosan concentration (mg/l)	Hairy root growth (mg dry wt.)	Diosgenin content (% dry wt.)
Control (no chitosan added)	249	0.040
10	237	0.064
20	240	0.053
30	240	0.102
40	234	0.125
50	259	0.094
60	261	0.077

supplemented with 3% sucrose, the best medium for the hairy root growth. The roots were cultured for 35 days. For the control, the same culture conditions were applied to hairy roots except that no chitosan was added to the culture medium. As can be seen from Table 1, the addition of chitosan had little effect on the hairy root growth. However, in the presence of 20 mg/l of chitosan, the diosgenin content reached 0.054% dry wt. This is three times the amount detected in the control roots.

In the second experiment, various concentrations of chitosan (10–60 mg/l) were added to half-strength WP liquid medium supplemented with 1% sucrose, the best medium for diosgenin yield, and the transformed roots were cultured for 35 days. Regardless of the chitosan concentration used (Table 2), no significant effect on the hairy root growth could be detected when compared with the control. However, the diosgenin content was in all cases significantly increased by the addition of chitosan. The maximum diosgenin content was obtained in cultures supplemented with 40 mg/l, in which the diosgenin reached 0.125% dry wt. This is three times the amount found in the untreated root cultures (0.040% dry wt) and five times the amount detected in the 8-month-old nontransformed roots (0.024% dry wt). These results demonstrated that an optimized chitosan concentration could greatly improve the diosgenin content in hairy roots of *T. foenum-graecum.*

2.3.7 Addition of Cholesterol

It is well-known that diosgenin arises via the mevalonic pathway (Dewick 1997), cholesterol being a key intermediate in its biosynthesis. This is why hairy root cultures were fed with cholesterol (30, 50 and 70 mg/100 ml) and allowed to grow for 35 days. Stock solution of cholesterol was prepared in ethanol and a known amount of this solution was added to the WP3 medium after autoclaving. A control medium (no cholesterol added) was used to determine whether ethanol had an inhibitory effect. After evaporation of ethanol, cholesterol precipitated in the medium. In all cases, hairy root growth was inhibited. To avoid the influence of ethanol and the precipitation of cholesterol in the culture medium, water-soluble cholesterol (Sigma, USA) encapsulated into the cavity of methyl β-cyclodextrin was tested at a concentration of 10 mg/100 ml. Preliminary experiments showed that methyl β-cyclodextrin alone (30 mg/100 ml) did not influence the hairy root growth. All of the three batches tested showed a large decrease in biomass, from 606 mg dry wt. in the control medium to 265, 281 and 220 mg dry wt., respectively (Table 3). The diosgenin content (0.002% dry wt.) was only about 11% of the control culture (0.018% dry wt). These results were not in line with those reported by Khanna et al. (1975a) for cell suspension cultures of *T. foenum-graecum* fed with cholesterol.

Table 3. Growth and diosgenin content of *Trigonella foenum-graecum* hairy roots, cultured for 35 days in WP 3 liquid medium supplemented with 10 mg/100 ml of cholesterol. Each value is the mean of three determinations. (Merkli et al. 1997)

Batch	Hairy root growth (mg dry wt.)	Diosgenin content (% dry wt.)
Control (no cholesterol added)	606	0.018
1	265	0.002
2	281	0.002
3	220	0.002

Table 4. Influence of pH on growth and diosgenin content of *Trigonella foenum-graecum* hairy roots, cultured for 35 days in WP 3 liquid medium. Each value is the mean of three determinations. (Merkli et al. 1997)

pH before autoclaving	pH after autoclaving	pH after 35 days of culture	Hairy root growth (mg dry wt.)	Diosgenin content (% dry wt.)
5.9	5.7	5.6	606	0.018
5.5	5.2	5.2	710	0.037
5.0	4.8	4.8	730	0.042

2.3.8 *Influence of pH*

Little is known about the effect of pH on growth and secondary metabolite production in hairy root cultures. Therefore, the effects of pH on hairy root growth and diosgenin content were measured. Transformed roots were cultured for 35 days in WP 3 liquid medium and the pH was adjusted to 5.0, 5.5 and 5.9 with 0.1 N NaOH before autoclaving. The biomass was not strongly affected and the roots were able to grow at all pH values (Table 4). Nevertheless, root growth was slightly inhibited at pH 5.9. The effect of pH was more pronounced on the diosgenin content. Highest diosgenin concentration (0.042% dry wt.) was recorded at pH 5.0. This is about three times the percentage obtained at pH 5.9. It is noteworthy that, in all cases, the pH remained remarkably stable between the beginning and the end of the experiment.

2.3.9 *Influence of Light*

The influence of light on the growth and steroidal saponin production in hairy root cultures of *Solanum aculeatissimum* was investigated by Ikenaga et al. (1995). It was demonstrated that the transformed roots grew better under continuous light and that steroidal saponins were biosynthesized in the light only.

Hairy roots of fenugreek were cultured in WP 3 liquid medium for 35 days on a rotary shaker (80 rpm) at 25 °C. One batch was kept in the dark and another batch under light, 16 h/day (Osram-L-Fluora 77 R, 700 lux). Dark-grown roots were pale brown in color and the biomass reached 606 mg dry wt. The diosgenin content was 0.018% dry wt., corresponding to a production of 108 μg/flask. On the other hand, light caused a significant inhibition of root growth (264 mg dry wt.) and indeed stimulated root structure disorganization by the formation of calli. However, light had no effect on the biosynthesis of diosgenin. Nevertheless, due to poor growth, the diosgenin production fell to 40 μg/flask.

2.3.10 Bioreactor Studies

The morphological and biochemical characteristics of transformed roots, as well as their genetic stability, make them extremely advantageous for large-scale production if an industrial process is to be attempted. However, due to the unusual rheological properties of hairy roots, mechanical agitation is not possible, shear stress causing disorganization and callus formation with consequently lowered productivity. Rodriguez-Mendiola et al. (1991) studied the growth characteristics and the diosgenin production of hairy roots of *Trigonella foenum-graecum* in two fermenter types, an airlift reactor and a column-mesh reactor, both of them having a 9 l nominal capacity. The bioreactors were aerated at 0.8 l/min and the experiments conducted under continuous subdued light. The results obtained, although very partial, indicated a GI of 500 after 60 days of culture in the airlift bioreactor and a diosgenin content reaching 0.025% dry wt.

2.3.10.1 Rotating-Drum Reactor

Based on the promising results obtained in Erlenmeyer flasks, we developed a root culture in a 7-l rotating-drum fermentor. For preculture, hairy roots were grown in WP3 liquid medium, as described above. For cultivation in the bioreactor, 1 g fresh wt. of fenugreek hairy roots, obtained from a 15-day flask culture, was transferred aseptically through an inoculation port. The configuration of the bioreactor used in this study is shown in Fig. 7 and consisted of

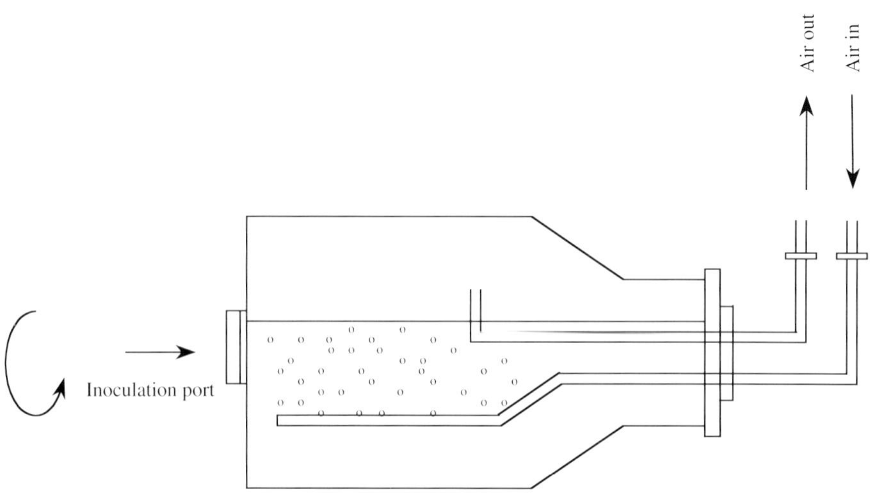

Fig. 7. Rotating-drum reactor for the hairy root culture

a cylindrical glass vessel, approximately 20 cm in diameter and 30 cm in length. It was filled with 2.5 l WP 3 liquid medium and rotated at 3 rpm on a Cellrotator (Shibata Hario Glass Co., Tokyo).

Humidified air was blown into the medium at a rate of 150 ml/min through a submerged nozzle. The culture was maintained in the dark in the above-mentioned conditions. The root tissue grew well, floated in the vessel following air flow and aggregated slightly. After 65 days of culture, 557 g fresh wt. was obtained. Figure 8 shows *T. foenum-graecum* hairy roots cultured in the rotating-drum fermenter. The average root growth rate was 3.4 $g l^{-1} day^{-1}$ and the diosgenin content reached 0.023% dry wt. The growth medium was also examined for the presence of diosgenin. However, no trace of the latter could be detected in the medium.

It is remarkable to note that in both studies, undertaken independently with two different bioreactors and different culture conditions, the root growth and the production of diosgenin are practically identical.

These preliminary studies demonstrate that further improvements through the development of a high-producing cell line may be possible. Operating fermenters with cell lines capable of exporting large amounts of sapogenins into the medium would offer the best operational system. A continuous flow system could then operate with the original biomass, using media capable of maintaining viability and synthetic capacity in the roots.

Fig. 8. Hairy root culture of *Trigonella foenum-graecum* after 65 days of growth in rotating-drum fermenter

3 Summary and Conclusion

Diosgenin is a natural sapogenin of prime importance for the pharmaceutical industry as starting material for the partial synthesis of oral contraceptives, sex hormones and other corticosteroids. Steroidal sapogenins have been reported in *Agave, Dioscorea, Costus, Solanum* and *Trigonella* species. *Trigonella* has drawn considerable attention as a source of diosgenin. In vitro callus and cell suspension cultures have been successfully established from seeds, roots and leaves. Optimization of the culture conditions for the growth and productivity of steroidal sapogenins and trigonelline has been achieved. It was demonstrated that the age of the culture, the nature and the dose of auxin and cytokinin and the addition of precursors can have marked effects on the growth and production of secondary metabolites. However, despite a substantial increase in the diosgenin production, the use of tissue culture as a commercial source of steroidal sapogenins does not seem to be competitive with their extraction from yams and sisal.

The *Agrobacterium*-mediated transformation of *Trigonella foenum-graecum* is a valuable method for obtaining hairy roots rapidly with a high growth rate and high yield of diosgenin. Compared with cell cultures, hairy roots are more easily established. After optimization of the culture conditions, WP medium was selected. This medium, supplemented with 3% sucrose (WP 3) was found the most appropriate for the root growth, whereas the same medium, diluted in half and supplemented with 1% sucrose (1/2 WP 1) was the most favorable for the production of diosgenin. The addition of 40 mg/l of chitosan to the 1/2 WP 1 liquid medium allowed the dry weight of diosgenin to reach 0.125% after 35 days of culture. This is three times the amount found in the untreated root cultures and five times the amount detected in the roots of 8-month-old nontransformed plants grown in the field. Cholesterol induced an important decrease in hairy root growth and inhibited the diosgenin production. Whereas light had no effect on the biosynthesis of diosgenin, decreasing the pH increased the sapogenin content significantly.

The growth and diosgenin production have been examined in a 7-l rotating-drum fermenter in batch fermentation on WP 3 culture medium. This study allowed us to obtain 557 g fresh wt. (about 28 g dry wt.) of hairy roots after 65 days, and a diosgenin content of 0.023% dry wt. Unfortunately, in the Erlenmeyer flasks as well as in the bioreactor, diosgenin was not released from the root cells into the medium. This result suggests that *T. foenum-graecum* hairy roots present very limited diosgenin excretion, which constitutes a major obstacle to the use of a bioreactor system. Much effort is necessary to improve the release of diosgenin into the culture medium. In particular, the use of permeabilizing agents may improve diosgenin production in hairy root culture of *Trigonella foenum-graecum*.

References

Allen ON, Allen EK (1981) The Leguminosae. a source book of characteristics, uses and nodulations. Macmillan, Madison, pp 666–667

Antony A, Gopinathan KP, Vaidyanathan CS (1975) Biosynthesis of trigonelline in root cultures of fenugreek (*Trigonella foenum-graecum* L.). Indian J Exp Biol 13:39–41

Bräutigam M, Franz G (1985) Versuche zur Gewebekultur von schleimbildenden pflanzlichen Geweben. Sci Pharm 53:237–246

Brain KR, Lockwood GB (1976) Hormonal control of steroid levels in tissue cultures from *Trigonella foenumgraecum*. Phytochemistry 15:1651–1654

Brain KR, Williams MH (1983) Evidence for an alternative route from sterol to sapogenin in suspension cultures from *Trigonella foenumgraecum*. Plant Cell Rep 2:7–10

Brummitt RK (1992) Vascular plant families and genera. Royal Botanic Gardens, Kew, pp 595–602

Cerdon C, Rahier A, Taton M, Sauvaire Y (1995) Effect of diniconazole on sterol composition of roots and cell suspension cultures of fenugreek. Phytochemistry 39:883–893

Culafic L, Savikin-Fodulovic K, Grubisic D, Neskovic M (1999) *Dioscorea balcanica* Kosanin and *D. caucasica* Lipsky: in vitro culture and production of diosgenin. In: Bajaj YPS (ed) Biotechnology in agriculture and forestry, vol 43. Medicinal and aromatic plants XI. Springer, Berlin Heidelberg New York, pp 85–104

Dewick PM (1997) Medicinal natural products. A biosynthetic approach. Wiley, Chichester, pp 152–269

Funk C, Brodelius P (1990) Influence of growth regulators and an elicitor on phenylpropanoid metabolism in suspension cultures of *Vanilla planifolia*. Phytochemistry 29:845–848

Gamborg OL (1970) The effects of amino-acids and ammonium on the growth of plant cells in suspension culture. Plant Physiol 45:372–375

Girardon P, Sauvaire Y, Baccou JC, Bessiere JM (1986) Identification of 3-hydroxy-4,5-dimethyl-2(5H)-furanone in aroma of fenugreek seeds (*Trigonella foenum-graecum*). Lebensm Wiss Technol 19:44–46

Gupta KC (1974) Influence of auxins on growth and cytohistology of fenugreek (*Trigonella foenum-graecum* L.) calli. Cytobios 9:103–107

Heywood VH (1971) The Leguminosae-A systematic purview. In: Harborne JB, Boulter D, Turner BL (eds) Chemotaxonomy of the Leguminosae. Academic Press, London, pp 1–29

Higgins JW (1976) A high-performance liquid chromatographic analysis of the benzoate esters of sapogenins isolated from *Agave*. J Chromatogr 121:329–334

Hooykaas PJJ, Klapwijk PM, Nuti MP, Schilperoort RA, Rörsch A (1977) Transfer of the *Agrobacterium tumefaciens* Ti plasmid to avirulent *Agrobacteria* and to *Rhizobium* ex planta. J Gen Microbiol 98:477–484

Ikenaga T, Oyama T, Muranaka T (1995) Growth and steroidal saponin production in hairy root cultures of *Solanum aculeatissimum*. Plant Cell Rep 14:413–417

Indrayanto G, Utami W, Syahrani A (1999) *Costus speciosus* (Koenig) J.E. Smith: in vitro cultures, micropropagation, and the production of diosgenin and other phytosteroids. In: Bajaj YPS (ed) Biotechnology in agriculture and forestry, vol 43. Medicinal and aromatic plants XI. Springer, Berlin Heidelberg New York, pp 57–77

Iribbarren AM, Pomilio AB (1983) Components of *Bauhinia candicans*. J Nat Prod 46:752–753

Jain SC, Nag TN, Mohan S, Khanna P (1975) Effect of ascorbic acid on and its estimation in plant tissue cultures. Sci Cult 41:292–293

Jain SC, Rosenberg H, Stohs SJ (1977) Steroidal constituents of *Trigonella occulta* tissue cultures. Planta Med 31:109–111

Joshi JG, Handler P (1960) Biosynthesis of trigonelline. J Biol Chem 235:2981–2983

Kamal R, Yadav R (1992) Diosgenin and tigogenin from *Trigonella polycerata* plant parts and callus cultures. Indian Drugs 29:360–361

Khanna P, Jain SC (1972) Effect of nicotinic acid on growth and production of trigonelline by *Trigonella foenum-graecum* L. tissue cultures. Indian J Exp Biol 10:248–249

Khanna P, Jain SC (1973) Diosgenin, gitogenin and tigogenin from *Trigonella foenum-graecum* tissue cultures. Lloydia 36:96–98

Khanna P, Nag TN (1973) Production of amino acids in vitro tissue culture. Indian J Exp Biol 11:310–311

Khanna P, Jain SC, Bansal R (1975a) Effect of cholesterol on growth and production of diosgenin, gitogenin, tigogenin and sterols in suspension cultures. Indian J Exp Biol 13:211–213

Khanna P, Bansal R, Jain SC (1975b) Effect of various hormones on production of sapogenins and sterols in *Trigonella foenum-graecum* L. suspension cultures. Indian J Exp Biol 13:582–583

Kuhn A, Gerhard H (1943) The trigonellin and nicotinic acid contents of semen foenugraeci. Arch Pharm (Weinheim, Germany) 281:378–379

Lloyd G, McCown B (1980) Commercially-feasible micropropagation of mountain laurel, *Kalmia latifolia*, by use of shoot-tip culture. Int Plant Prop Soc Proc 30:421–427

Lockwood GB, Brain KR (1976) Influence of hormonal supplementation on steroid levels during callus induction from seeds of *Trigonella foenumgraecum*. Phytochemistry 15:1655–1660

Mehra P. Yadav R, Kamal R (1996) Influence of nicotinic acid on production of trigonelline from *Trigonella polycerata* tissue culture. Indian J Exp Biol 34:1147–1149

Merkli A (1996) Production de diosgénine par les racines transformées (hairy roots) de *Trigonella foenum-graecum* L. (Fabaceae). PhD Thesis, Univ Geneva, Switzerland

Merkli A, Christen P, Kapetanidis I (1997) Production of diosgenin by hairy root cultures of *Trigonella foenum-graecum* L. Plant Cell Rep 16:632–636

Murashige T, Skoog F (1962) A revised medium for rapid growth and bioassays with tobacco tissue cultures. Physiol Plant 15:473–497

Oncina R, Botia JM, Del Rio JA, Ortuño A (2000) Bioproduction of diosgenin in callus cultures of *Trigonella foenum-graecum* L. Food Chem 70:489–492

Ortuño A, Oncina R, Botia JM, Del Rio JA (1998) Distribution and changes of diosgenin during development of *Trigonella foenum-graecum* plants. Modulation by benzylaminopurine. Food Chem 63:51–54

Petit A, David C, Dahl DA, Ellis JG, Guyon P, Casse-Delbart F, Tempé J (1983) Further extension of the opine concept: plasmids in *Agrobacterium rhizogenes* cooperate for opine degradation. Mol Gen Genet 190:204–214

Radwan SS, Kokate CK (1980) Production of higher levels of trigonelline by cell cultures of *Trigonella foenum-graecum* than by the differentiated plant. Planta 147:340–344

Rodriguez-Mendiola MA, Stafford A, Cresswell R, Arias-Castro C (1991) Bioreactors for growth of plant roots. Enzyme Microb Technol 13:696–702

Sauvaire Y, Baccou JC (1978) L'obtention de la diosgénine, (25*R*)-spirost-5-ène-3β-ol; problèmes de l'hydrolyse acide des saponines. Lloydia 41:247–256

Sen B, Gupta S (1979) Differentiation in callus cultures of leaf of two species of *Trigonella*. Physiol Plant 45:425–428

Sharma AUGL, Khanna P (1977) Flavonoids from in vitro seedling callus culture of *Trigonella foenum-graecum* Linn. Indian J Pharm 39:142–143

Shekhawat NS, Galston AW (1983) Mesophyll protoplasts of fenugreek (*Trigonella foenum-graecum*): isolation, culture and shoot regeneration. Plant Cell Rep. 2:119–121

Singh NN, Kokate CK, Tipnis HP (1981) A note on development of callus cultures of *Trigonella foenum-graecum* for diosgenin bioproduction. Indian Drugs 19:25–26, 33

Trevelyan WE, Procter DP, Harrison JS (1950) Detection of sugars on paper chromatograms. Nature 166:444–445

Trisonthi P, Baccou J-C, Sauvaire Y (1980) Essai d'amélioration de la production de sapogénines stéroïdiques par les tissus de fenugrec (*Trigonella foenum-graecum*) cultivés in vitro. CR Acad Sci Paris Ser D 291:357–360

Tutin TG, Heywood VH, Burger NA, Moore DM, Valentine DH, Walters SM, Webb DA (eds) (1968) Flora Europaea, vol 2. Cambridge University Press, Cambridge, 152 pp

Willis JC (1973) A dictionary of the flowering plants and ferns, 8th edn. Cambridge University Press, Cambridge, pp 1174–1175

White PR (1943) A handbook of plant tissue culture. Jaques Cattell Press, Lancaster, Pennsylvania

Xu Z-H, Davey MR, Cocking EC (1982) Organogenesis form root protoplasts of the forage legumes *Medicago sativa* and *Trigonella foenum-graecum*. Z Pflanzenphysiol 107:231–235

18 *Vaccinium* Species (Small-Fruited Berries): In Vitro Culture and the Production of Food Colorants and Phytochemicals

M.A.L. SMITH[1]

1 Introduction

1.1 Distribution and Importance of the Plant

The genus *Vaccinium*, from the heath family (Ericaceae), includes a wide range of popular berry species of economic importance, including the American cranberry (*V. macrocarpon* Ait.), the wild lowbush blueberry (*V. angustifolium* Ait.), cultivated highbush and rabbiteye blueberries (*V. corymbosum* L. and *V. ashei* Reade), bilberry (*V. myrtillus* L.) and lingonberry (*V. vitis-idaea* L.). While these crops are well known throughout the world, in many cases, their individual distributions are quite narrow (Lyrene and Perry 1988; Scorza and Welker 1988; Kalt and Dufour 1997). Wild lowbush blueberry, for example, is localized in the extreme northeastern United States and maritime provinces of Canada (Fig. 1A,B); bilberry is grown only in a few European countries with an isolated pocket of distribution in the Rocky Mountain region of the USA, and cranberry production, which until recently was confined to the eastern and western coasts of the USA, has recently expanded into higher elevations in South America. The harvested berries are marketed fresh, frozen, and in some cases, sweetened and dried (personal communication, D. Nolte, Decas Cranberry Co.). They are also popular components in bakery items, dried cereals, jams, juices, and numerous related edible products.

Less well-known species of *Vaccinium*, including wild plantings, are limitedly distributed in pockets throughout the world. *V. pahalae* Skottsb., common name ohelo, is typical of this latter category. Ohelo is a creeping shrub which is only known to grow on the main island of Hawaii, in bog-type areas, weathered lava flows, and on mountain slopes. It is expected to hybridize freely with other wild *Vaccinium* species which colonize this precarious environment. The edible berries range from dark red to pale yellow, and are eaten raw or used in baking. This particular plant is the focus of a rich island folklore, and was considered to be sacred to the volcano goddess Pélé. In fact, a legend states that ohelo berries were customarily thrown into the center of volcanic

[1]University of Illinois, Department of Natural Resources and Environmental Sciences, Urbana, Illinois 61801, USA

Biotechnology in Agriculture and Forestry, Vol. 51
Medicinal and Aromatic Plants XII (ed. by T. Nagata and Y. Ebizuka)
© Springer-Verlag Berlin Heidelberg 2002

Fig. 1. Wild blueberry (*Vaccinium angustifolium*), one of the prominent members of the genus with known bioactive properties. **A** The low-growing crop spreads over acreages in the maritime provinces of Canada, and the extreme northeastern USA. **B** Prolific berries are available for harvest within a narrow (2–3 week) window of time in the late summer

activity to allay the goddess's wrath; only after this offering was it considered safe to eat the berries (Degener 1984).

1.2 Production of Bioactive Compounds in *Vaccinium* Fruits

Members of the genus *Vaccinium* are primarily grown as a source of edible fruit, but also as landscape ornamentals, and are used to colonize wild areas. In most cases, plants from this genus prefer acid soils and require some inactive dormant cold period in order to successfully produce flowers and fruits. Bog-type environments favor production of some of these fruits (e.g., cranberry and ohelo), and in some cases, production is largely limited to established wild stands, since propagation and new plantings are not easily initiated (wild blueberry). The berries are extracted and marketed as a valuable source of natural anthocyanin pigments for the food-processing industry, as replacements for potentially harmful synthetic pigments and dyes (P. Hereld, the Hereld Organization, pers. comm.). In cranberry, for example, the waste product of the juicing process (the pomace) can be further extracted as a source of the natural anthocyanin pigments (Francis 1989). The flavonoid content (pigments and other polyphenolic compounds) from different species and cultivars has been extensively investigated in an effort to identify donors for valuable extracts (Fuleki and Francis 1967; Wang et al. 1978). More recently, consumer interest in these species has risen due to increasing evidence for the health protective properties of the berries.

While anthocyanins have most frequently been cited as the bioactive constituents of *Vaccinium* extracts, considerable evidence has also indicated that the proanthocyanidin fractions possess considerable anticancer and antiradical activity. Mixtures of flavonoids derived from *Vaccinium* berries may have additive or synergistic biological benefits that cannot be realized from one isolated compound; hence, some of the products derived from the berries are rich in mixtures of proanthocyanidins and anthocyanins (Table 1).

Health benefits associated with consumption of some *Vaccinium* berries have been recognized for centuries, but it is just within the past decade that potent antioxidant, anticancer, cardioprotective, and other bioactive properties have been scientifically demonstrated and widely appreciated. The anthocyanins, the proanthocyanidins, or combinations of these phytochemicals have been reported as the active ingredients from foods and supplements containing *Vaccinium* spp. (Fig. 2). The anthocyanin composition of some of the most important *Vaccinium* members is given in Table 2. The exact proanthocyanidin composition has not been elucidated for these species. Both leaves and fruits of the bilberry have been known as ingredients of traditional European medicines since the 16th century. Bilberry extracts are rich in up to 15 free anthocyanins (3-*O*-arabinosides, 3-*O*-glucosides, and 3-*O*-galactosides of cyanidin, delphinidin, peonidin, petunidin, and malvidin) as well as associated

Table 1. Summary of in vivo/in vitro production and harvest of secondary metabolites in the genus *Vaccinium*

Fruits	Secondary metabolite harvested as pharmaceutical or food supplement	In vivo/ in vitro	Sources or references
Lowbush blueberry (*V. angustifolium*)	Anthocyanins	In vivo	Francis et al. (1966)
Rabbiteye blueberry (*V. ashei*)	Anthocyanins, proanthocyanidins, and carotenoids	In vitro	Nawa et al. (1993)
Bilberry (*V. myrtillus*)	Anthocyanin and proanthocyanidin extract	in vivo	Bettini et al. (1991); Indena Corp. ESA Labs
	Anthocyanin and proanthocyanidin extract; carotenoids	In vitro	Madhavi et al. (1998)
American cranberry (*V. macrocarpon*)	Anthocyanin-rich powdered extract	In vivo	Fuleki and Francis (1967); Decas Cranberry Co.; Ocean Spray Cranberries; Cape Cod Biolab Corp.
	Anthocyanins	in vitro	Madhavi et al. (1995)

Indena Corp. is based in Italy. Decas Cranberry Co, ESA Labs, Ocean Spray Cranberries, Inc., and Cape Cod Biolab Corp. are based in the USA.

proanthocyanidins, and are currently marketed through pharmacies in Europe. The antioxidant capacity of the rich flavonoid complement in bilberry extracts effectively inhibits the oxidative modifications of low density lipoproteins (LDL). Because of this demonstrated capacity, bilberry extracts are consumed in part to inhibit the formation of atherosclerotic plaques in the arterial wall (Laplaud et al. 1997). Research has cited the ability of bilberry anthocyanins to relax coronary artery segments (antagonize contractile responses) as a potential inhibitor of heart disease (Bettini et al. 1991), and to enhance microvascular blood flow (Colantuoni et al. 1991). These same extracts have been lauded for antiulcer and therapeutic activity (Cristoni and Magistretti 1987), and for clinical applications in ophthalmology, including the ability to increase night vision acuity (Morazzoni and Bombardelli 1996). In all these applications, the anthocyanin pigments are designated as the active principles responsible for bioactivity (Kalt and Dufour, 1997). Related fruits from other members of the genus are recognized for other miscellaneous health benefits: blueberries and cranberries as sources of antiadhesins (effective in treatment of urinary tract infections); blueberries as a source of cough suppressants and

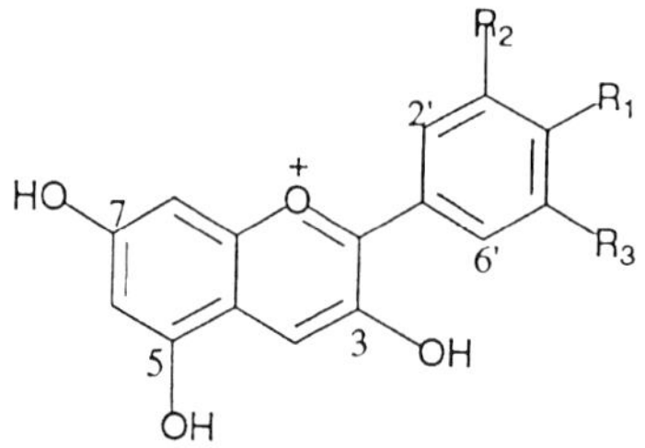

The structure of anthocyanidin.

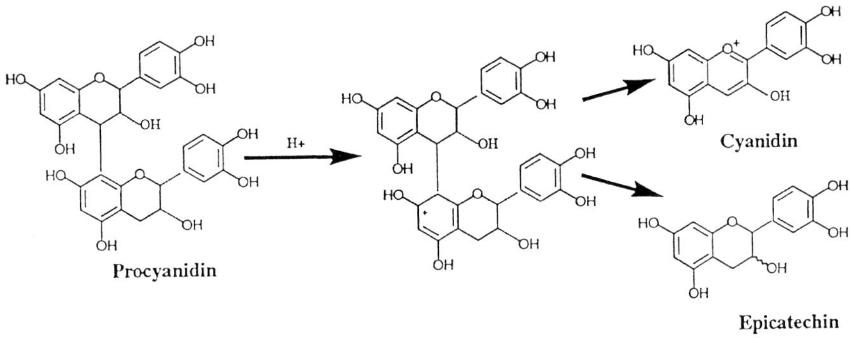

Hydrolysis of proanthocyanidins in acid. Proanthocyanidins are encountered in the form of procyanidins.

Fig. 2. Anthocyanins and proanthocyanidins are both recognized as biologically active phytochemicals in the genus *Vaccinium* and are responsible for antioxidant, cardioprotective, and anticancer properties associated with consumption of the fruits

as a diarrhea remedy; cranberries as donors of cardio-protective phytochemicals (Ofek et al. 1991; Avorn et al. 1994; Kalt and Dufour 1997; Wilson et al. 1998).

While crude extracts from wild blueberry, cranberry and lingonberry fruits each demonstrated inhibition of the promotion stage of chemically induced carcinogenesis, it was the proanthocyanidin-rich fraction which proved to be most highly anticarcinogenic (Bomser et al. 1996; Smith et al. 2000). Proanthocyanidins are one of the major flavonoid classes found in these fruits, and are associated with, but separate from, the anthocyanin pigments. Further research has recently shown that blueberries, on a fresh weight basis, have the highest antioxidant capacity of all previously tested fruits and vegetables in the typical diet, which is indicative of the ability of these fruits to protect against conditions of neurodegenerative changes with aging, cardiovascular disease, and carcinogenesis (Prior et al. 1998). In these tests, the anthocyanin pigments themselves are cited as the active components responsible for antioxidant capacity.

Table 2. The anthocyanins in key members of the genus *Vaccinium*

Fruits	Anthocyanins[a]	References
Highbush blueberry (*V. corymbosum*)	Cy 3-glc, 3-gal, 3-ara; Dp 3-glc, 3-gal, 3-ara; Mv 3-glc, 3-gal, 3-ara; Pt3-glc, 3-gal, 3-ara; Pn 3-glc, 3-gal, 3-ara	Ballinger et al. (1970)
Lowbush blueberry (*V. angustifolium*)	Cy 3-glc, 3-gal, 3-ara; Dp 3-glc, 3-gal, 3-ara; Mv 3-glc, 3-gal, 3-ara; Pt 3-glc, 3-gal, 3-ara; Pn 3-glc, 3-gal, 3-ara	Francis et al. (1966)
Rabbiteye blueberry (*V. ashei*)	Cy 3-glc, 3-gal, 3-ara; Dp 3-glc, 3-gal, 3-ara; Mv 3-glc, 3-gal, 3-ara; Pt 3-glc, 3-gal, 3-ara; Pn 3-glc, 3-gal, 3-ara	Ballington et al. (1987)
Lingonberry (*V. vitis-idaea*)	Cy 3-gal	Swain (1962)
Bilberry (*V. myrtillus*)	Cy 3-glc, 3-gal, 3-ara; Dp 3-glc, 3-gal, 3-ara; Mv 3-glc, 3-gal, 3-ara; Pt 3-glc, 3-gal, 3-ara; Pn 3-glc, 3-gal, 3-ara	Baj et al. (1983)
American cranberry (*V. macrocarpon*)	Cy 3-gal, 3-ara; Pn 3-gal, 3-ara	Zapsalis and Francis (1965)
European cranberry (*V. oxycoccus*)	Cy 3-gal, 3-glc, 3-ara; Pn 3-gal, 3-glc, 3-ara; Dp 3-glc, Pt 3-glc, My 3-glc	Andersen (1989)

[a] Cy, Cyanidin; Dp, delphinidin; Pt, petunidin; Pn, peonidin; Mv, malvidin; Pl, pelargonidin; Pg, pelargonidin; glc-glucose; gal, galactose; ara, arabinose

The continuing studies and promising research results on bioactive principles contained in fruits of *Vaccinium* are expected to further enhance the global market for these berries. Because investigations have elucidated, in part, the identity of chemopreventive compounds such as the flavonoids (in particular, the anthocyanin pigment complex and proanthocyanidins), a new market has developed for extracts in supplement (pill) form. In the US, these supplements are currently marketed through health food stores and groceries, whereas in Europe and Asia, a large market for flavonoid extracts as pharmaceuticals and dietary supplements already exists (Kalt and Dufour 1997).

2 In Vitro Approaches

2.1 Tissue Culture and Biotechnology of *Vaccinium*

Several members of the genus have been adapted to in vitro culture, as shoot cultures, in suspension, and as tissue culture targets for gene transfer research. Micropropagation has been reported in cranberry, lingonberry, and blueberry as a clonal propagation technique (Lyrene 1980; Scorza and Welker 1988;

Marcotrigiano and McGlew 1991; Reed and Abdelnour-Esquivel 1991), but shoot cultures and callus cultures have also been used as a vehicle to facilitate in-depth investigations on stress physiology for these crops (Muralitharan et al. 1990, 1992). Improvements in the in vitro cultivation of bilberry and ohelo microshoots were recently achieved through manipulation of the cytokinin and iron sources in the proliferation medium (Shibli et al. 1997b). The direct regeneration of shoots and roots from leaf tissue from these same two *Vaccinium* species was accomplished as a prelude to use of these tissues in transformation research, aimed at intensifying the levels of phytochemicals in cell cultures (Shibli and Smith 1996). In these studies, regeneration was accomplished on thidiazuron-supplemented medium (0.9–2.7 μM) from whole leaves and leaf sections of ohelo (Figs. 3, 4); similar results were obtained from bilberry leaves and hypocotyl segments. Tissue cultures of *Vaccinium* species have been the target for in vitro induction of tetraploidy using colchicine (Perry and Lyrene 1984; Lyrene and Perry 1988). Genetic engineering using particle bombardment was reported for cranberry germ plasm, which resulted in transformation of the plant (Serres et al. 1992).

Anthocyanins have been expressed and manipulated in cell cultures (callus and suspensions) of rabbiteye blueberry, cranberry, and ohelo (Nawa et al. 1993; Madhavi et al. 1995; Smith et al. 1997; Kandil et al. 2000). Both chemical and physical microenvironmental manipulation of the cell cultures has resulted in dramatically enhanced product recovery, continuous production, and scale-up of cultures to bioreactors (Shibli et al. 1997a; Smith 1998; Pépin et al. 1999). Cell cultures from key genotypes of *Vaccinium*, including the bilberry and the ohelo, have recently been identified as sources of bioactive compounds, with antioxidant and anticancer potential (Kandil et al. 2000). Bilberry cell cultures demonstrate high levels of peroxidase enzyme activity in the culture medium, and have been used as a model system for in depth investigation of two cationic peroxidases produced in vitro (Melo et al. 1995, 1997).

The natural chemopreventive properties of bilberry cell culture extracts were further investigated by fractionating culture extracts into flavonoids (proanthocyanidin and anthocyanin pigments) and a hexane extract containing carotenoids, sterols, and lipids (Madhavi et al. 1998). The extract was screened for anticancer activity by in vitro testing of the ability to induce phase II xenobiotic detoxification enzyme quinone reductase (QR). Fractions from the in vitro cell cultures provided a clear, significant induction of QR activity, indicative of inhibition of the initiation stage of carcinogenesis. A twofold stimulation of QR activity was observed with extracts added at approximately 9 μg/150 μl per well, which indicates that one or more components possess substantial QR-inducing capacity, comparable to known QR inducers such as furanon and catechol. The extracts were not cytotoxic at the dose levels tested; so the strong induction of quinone reductase by the fraction is indicative of potent anticancer activity, with an efficacy comparable to several active dietary constituents recently identified. These extracts also exhibited a growth

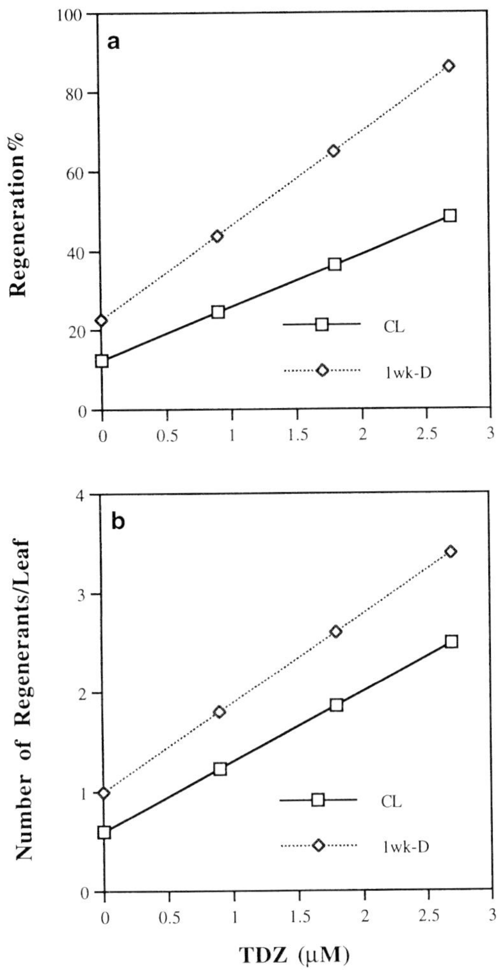

Fig. 3. Regeneration from whole leaves of ohelo on TDZ-supplemented medium. *CL* Explants held in continuous light; *1 wk-D* explants held 1 week in darkness prior to exposure to light. (Shibli and Smith 1996)

suppressing effect (antiproliferative capacity) against an estrogen receptor-positive human breast cancer line, and an estrogen receptor-negative human breast cancer cell (Madhavi et al. 1998).

Extracts from ohelo cell suspension cultures demonstrated parallel chemo-preventive bioactivity in anticancer screening (Smith et al. 1997; Kandil et al. 2000). In addition, a galvinoxyl free radical assay was adapted to assess the antioxidant capacity of extracts from ohelo cell cultures. Galvinoxyl is a stable free radical which reacts slowly with oxygen or quickly with compounds that are good hydrogen atom/electron donors (therefore good antioxidants). This

Fig. 4. Regeneration from whole leaves of ohelo on TDZ-supplemented medium. (Shibli and Smith 1996)

assay was adapted as a rapid, inexpensive means to assess antioxidant capacity of single, pure flavonoid compounds as well as complex mixtures in an extract. In preliminary tests, a 70% acetone extract of ohelo cell cultures demonstrated antioxidant capacity comparable to well-known antioxidant fruits, but a specific subfraction rich in acylated anthocyanins and proanthocyandins provided a fivefold increase in antioxidant capacity, indicating that these phytochemicals are key players in the bioactivity.

The ohelo cell suspension system was subsequently scaled-up to 12 l bioreactor volume (Fig. 5). Analysis of the cell culture extracts revealed a strong bioflavonoid profile, including a rich proanthocyanidin fraction and associated anthocyanin pigments (Smith 1998; Pépin et al. 1999).

2.2 Establishment of Cell Suspensions

Microcultures of *Vaccinium pahalae* were maintained by subculturing microshoots (3.0–4.0 mm) to a woody plant medium (Lloyd and McCown 1981) supplemented with 18 μM zeatin at 7-week intervals. Cultures were grown at a reduced irradiance level (20–25 $\mu mol\, m^{-2} s^{-1}$), because microplants exhibited slow growth and declined at higher light intensity (Shibli and Smith 1996).

Callus was induced in the dark from excised microshoot leaf and stem segments approx. 2 mm long by explanting onto the surface of 40 ml of initiation media containing 1.0 mM $CaCl_2(2H_2O)$, 1.0 mM $MgSO_4$, 1.25 mM NaH_2PO_4, 8.0 mM KNO_3, 20.4 mM $(NH_4)_2SO_4$, and B5 vitamins and micronutrients (Gamborg et al. 1968), supplemented with 5.4 μM NAA, 0.45 μM 2,4-D, 2.3 μM

Fig. 5. Ohelo cell suspensions have been scaled-up to production in 2 or 12-l bioreactors

kinetin, 2% sucrose, and 6% agar in GA7 vessels (Magenta Crop., Chicago, IL). However, initial calluses tended to be slow-growing and not friable enough to facilitate bulk-up and eventual transfer to suspension culture medium, even after multiple subcultures. In an effort to increase the friability of callus (to make it more amenable to subsequent suspension culture), subcultures were made to woody plant medium with elevated iron chelate; ferric disodium ethylenediamine tetraacetate (FeEDTA at 500–800 μM) for two subculture periods. Data on fresh weight and observations of callus friability were made at the end of the second subculture on high iron-containing medium treatments. Callus was then transferred to liquid suspension culture medium with standard (100 μM) iron concentration, and evaluated after the first cell cycle in terms of anthocyanin concentration and fresh weight (Table 3).

When calluses from each elevated iron treatment were transferred to a liquid suspension media with standard iron chelate level (100 μM FeEDTA), friability was maintained, and a fine uniform suspension was achieved from the

Table 3. The influence of elevated FeEDTA (between 500–800 μM) in solid WPM on growth and anthocyanin production of ohelo callus and suspension cultures

	FeEDTA levels in solid WPM (μM)			
	500	600	700	800
Solid culture Fresh wt. (g)[a]	4.28 a	3.95 b	3.74 b	3.16 c
Suspension culture[b] Fresh wt. (g)	5.44 a	5.59 a	5.67 a	5.70 a
Anthocyanin content [mg (g fresh wt.)$^{-1}$]	0.30 a	0.32 a	0.26 b	0.25 b

[a] Means within rows having different letters are significantly different according to LSD at the 0.05 level of probability ($n = 10$).
[b] Calluses were cultured in indicated FeEDTA levels in solid WPM, then were transferred to liquid WPM with standard FeEDTA level (100 μM).

500 and 600 μM solid culture pretreatments. Anthocyanin concentration was also enhanced by "pretreating" calluses on solid medium with up to 600 μM FeEDTA prior to introduction to liquid suspension media (Table 3). A high FeEDTA level in solid medium used once or twice as "pulse" was sufficient to induce sustained callus friability and vigorous growth. In all subsequent research on this cell culture system, once friable calluses were obtained, they were consistently maintained on standard solid medium with 100 μM FeEDTA, and friability was retained. Therefore, solid medium with 600 μM FeEDTA was used routinely to obtain friable calluses shortly after callus initiation from leaves, and friable and rapidly growing calluses were selected and subcultured to standard medium for continuous callus maintenance before initiation of suspension cultures.

Suspension cultures were initiated by transferring about 1.0 g fresh weight (fresh wt.) of friable callus into 40 ml of a standard liquid medium that contained woody plant medium salts, the same auxin composition, 20 μM BA, and 50 g l^{-1} sucrose in 125-ml flasks. Cultures were incubated on a rotary shaker (150 rpm) under 90 μmol m^{-2} s^{-1} continuous fluorescent light at 25 °C. After 14–20 days, any cell aggregates larger than 2.0 mm were discarded, and only fine pigmented cells and small clusters were subcultured to fresh medium. Subsequent subcultures were performed at 14-day intervals by transferring 2.0 ml settled cell volumes (SCV), about 1.0 g fresh wt., to fresh medium.

The biomass and anthocyanin productivity of suspension cultures was further enhanced by elicitation of the cell cultures with biotic and abiotic agents. Specifically, iron chelate in the form of sodium ferric ethylenediamine di-(*o*-hydroxyphenylacetate) [FeEDDHA] at 90 μM as an abiotic agent and β-glucan at 10 μM as a biotic elicitor provoked significantly higher anthocyanin

Table 4. The influence of FeEDDHA (as an abiotic elicitor), beta-glucan (as a biotic elicitor), and methyl jasmonate (as a signaling compound) on anthocyanin production in ohelo suspension cultures

	Treatment			
	Control	FeEDDHA (90 μM)	Beta-glucan (10 μM)	Methyl jasmonate (0.5 μM)
Fresh wt. ($g\,l^{-1}$)[a]	159 c	173 b	191 a	194 a
Anthocyanin yield ($mg\,l^{-1}$)	82 d	129 c	145 b	164 a

[a] Means within rows having different letters are significantly different according to LSD at the 0.05 level of probability ($n = 10$).

accumulation levels as compared to control cultures (Table 4). Addition of 0.5 μM methyl jasmonate alone (a signaling compound) provoked a twofold increase in anthocyanin production over the control (Table 4); however, no additive effect on anthocyanin production occurred.

Deliberate manipulation of physical microenvironmental variables, such as culture surface-to-volume ratios, irradiance wavelength, and subculture timing have resulted in dramatic shifts in flavonoid productivity in vitro. For example, a preliminary experiment performed with ohelo suspension-cultured cells exposed for 2 h daily to 300 $\mu mol\,m^{-2}\,s^{-1}$ (=65 $W\,m^{-2}$, with 700 $mW\,m^{-2}$ UV-B) provided by a xenon lamp elicited a higher anthocyanin production response (200 $mg\,l^{-1}$) than the control cultures (120 $mg\,l^{-1}$) grown under the usual white fluorescent culture room lighting (70 $\mu mol\,m^{-2}\,s^{-1}$ (=15 $W\,m^{-2}$) for 24 h. Subsequently, three distinct treatments were set up under the xenon lamp irradiance: (1) cells held in standard plastic multi-well plates with lids, (2) cells held in multiwell plates, but covered with a plastic film permeable to UV-B wavelengths, and (3) cells held in the same environment as 2, but ventilated. As shown in Fig. 6, the cell culture productivity was hastened and intensified by provision of UV light, and by increasing the available oxygen to the cells in culture. Given these results, elevated dissolved oxygen concentrations (50% DO) have been implemented in standard 12-l bioreactor setups to support flavonoid accumulation.

2.3 Extraction and Analysis of Flavonoids from *Vaccinium* Cell Cultures

Callus and suspension culture evaluations routinely assess biomass fresh wt. ($g\,l^{-1}$), anthocyanin content ($mg\,[g\,\text{fresh wt.}]^{-1}$), and total anthocyanin yield ($mg\,l^{-1}$). Fresh weight was recorded upon harvest for callus cultures; for suspension cultures, cells were separated from cell suspension medium by vacuum filtration through Whatman #4 filter paper, just until the point when free liquid is no longer expressed, and fresh wt. was then immediately recorded.

Biomass was extracted overnight in 1.0% HCl-MeOH (1:50 w/v) at 4 °C for anthocyanin measurement. For calculations of anthocyanin content and

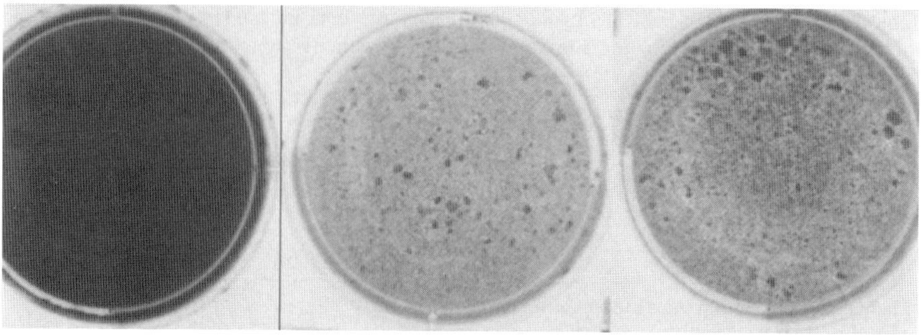

Fig. 6. Effects of irradiation and oxygen availability on flavonoid accumulation in ohelo cell cultures. Suspension-cultured cells held in multiwell plates were exposed for 2 h daily to 300 μmol m^{-2} s^{-1} (=65 W m^{-2}, with 700 mW m^{-2} UV-B) provided by a xenon lamp. After 3 days, cells with lids permeable to UV-B wavelengths (*right*) accumulated more flavonoids than cultures with standard polypropylene lids (*center*). Cell cultures which had the UV-permeable lids and also received additional aeration (*left*) exhibited the most intense pigment accumulation, more rapidly than other treatments

anthocyanin yield, extract was filtered through Whatman #4 filter paper, and anthocyanin was calculated from absorbance measurements made at 535 nm using a molar extinction coefficient of 46,200 cm^{-1} M^{-1} (Smith et al. 1997), using a DU-65 spectrophotometer (Beckman Instruments, Inc., Fullerton, CA). For further analysis, extracts after filtration were evaporated to dryness at 30 °C with a Buchi rotary evaporator (Switzerland). The residue was dissolved in water and extracted with ethyl acetate (13 v/v) three times. The aqueous fraction was evaporated to remove residual ethyl acetate and adsorbed onto an activated OnGuard-RP Sep-Pak cartridge (Dionex, CA). The cartridge was washed with water and anthocyanins were eluted with 0.01% HCl in methanol. Peak identification used authentic standards isolated from cranberry fruit and other known sources.

Anthocyanins from callus cultures of bilberry, cranberry, and ohelo were subjected to acid hydrolysis in 2 N HCL:methanol (1:1) for 30 min. Anthocyanidins derived from acid hydrolysis were analyzed using a Whatman Partisil ODS-3 (250◊4.6 mm, 5mu) column connected to a YMC-Pack S5 120 A ODS-AM guard column. Solvents were 10% formic acid (A) and 100% acetonitrile (B). Separation was obtained by an isocratic elution of 20% B in A at 1 ml/min. Electrospray (ES) mass spectrometry (MS) of the anthocyanins was accomplished on a VG Quattro mass spectrometer, and the spectra were obtained in a positive ion mode. The ES voltage was +4 kV. Tandem MS was performed on selected [M+] ions. The total anthocyanins were purified by preparative paper chromatography in butanol:acetic acid:water (4:1:5) for MS analysis.

Proanthocyanidins were estimated as anthocyanidins by the method of Nawa et al. (1993). Samples were repeatedly extracted in cold (4 °C) 1% HCl methanol to remove all pigments. The residue after pigment extraction was suspended in 1% HCl methanol (1:10 w/v), incubated overnight at 50 °C and then filtered. Absorbance was measured at 535 nm and expressed as absorbance units (A.U.)/g fresh wt.

Whereas reverse phase HPLC chromatograms of the anthocyanins from cranberry fruits showed four major pigments comprised of cyanidin and peonidin 3-galactosides and 3-arabinosides, the same analysis showed that simpler profiles consisting mainly of cyanidin-3-galactoside, cyanidin-3-glucoside, and cyanidin 3-arabinoside were produced in pigmented cranberry callus cultures. These same major pigments were observed in callus cultures from bilberry. Anthocyanins were not detected in dark-grown callus cultures. On acid hydrolysis of the anthocyanins from cultures, cyanidin and peonidin were the aglycones.

The major anthocyanin in ohelo cell suspension culture was cyanidin-3-galactoside. ES mass spectrometry confirmed the presence of cyanidin and peonidin glycosides in cell cultures. In low resolution MS, molecular ions were observed at m/z 449, 419, and 463 corresponding to cyanidin galactoside, cyandin arabinoside, and peonidin galactoside respectively. When the molecular ions corresponding to m/z 449 and 419 were subjected to tandem MS, a major fragment ion was observed at m/z 287 in both cases, indicating that they are cyanidin derivatives. Tandem MS of the molecular ion atm/z 463 revealed a major fragment at m/z 301, indicative of peonidin glycoside. Further analysis revealed that cell culture extracts had a very high absorption due to proanthocyanidins, as compared to fruit (Madhavi et al. 1995). Callus cultures in the dark accumulated proanthocyanidins at comparable levels to cultures under light.

3 Summary, Conclusions, and Prospects

At present, the *Vaccinium* cell culture systems provide useful models for elucidation of the complex anthocyanin pathway, and have potential as vehicles for gene transfer research. Callus and suspension culture systems for flavonoid production were developed for three species (cranberry, ohelo, and bilberry), but only the ohelo system has been scaled-up to 12 l bioreactor volumes. Given the increasing public interest in anthocyanins as a source of biomedicinal properties, either in foods or food supplements, these models will be used to characterize the particular species of pigment that confer activity. In particular, issues related to possible additive or synergistic interactions between related phytochemicals (e.g., anthocyanins and associated proanthocyanidins or other flavonoids), which may work in tandem to protect human health, have yet to be fully understood. The simple, predictable, and highly repeatable cell culture production systems for bioactive flavonoid production offer a useful strategy for in depth study of these areas in pharmacological research.

4 Protocol

1. Initial explants: nodal segments and shoot tips (for shoot culture production); leaf and internode segments (for callus initiation).
2. Culture conditions for suspension cultures: agitation by orbital motion at 150 rpm; 90 μmol m^{-2} s^{-1} continuous fluorescent light at 25 °C.
3. Suspension culture medium: liquid woody plant medium (Lloyd and McCown, 1981) with 5.4 μM NAA, 0.45 μM 2,4-D, 20 μM BA, and 50 g l^{-1} sucrose; 40 ml in 125-ml flasks.
4. Maintenance: subculture every 14 days by transferring 2.0 ml settled cell volume to fresh medium.
5. Anthocyanin analysis: absorbance measurements at 535 nm.

References

Andersen O (1989) Anthocyanins in fruits of *Vaccinium oxycoccus* L. (small cranberry). J Food Sci 54:383–384

Avorn J, Monan M, Gurwitz J, Glynn R, Choodnovskiy I, Lipsitz L (1994) Reduction of bacteriuria and pyuria after ingestion of cranberry juice. JAMA 271:751–754

Baj A, Bombardelli E, Gabetta B, Martinelli E (1983) Qualitative and quantitative evaluation of *Vaccinium myrtillus* anthocyanins by high-resolution gas chromatography and high-performance liquid chromatography. J Chromatogr 279:365–372

Ballinger W, Maness E, Kushman L (1970) Anthocyanin in ripe fruit of highbush blueberry, *V accinium corymbosum* L. J Am Soc Hortic Sci 95:283–285

Ballington J, Ballinger W, Maness E (1987) Interspecific differences in the percentage of anthocyanins aglycones, and aglycone-sugars in the fruit of seven species of blueberries. J Am Soc Hortic Sci 112:859–864

Bettini V, Aragno R, Bettini M, Braggion G, Calore L, Concoloato M, Favaro P, Penada G, Montin S (1991) Vasodilator and inhibitory effects of *Vaccinium myrtillus* anthocyanosides on the contractile responses of coronary artery segments to acetylcholine. Role of the prostacyclins and of the endothelium-derived relaxing factor. Fitoterapia 62:15–28

Bomser J, Madhavi D, Singletary K, Smith M (1996) In vitro anticancer activity of fruit extracts from *Vaccinium* species. Planta Med 62:212–216

Colantuoni A, Bertuglia S, Magistretti M, Donato L (1991) Effects of *Vaccinium myrtillus* anthocyanosides on arterial vasomotion. Arzneimittelforschung 41:905–909

Cristoni A, Magistretti M (1987) Antiulcer and healing activity of *Vaccinium myrtillus* anthocyanosides. Farmaco 42:29–43

Degener O (1984) Plants of Hawaii National Park illustrative of plants and customs of the South Seas. Braun-Brumfield Inc, Ann Arbor, MI, pp 240–245

Francis F (1989) Food colorants: anthocyanins. Crit Rev Food Nutr 28:273–314

Francis F, Harborne J, Barker W (1966) Anthocyanins in the lowbush blueberry, *Vaccinium angustifolium*. J Food Sci 31:583–587

Fuleki T, Francis F (1967) The co-occurrence of monoglucosides and monogalactosides of cyanidin and peonidin in the American cranberry, *Vaccinium macrocarpon*. Phytochemistry 6:1705–1708

Gamborg O, Miller R, Ojima K (1968) Nutrient requirements of suspension cultures of soybean root cells. Exp Cell Res 50:151–158

Kalt W, Dufour D (1997) Health functionality of blueberries. HortTechnology 7:216–221

Kandil F, Song L, Pezzuto J, Seigler D, Smith MAL (2000). Isolation of oligomeric proanthocyanidins from flavonoid-producing cell cultures. In Vitro Cell and Dev Biol Plant 36:492–500

Laplaud P, Lelubre A, Chapman M (1997) Antioxidant action of *Vaccinium myrtillus* extract on human low density lipoproteins in vitro: initial observations. Fund Clin Pharmacol 11: 35–40

Lloyd G, McCown B (1981) Commercially feasible micropropagation of mountain laurel, *Kalmia latifolia*, by use of shoot tip culture. Int Plant Prop Soc Proc 30:421–427

Lyrene P (1980) Micropropagation of rabbit-eye blueberries (*V. ashei*). HortScience 15:80–81

Lyrene P, Perry J (1988) Blueberries (*Vaccinium* spp.). In: Bajaj YPS (ed) Biotechnology in agriculture and forestry, vol 6. Crops II. Springer, Berlin Heidelberg New York, pp 181–198

Madhavi D, Smith M, Berber-Jiménez M (1995) Expression of anthocyanins in callus cultures of cranberry (*Vaccinium macrocarpon* Ait). J Food Sci 60:351–355

Madhavi D, Bomser J, Smith M, Singletary K (1998) Isolation of bioactive constituents from *Vaccinium myrtillus* (bilberry) fruits and cell cultures. Plant Sci 95–103

Marcotrigiano M, McGlew S (1991) A two-stage micropropagation system for cranberries. J Am Soc Hortic Sci 116:911–916

Melo N, Cabral J, Fevereiro M (1995) Extracellular peroxidases from cell suspension cultures of *Vaccinium myrtillus*. Purification and characterization of two cationic enzymes. Plant Sci 106:177–184

Melo N, Larson E, Welinder K, Fevereiro P (1997) Characterization of two major cationic peroxidases from cell suspension cultures of *Vaccinium myrtillus*. Plant Sci 122:1–10

Morazzoni P, Bombardelli E (1996) *Vaccinium myrtillus* L. Fitoterapia 47:3–29

Muralitharan M, Van Stevenick R, Chandler S (1990) Growth characteristics and ion contents of non-selected and salt-selected callus lines of highbush blueberry (*Vaccinium corymbosum*) cultivars Blue Crop and Denise Blue. Plant Cell Rep 9:151–155

Muralitharan M, Chandler S, Van Stevenick R (1992) Effects of Na_2SO_4, K_2SO_4, and KCl on growth and ion uptake of callus cultures of *Vaccinium corymbosum* L. cv. Blue Crop. Ann Bot 69:459–465

Nawa Y, Asano S, Motoori S, Ohtani T (1993) Production of anthocyanins, carotenoids, and proanthocyanidins by cultured cells of rabbiteye blueberry (*Vaccinium ashei* Reade). Biosci Biotech Biochem 57:770–774

Ofek I, Goldhur J, Zafriri D, Lis H, Adar R, Sharon N (1991) Anti *Escherichia* adhesion activity of cranberry and blueberry juices. N Engl J Med 324:1599

Pépin M-F, Smith MAL, Reid J (1999). Application of imaging tools to plant cell culture: relationship between plant cell aggregation and flavonoid production. In Vitro Cell and Dev Biol Plant 35:290–295

Perry J, Lyrene P (1984) In vitro induction of tetraploidy in *Vaccinium darrowi*, *V. elliottii*, and *V. darrowi* ◊ *V. elliotti* with colchicine treatment. J Am Soc Hortic Sci 109:4–6

Prior R, Cao G, Martin A, Sofic E, McEwen J, O'Brien C, Lischner N, Ehlenfeldt M, Kalt W, Krewer G, Mainland C (1998) Antioxidant capacity as influenced by total phenolic and anthocyanin content, maturity, and variety of *Vaccinium* species. J Agric Food Chem 46:2686–2693

Reed B, Abdelnour-Esquivel A (1991) The use of zeatin to initiate in vitro cultures of *Vaccinium* species and cultivars. HortScience 26:1320–1322

Scorza R, Welker W (1988) Cranberries (*Vaccinium macrocarpon* Ait.) In: YPS Bajaj (ed) Biotechnology in agriculture and forestry vol 6. Crops II. Springer, Berlin Heidelberg New York, pp 181–198

Serres R, Stang D, McCabe D, Russel D, Mahr D, McCown B (1992) Gene transfer using electric discharge particle bombardment and recovery of transformed cranberry plants. J Am Soc Hortic Sci 117:174–180

Shibli R, Smith M (1996) Direct shoot regeneration from *Vaccinium pahalae* (ohelo) and *V. myrtillus* (bilberry) leaf explants. HortScience 13:1225–1228

Shibli R, Smith M, Kushad M (1997a). Headspace ethylene accumulation reduced secondary metabolite production in *Vaccinium pahalae* cell culture. Plant Growth Regul 23:201–205

Shibli R, Smith M, Nasr R (1997b) Iron source and cytokinin mitigate the incidence of chlorosis and hyperhydration in vitro. J Plant Nutr 20(6):773–781
Smith M (1998) Bioreactor-based production of natural anthocyanins and related chemopreventive compounds. Proceedings of INFCOL III, the 3rd International Symposium on Natural colorants for food, nutraceuticals, beverages, confectionery, and cosmetics, Princeton, NJ, pp 344–350
Smith M, Madhavi D, Fang Y, Tomczak M (1997) Continuous cell culture and product recovery from wild *Vaccinium pahalae* germplasm. J Plant Physiol 150:462–466
Smith MAL, Marley K, Seigler D, Singletary K, Meline B (2000) Bioactive properties of wild blueberry fruits. J Food Sci 65:352–356
Swain T (1962) The chemistry of flavonoid compounds. Pergamon Press, New York
Wang P, Du C, Francis F (1978) Isolation and characterization of polyphenolic compounds in cranberries. J Food Sci 43:1402–1404
Wilson T, Porcari J, Harbin D (1998) Cranberry extract inhibits low density lipoprotein oxidation. Life Sci 62:381–386
Zapsalis C, Francis F (1965) Cranberry anthocyanins. J Food Sci 30:396–399

Subject Index

Printing (Computer to Film): Saladruck Berlin
Binding: Stürtz AG, Würzburg